THE HUNT FOR THE DAWN MONKEY

The Hunt for the Dawn Monkey

Unearthing the Origins of
Monkeys, Apes, and Humans

CHRIS BEARD

ILLUSTRATIONS BY MARK KLINGLER

UNIVERSITY OF CALIFORNIA PRESS Berkeley Los Angeles London

University of California Press
Berkeley and Los Angeles, California

University of California Press, Ltd.
London, England

© 2004 by the Regents of the University of California

Library of Congress Cataloging-in-Publication Data

Beard, K. Chris.
    The hunt for the dawn monkey : unearthing the
origins of monkeys, apes, and humans / Chris Beard;
illustrations by Mark Klingler.
        p.   cm.
    Includes bibliographical references and index.
    ISBN 0–520–23369–7 (cloth : alk. paper)
    1. Primates, Fossil.   2. Monkeys, Fossil.
3. Fossil hominids.   4. Human beings—Origin.
5. Paleoanthropology.   I. Title.
QE882.P7B35   2004
569'.8—dc22   2004001403

Manufactured in the United States of America
13   12   11   10   09   08   07   06   05   04
10   9   8   7   6   5   4   3   2   1

For Sandi
Amor vitae supervivit

Contents

Illustrations

PLATES

Preface

The story of human evolution has been told many times before, and it will no doubt continue to be revised and updated as new fossils are discovered. My goal in writing this book has been to add a much-needed prologue to what is now a familiar tale. If the major outlines of human origins are settled, the search for anthropoid origins remains scientifically in its infancy. Great strides have been made over the past two centuries, but we remain fairly ignorant of such basic questions as when, where, how, and why our earliest anthropoid ancestors evolved. This appraisal is not meant as a critique. Ignorance is to science as economic opportunity is to capitalism. It is more rewarding to toil in earnest on an unsettled issue than to tinker at the margins of a topic that is largely known. The story of anthropoid origins is fascinating precisely because so much of it remains in flux. At the same time, it is a story that has never been made available to a wide audience, one that extends beyond the narrow group of academic specialists who have devoted much of their professional lives to solving its mysteries.

Teamwork plays a vital role in paleontology, because scientific advances in this field hinge on isolated discoveries that reach across vast swaths of space and time. Over the past decade or so, I have had the privilege of working with some of the finest and most accomplished paleontologists in the world, in places that few ever get to visit. I have no doubt that I have learned more from my colleagues than vice versa. Throughout, we have been united by our mutual goal of illuminating the remote ancestry that we humans share with other anthropoid or "higher" primates.

Paleontology is one of the few academic disciplines in which field exploration remains a fundamental part of the quest to expand knowledge and understanding. This unique combination of the possibility for personal adventure and intellectual fulfillment is what attracted me to paleontology in the first place. I hope that I am able to impart a fraction of what I have experienced and learned during these past few years in this book.

My role in this story would not have been possible without the support and cooperation of a large number of individuals and institutions. It gives me great pleasure to thank my colleagues at the Carnegie Museum of Natural History, Mary Dawson and Luo Zhexi, who have often ventured into the field with me and who have served as frequent sounding boards for my ideas, while constantly providing me with their own unique expertise. Equally important have been a number of other colleagues who have worked alongside me in the field in China: Dan Gebo, Marc Godinot, Wulf Gose, John Kappelman, Leonard Krishtalka, Ross MacPhee, Jay Norejko, Tim Ryan, and Alan Tabrum. I am also deeply indebted to my friends and colleagues at the Institute of Vertebrate Paleontology and Paleoanthropology in Beijing: Qi Tao, Wang Banyue, Li Chuankuei, Wang Yuanqing, Tong Yongsheng, Wang Jingwen, Huang Xueshi, and Guo Jianwei. For years, these world-class scholars and indomitable scientists have welcomed my American colleagues and me into their country and into their homes. During our joint expeditions to various parts of China, they have imparted their knowledge and perseverance along with their unmatched organizational skills, without which none of the expeditions could ever have been launched. More important, they have extended a hand of friendship to me and many other *wei guo ren* that will always be cherished. I also want to thank some of the scientists who played such critical roles in my formal education and professional training, and who have enlightened and inspired me through the years, among them Rich Kay, Elwyn Simons, Ken Rose, Tom Bown, and Alan Walker.

Fieldwork in distant locales can be expensive, and none of this research could have been conducted without the financial support of various institutions, including the Carnegie Museum of Natural History, the

Leakey Foundation, the National Geographic Society, and the Physical Anthropology Program at the National Science Foundation. A fellowship from the John D. and Catherine T. MacArthur Foundation gave me the flexibility to take on this challenge. In writing the manuscript, I have benefited from the wisdom and insight of numerous friends and colleagues, including Mary Dawson, Dan Gebo, Bert Covert, Ken Rose, Jean-Jacques Jaeger, Marc Godinot, and Hans Sues. The original artwork in this book is due to the talent and creativity of Mark Klingler, scientific illustrator at the Carnegie Museum of Natural History. Original photographs have generously been provided by Patrick Aventurier and the Gamma Agency, Bert Covert, Marc Godinot, David Haring, Rich Kay, Mohamed Mahboubi, Elwyn Simons, and Fred Szalay. Archival photographs have been provided by the American Museum of Natural History, the Carnegie Museum of Natural History, the Muséum National d'Histoire Naturelle, and Thierry Pélissié (on behalf of Phosphatières du Quercy). I also gratefully acknowledge the skill and patience of my editor at the University of California Press, Blake Edgar. On a personal level, I owe the greatest debt of all to my parents, Kenneth and Nancy Beard, who instigated my sense of curiosity at an early age, and especially to my wife, Sandra Beard, whose love and support make it all worthwhile.

1

Missing Links and Dawn Monkeys

In rural China, the highest compliment you can get is not that you're attractive or smart. It's that you work really hard. As I shift to stay in the scant midday shade offered by a deep ravine on the northern bank of the Yellow River, this proletarian attitude makes a lot of sense. When I left the United States earlier this month, spring had barely begun. Checking the calendar in my field notebook, I see that it's only mid May—too early in the season for a heat wave. Yet for the past few days, my team has endured triple digit temperatures. Each of us sports a tan several shades deeper than our normal hue. A few yards away, where he chips at a piece of freshwater limestone that just might contain a fossil, my colleague Wang Jingwen is beginning to live up to his nickname, which translates roughly as "black donkey." I'm told that the local villagers have been praising our work ethic, because when it gets this hot, even the peasants take a siesta under a shade tree.

We have no choice but to tolerate the heat of the noon sun, because it provides the best lighting conditions for finding fossils. At this time of day, there are no shadows to hide the small jaws and limb bones that have been entombed in these rock strata for the past forty thousand millennia or so. Having traversed twelve time zones to get here, I'm not about to forgo the chance to find an important specimen merely because of the oppressive heat. My persistence is rewarded when I split apart another block of greenish-gray limestone. Inside I find a nearly complete maxilla, or upper jaw, of a small rodent, replete with three black teeth that glisten like fresh obsidian in the sunlight. Peering through a hand lens that I keep tied to a leather thong draped like a necklace under my tee shirt, the diagnostic pattern of cusps and crests on the fossilized teeth readily identifies the creature as *Pappocricetodon schaubi*. A primitive progenitor of modern mice, rats, and gerbils, *Pappocricetodon* is the most abundant fossil mammal known from this site.[1] Though it's not exactly the pivotal discovery I had hoped for, finding the mortal remains of any animal that lived millions of years ago invigorates the mind. I begin to contemplate the weighty scientific issues that have led me to travel halfway around the world, to this remote part of central China's Shanxi Province.

My particular area of scientific expertise, vertebrate paleontology, is in the midst of a sea change. Much of what I learned as a graduate student is being challenged by provocative new fossils and new methods of interpreting them, if not discarded altogether. Increasing globalization and the collapse of the Soviet Union and its satellite states have opened up most of the world to paleontological exploration, including places that, only a few years earlier, I never dreamed of being able to visit in search of fossils. On a separate front, molecular biologists are sequencing the DNA of various organisms at an increasingly frenetic pace, churning out megabytes of raw data that are being used to test old ideas, and to propose new ones, about the evolutionary relationships of living plants and animals. All in all, it feels like a unique moment in history and a great time to be a paleontologist, especially when you're involved in one of the most exciting debates to hit the field of paleoanthropology in many years.

Paleoanthropology is the scientific study of human origins. In the strictest sense, paleoanthropologists seek to illuminate the evolutionary history of the human lineage as it evolved from our more apelike ancestors. Fossil hominids are the crown jewels of paleoanthropology. Without them, theories about when, where, and how our species evolved would be helter-skelter, unconstrained by hard data. One of the great triumphs of twentieth century science has been the recovery of an amazing diversity of hominid fossils, mainly from eastern and southern Africa, but also from various parts of Eurasia, ranging from France and Spain to China and Indonesia. Discoveries of new fossil hominids continue unabated. Considered as a whole, the fossil record of early humans is now complete enough that, at least in broad strokes, we know how humans evolved from more apelike precursors. Virtually all paleoanthropologists agree, for example, that the human lineage originated sometime between five and seven million years ago in Africa, and that early humans acquired the ability to walk upright on two legs millions of years before their brains enlarged much beyond those of chimpanzees.[2]

A fuller consideration of human origins requires us to place our own evolutionary history within a broader context. Did humans take longer to evolve our unique characteristics than other living primates, or did our ancestors simply experience unusually high rates of evolution? For that matter, how unique are humans with respect to other primates anyway? Which seemingly "human" traits are ours alone, and which are shared with various primate relatives? Where do humans lie on the family tree of all primates, and what does that tree look like? Where do primates lie on the larger family tree of all mammals? Were there particularly critical events during the earlier phases of our evolutionary history, before our own lineage branched away from those leading to chimpanzees and other living primates? Today, these questions pose far greater scientific challenges than simply filling in the constantly shrinking gaps in the human fossil record. Yet, ironically, when most people hear the term "missing link," they think of a gap in the fossil record that supposedly fails to link modern humans with our apelike ancestors. The dirty little secret of paleoanthropology is that, while there are plenty of missing links, they don't occur where most people think they do. They exist farther back in deep time. Ultimately, this is why I'm at the bottom of a ravine on the banks of the Yellow River.

The ravine itself is a natural erosional feature, an ephemeral drainage flowing into the Yellow River from the north. It dissects a relatively flat plateau, which—like most rural parts of central China—is now under intensive wheat cultivation. Standing on top of the plateau at the head of the ravine offers a panoramic view of the surrounding terrain. To the south, on the far side of the Yellow River in Henan Province, lie rugged mountains composed primarily of limestone of Ordovician age. Some 450 million years ago—about twice the age of the earliest known dinosaurs—the rock now forming the crest of this range was deposited in a warm, shallow sea not unlike that surrounding the modern Bahamas.

To the north and east, wheat fields extend across the plateau as far as the eye can see. Immediately west of the ravine, the sleepy village of Zhaili shelters the peasant farmers who tend the surrounding fields. A narrow path, hewn into the western wall of the ravine, provides access to the bottom some 150 feet below for the villagers and their sheep and goats. Walking down this path, you can't help but notice the peculiar nature of the nearly vertical walls of the ravine. The rock defining both sides of the ravine is soft and pliable, so easy to work that many people in this part of China actually carve small caves into it, which function as storage rooms or even small homes. Geologically, this type of rock is known as loess. It is composed of wind-blown sediment laid down by countless dust storms that swept across this part of China during the Pleistocene Epoch, when vast ice sheets were expanding and contracting farther north in Siberia.

What is unique about this particular ravine, though, is not the loess. In this part of Shanxi Province, loess is ubiquitous, draping over older geological features like autumn leaves covering a well-kept lawn. But here, as the ravine approaches the Yellow River, it cuts deep into the loess. For the last fifty yards or so of its existence, the ravine finally succeeds in breaking through the loess altogether to expose the much older underlying strata. Even to the untrained eye, it is clear that these rocks are different, in terms of both their composition and their segregation into different layers or beds. They consist of alternating bands of blue-green mudstone, pale yellow and white limestone, and thick gray sandstones, the last of which show internal evidence of stratification in the form of minute swales of sand grains known as cross-bedding. The fossils we seek are concentrated in the layers of mudstone and limestone. They are roughly forty million years old, about six times older than the earliest putative hominids ever discovered. They pertain to an interval of Earth history known as the Eocene, the Greek roots of which translate more or less as "dawn of recent [life]."

As its etymology suggests, the Eocene was a pivotal period in the history of life on Earth—a time of transition from ancient to modern. The earliest members of most living orders of mammals first appeared and became geographically widespread, replacing more archaic forms that left no living descendants. Such distinctive and highly specialized types of modern mammals as bats and whales first showed up in the Eocene, together with the earliest odd-toed ungulates (horses, rhinos, and tapirs), even-toed ungulates (pigs, camels, and primitive relatives of deer and antelopes), and others. The order of mammals to which we belong, the Primates, also first became geographically widespread and ecologically prominent at the beginning of the Eocene, although a few scattered fossils hint that primates are somewhat older yet. At the same time, the Eocene witnessed the decline and extinction of many groups of mammals that first evolved alongside the dinosaurs, or immediately following their demise. Examples include the vaguely rodentlike multituberculates, the raccoon- or bearlike arctocyonids, and the large herbivores known as pantodonts and uintatheres. The Eocene also witnessed a great evolutionary diversification of flowering plants, together with the insects that feed on them.[3]

In terms of its prevailing climate, the Eocene was virtually a mirror image of the Pleistocene or "Ice Ages," when much of human evolution transpired. It began with a pronounced episode of global warming some fifty-five million years ago. Such optimal conditions allowed tropical and subtropical forests—and the animals that inhabit them—to occur at much higher latitudes than they do today. Because primates have always prospered in these warm forest habitats, the Eocene was truly a heyday for primate evolution. Among their other accomplishments, Eocene primates extended their geographic range far beyond its current limits. Fossils of Eocene primates have been found as far north as Saskatchewan in North America, England and Germany in Europe, and Mongolia in Asia. As I discuss in greater detail in subsequent chapters, the fossil record shows that during the Eocene, even these northern continental regions supported diverse evolutionary radiations of primates. After enduring for more than twenty million years, the greenhouse world of the Eocene ended thirty-four million years ago, when the Earth's climate once again became cooler and drier. It is unlikely to be a coincidence that this severe climatic deterioration witnessed the extinction of primates in North America and Europe, where tropical and subtropical habitats disappeared.

The vast majority of the fossil primates known from the Eocene resemble the most primitive primates alive today. These animals, collectively known as prosimians, include the diverse radiation of lemurs native to Madagascar, the bushbabies of continental Africa, the lorises of Africa and southern Asia, and, perhaps strangest of all, the tarsiers of Southeast Asian islands. Prosimians resemble other primates, including humans, in possessing nails rather than claws on most digits of their hands and feet, and in having eyes that face forward to allow for enhanced, "stereoscopic" vision. Like all primates aside from humans, prosimians have a grasping big toe, functionally akin to the human thumb. Yet prosimians also differ from humans and our nearest primate relatives, the monkeys and apes, in many aspects of their anatomy, physiology, and behavior.

Monkeys, apes, and humans are collectively known as anthropoids or "higher primates." Compared to prosimians, living anthropoids possess larger brains, eye sockets that are almost completely surrounded by bone, a single lower jaw bone (or mandible) formed by the fusion of two separate bones at the chin, and many other anatomically advanced features. In terms of their behavior, anthropoids again differ from most prosimians, although there is some overlap between species of each group. In general, anthropoids tend to live in complex groups characterized by intricate social interactions among individual members. Some prosimian species, in contrast, live quite solitary lives. All anthropoids aside from the South American owl monkey *(Aotus)* are diurnal—that is, they are mainly active during daytime. Many prosimians, notably tarsiers, bushbabies, lorises, and some lemurs, strongly prefer to move about and feed at night. These profound differences between prosimians and anthropoids extend to the molecular level. Analyses of long sequences of the DNA of various species of monkeys, apes, and humans show that all of these species are far more similar to one another than any of them are to prosimians. In an evolutionary context, this means that, whether we analyze anatomy, behavior, or DNA, the conclusion remains inescapable. We humans are much more closely related to monkeys and apes than we are to lemurs or tarsiers. Put slightly differently, monkeys share a more recent common ancestor with us than they do with prosimians.

Despite unanimous scientific agreement that humans share a close common ancestry with monkeys and apes, one of the most controversial issues in paleoanthropology today is how, when, and where the first anthropoids—the common ancestors of monkeys, apes, and people—evolved. In stark contrast to the relatively abundant fossil record for early humans, the fossil record for anthropoid origins is spotty, incomplete, and seemingly incoherent. Paleontology, like other branches of science, abhors such a vacuum. The main purpose of our expedition is to help flesh out this distant phase of our evolutionary history. Yet the simple fact that our team is searching for fossils of early anthropoid primates in Eocene rocks in central China is, in several respects, unorthodox—if not downright heretical.

Our goal is to test a bold new hypothesis about anthropoid origins—one that moves the birthplace of these remote human ancestors from Africa to Asia while it ruptures the established evolutionary timetable by tens of millions of years. This sweeping idea rests on the wobbly foundation provided by some fragmentary fossils from another Chinese site, known as Shanghuang, that I had recently named *Eosimias* ("dawn monkey" in Latin and Greek). If we are to have any hope of gaining scientific traction, we must find better fossils of *Eosimias* and animals like it. The

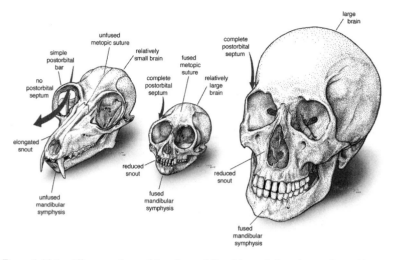

Figure 1. Major differences in cranial anatomy distinguish prosimians from anthropoids. Illustrated here (from left to right) are skulls of a ruffed lemur *(Varecia variegata),* a South American squirrel monkey *(Saimiri sciureus),* and a human *(Homo sapiens).* Note the basic similarity in skull form in the two anthropoids, which differ from the lemur in having a relatively larger brain, a reduced snout, fused mandibular symphysis and metopic suture, and a complete postorbital septum. Original art by Mark Klingler, copyright Carnegie Museum of Natural History.

bottom of the ravine on the northern bank of the Yellow River seems like a promising place to start.

To search for such elusive fossils, a highly interdisciplinary and international team of scientists has converged on this remote corner of central China. Each member brings a unique set of skills and knowledge to the table. On the Chinese side are four scientists from the Institute of Vertebrate Paleontology and Paleoanthropology (or IVPP), a branch of the Chinese Academy of Sciences. Tong Yongsheng, a veteran of numerous field campaigns all over the People's Republic, originally hails from Zhejiang Province, along China's southern coastline. A muscular man of medium build, Tong specializes in small mammals from the Eocene, especially rodents and insectivores (shrews, hedgehogs, and the like). Wang Jingwen, who grew up in Beijing, primarily studies ungulates, or hooved mammals, from the Eocene. Lately, though, Wang has developed an abiding interest in early primates, which allows the two of us to collaborate closely on joint research projects. Huang Xueshi boasts the most eclectic interests of any member of our team, having worked on fossils ranging in age from Paleocene to Oligocene. Huang's excellent mastery of English, combined with his strong local dialect, makes him the object of the occasional joke. Other Chinese sometimes ask him to speak to them in English so that they can better understand him! Guo Jianwei, the youngest Chinese member of the team, focuses on the evolution of ruminant artiodactyls—the large group of even-toed ungulates that includes living deer, giraffes, antelopes, goats, and cattle.

The American members of the team include both paleontologists and geologists. Mary Dawson, my colleague at the Carnegie Museum of Natural History, specializes in the early evolution of rodents, rabbits, and their kin. Her role in the discovery of the first Eocene vertebrates north of the Arctic Circle, on Ellesmere Island in the Canadian Arctic archipelago, has won her widespread acclaim. John Kappelman, an anthropologist from the University of Texas, is a leading expert on the later phases of higher primate evolution, especially the evolution of apes during the Miocene Epoch. John's role in our expedition relates to his other professional hat, that of paleomagnetic stratigrapher. Together with Wulf Gose, a geologist from the University of Texas, and Tim Ryan, his graduate student, Kappelman hopes to determine the age of the fossils we find, using the episodic reversals in the Earth's magnetic field as a guide.

Wen Chaohua, a peasant farmer from the neighboring village of Zhaili, rounds out our field crew. I first met Mr. Wen the previous year, when we hired him as a manual laborer. Slight of build but surprisingly strong, Wen rapidly earned a spot on our team because of his solid work habits, his quick smile, and his unbridled enthusiasm for finding fossils. Though Wen has only the minimal educational background typical of rural Chinese of his generation, he shows plenty of raw intelligence. Had he been fortunate enough to grow up under different circumstances, I'm sure Wen could have been successful in almost any endeavor he chose to pursue. This year, Wen looks positively professorial wearing his new eyeglasses, which correct a minor astigmatism that had bothered him last year. Like me, Wen sports a small hand lens tied around his neck, which he uses to examine small fossils up close. In recognition of his hard work, Mary Dawson gave Wen her own hand lens at the conclusion of our previous field season. Now that he has the standard tools of the trade, Wen takes even greater pride in his work. Our reward is a steady stream of fossils.

Wen's role on our field crew is simply to extract large blocks of fossil-bearing rock from the bottom of the ravine. Other members of the team then break each block down more finely in search of any fossils that might lie inside. Wen's tool of choice for this enterprise is a large steel rock pick hafted onto a stout wooden handle. This Wen wields with all of the exuberance of a forty-niner searching for a vein of gold. Invariably, Wen himself uncovers many fossils, simply because he exposes so much fresh fossil-bearing rock with each powerful swing of his pick. At first, it was hard to restrain Wen from attempting to extricate the fossils he encountered during his daily assault on the layers of limestone and mudstone. Now, with a field season of experience under his belt, Wen understands that whenever he happens across a fossil, he must stop his work and alert the rest of the team.

I find that fieldwork in almost any locale quickly settles down into a daily routine. The work itself is often repetitive, even though the scientific results can vary dramatically from day to day. Our days in the bottom of the ravine by the Yellow River consist mostly of reducing large blocks of fossil-bearing rock to smaller ones, a process that is randomly punctuated by Wen's standard victory call—"You yige ya-chuang! You yige ya-chuang!" (I've got a jaw!)—whenever he finds something he thinks is interesting. Wen himself is particularly fond of large fossils, possibly because of his culture's long-standing fascination with "dragon bones." Usually, I know that Wen's most agitated cries mean that he has stumbled across the limb bones or jaws of the hippolike animal known as *Anthracokeryx*, the most common large mammal found at this locality. But Wen appreciates that the rest of us become more excited by relatively complete specimens of smaller mammals.

Today, Wen is in particularly fine form, whacking away at the freshwater limestone with gusto. It is May 21, 1995, and Wen knows that the field season is scheduled to end within the week, so that our team can return to Beijing in time to plan the logistics of future research before the American members have to catch their return flights home. The end of the field season means big changes in all of our daily lives. Most of us will return to our academic lifestyles, writing grant proposals and technical articles, preparing lectures, and attending administrative meetings. Wen will go back to being a farmer in the village of Zhaili. Maybe it's the thought of the upcoming changes that spurs Wen onward. In any case, he seems determined to find something important today. Looking back on it now, I doubt that Wen could possibly have dreamed of making such a momentous discovery as he hoisted his pick once more.

I can still hear the distinct thump of Wen's rock pick striking that fateful blow. Immediately, Wen's excited chatter makes me drop whatever I'm working on to see what all the fuss is about. Wen shouts: "Yige xiao ya-chuang, heng piao-liang! Ni kan-kan!" (A small jaw—very beautiful. You must see it for yourself!). As soon as I see what Wen's hefty pick has revealed, my heart begins to race. A large block of freshwater limestone

Figure 2. The author and Wen Chaohua at Locality 1 in the Yuanqu Basin of central China, where Mr. Wen discovered the complete lower dentition of *Eosimias centennicus* in 1995.

has been split cleanly in two by a single blow from Wen's pick. Through sheer luck, the plane in which the block has fractured corresponds exactly with the bedding plane on which both halves of an *Eosimias* lower jaw were entombed some forty million years ago. Unlike the fragmentary jaws of *Eosimias* we had collected at Shanghuang, this specimen is virtually complete, with all of the teeth intact and well preserved. The region near the chin makes it immediately apparent that the two halves of the lower jaw of *Eosimias* are not fused as they are in modern monkeys, apes, and humans. Despite the presence of this prosimianlike condition, I can also make out the remarkably anthropoidlike front teeth of *Eosimias*. Here, in a single specimen, lies compelling evidence that *Eosimias* occupies a critical position on the evolutionary tree of primates—one intermediate between living prosimians and anthropoids. This precious fossil is exactly what we've been looking for—the pot of gold at the end of the rainbow!

Still reeling from the excitement of Wen's discovery, I realize that other members of the crew are crowding anxiously around me, waiting to learn what is so interesting. Mary Dawson approaches to peer at the block of stone in my hands. As soon as she sees the dual rows of teeth lying on the limestone slab like an exquisite string of black pearls, she exclaims, "Chris, this specimen is going to confirm what we've thought all along! *Eosimias* is a primitive little monkey after all! No one will be able to complain about the Shanghuang specimens anymore." I grin and agree wholeheartedly. Tong Yongsheng and Wang Jingwen then come closer, converse briefly with Wen in Chinese, and begin to examine the amazing specimen for themselves. After a minute or so, they too look up at me with glints in their eyes. "This fossil is very important," intones Tong seriously. "Maybe it proves that all anthropoids began in China." Wang agrees, then adds, "Chris, you are very lucky! Everywhere you go in China you find interesting primates. Maybe it's because of your nickname." My Chinese nickname, *xiao hou-ze,* means "little monkey," in recognition of my favorite fossils.

As far as my new hypothesis about anthropoid origins was concerned, Wen's pivotal discovery couldn't have come at a better time. Ever since I had introduced *Eosimias* as a critical new link in the search for anthropoid origins, both the tiny fossil and I had been at the center of a controversy of monumental proportions, and I could not afford the luxury of ignoring the academic brouhaha. My career had barely begun, yet my scientific reputation was under assault. I needed fresh and compelling evidence if my new interpretation of anthropoid origins was to gain ground, and Wen's remarkable breakthrough promised to provide exactly that. Novel ideas in paleontology depend heavily on the fossils that support them. Until now, however, my biggest challenge had been that most of the fossil record seemed to be stacked against me.

For the past several decades, all undisputed early anthropoids had been discovered in Africa, mainly at a series of sites in the Fayum region of northern Egypt being excavated by Elwyn Simons of Duke University and his students and colleagues. This African dominance of the early fossil record of anthropoids dovetailed nicely with the broad consensus that later stages of anthropoid evolution, especially the origins of apes and humans, were confined to that continent. Yet I doubted that the geographical component of primate evolution could be as simple as this "Out of Africa" theory implied. Did most, or even all, of the major evolutionary transitions in primate and human evolution occur in Africa? For me, Asia is a far more likely birthplace for the lineage we share with living apes and monkeys. Yet my views lie distinctly in the minority at present.

Despite Africa's legitimate claim as a potential birthplace for the earliest anthropoids, three lines of evidence have persuaded me to focus my efforts on Asia. These include: (1) the geographic distribution of tarsiers, the group of prosimians that seems to be the nearest evolutionary cousins of anthropoids; (2) some fragmentary fossils from Myanmar (a nation formerly known as Burma), discovered decades ago, that appear to document the presence of early—and anatomically primitive—anthropoids in Southeast Asia; and (3) results from my own earlier expeditions to China, which yielded the contentious fossils that had ignited the paleoanthropological firestorm in the first place.

The first important hint that Asia may have been the birthplace of all anthropoids comes from the geographic distribution of tarsiers, which live only on various offshore islands in Southeast Asia. By any objective standard, tarsiers are among the strangest primates that have ever lived. Tarsiers are the only primates that eat nothing but live animal prey—mainly insects, but also small vertebrates such as lizards, snakes, and even birds, which tarsiers have been reported to catch in midflight.[4] In contrast, most other primates tend to be vegetarians; yet others, like most humans, consume lots of vegetables along with their meat. Although tarsiers are not habitual bipeds like us, their own special way of moving about is at least as distinctive. The hindlimbs of tarsiers are extremely long and muscular, allowing them to leap across distances many times their own body length. Finally, tarsiers resemble many other prosimians in that they are most active at night. Yet tarsiers lack the familiar "glow-in-the-dark" structure in the back of their eyes (technically known as the tapetum lucidum) that concentrates diffuse nighttime light in the eyes of lemurs, cats, and many other mammals. To compensate for this anatomical deficiency, tarsiers have evolved the largest eyes of any living primate. Indeed, the volume of a tarsier eyeball more or less equals that of a tarsier brain![5]

Despite the generally odd biology of tarsiers, a great deal of evidence suggests that these animals are the nearest living relatives of anthropoids. For example, the noses of tarsiers resemble those of humans and other anthropoids in lacking the moist, hairless region between the nostrils, known as the rhinarium, that creates the familiar "wet nose" of dogs, lemurs, and many other mammals. Like those of anthropoids, the eye sockets of tarsiers are almost completely enclosed by bone. In contrast, lemurs have much simpler eye sockets, in which the outer margin is defined by a simple, rodlike strut of bone. Although the hindlimbs of tarsiers are highly specialized and differ from those of anthropoids, some of the individual bones (especially the talus—the ankle bone that articulates with the bones of the lower leg) closely resemble those of certain monkeys. Lemurs differ appreciably from both tarsiers and anthropoids in these respects. Both tarsiers and anthropoids lack the tapetum lucidum layer in the back of the eyeball, while lemurs still retain this ancient mammalian structure. Evidence from physiology and molecular biology likewise indicates that tarsiers and anthropoids are more closely related to one another than either group is to lemurs. For example, in contrast to lemurs and most other mammals, neither tarsiers nor anthropoids have the ability to synthesize vitamin C. Like humans, tarsiers must therefore ingest sufficient quantities of this compound to meet their daily nutritional requirements.[6] Similarly, DNA sequencing has shown that the genomes of tarsiers and anthropoids have been modified from the ancestral primate condition in exactly the same way, by having chunks of extraneous DNA included in their genomes in precisely the same locations.[7] Although some of these similarities between tarsiers and anthropoids may be spurious (caused by convergent evolution from different ancestral conditions), it seems very unlikely that all of them are. Instead, the simplest hypothesis requires us to view tarsiers and anthropoids as descendants of a common ancestor—one that possessed most, if not all, of the preceding biological traits. This common ancestry shared by tarsiers and anthropoids existed for some unknown length of time after the evolutionary schism that produced the ancestors of all other living primate lineages (lemurs, lorises, and bushbabies).

Accepting a unique common ancestry between tarsiers and anthropoids has significant implications for reconstructing the geography of anthropoid origins. By definition, the anthropoid lineage originated when ancestral tarsiers first diverged from ancestral anthropoids. Ultimately, this evolutionary divergence between tarsiers and anthropoids corresponded to a single episode of speciation. Documenting such a geologically brief event typically lies beyond the power of resolution of paleontology. However, from everything we currently know about speciation, it occurs on a local, rather than global, scale. Accordingly, the pivotal speciation event that gave rise to the tarsier and the anthropoid lineages must have occurred at a unique (if currently unknown) point in space and time. Once we conclude that tarsiers and anthropoids are each other's

nearest evolutionary cousins, we must also assume that both lineages originated in the same place (since speciation, like politics, is local). As it happens, ascertaining the birthplace of tarsiers is more straightforward than
doing the same for anthropoids.

Today, tarsiers are found only on the Indonesian islands of Sumatra,
Borneo, and Sulawesi, some of the more southerly islands of the Philip-
pine archipelago, and small satellite islands nearby. Undoubted fossil tarsiers are rare, and individual specimens are highly fragmentary, but these
too have only been found in Asia.[8] Fossils pertaining to extinct prosimians that may be closely related to tarsiers have been found in North America, Europe, and Asia (these animals will be explored more fully in chapter 3). Significantly, fossil tarsiers—or even plausible fossil relatives of
tarsiers—have never been found in Africa.[9] The narrow geographic range
of tarsiers throughout their evolutionary history therefore provides an
important guide to where tarsiers and anthropoids first diverged, with
the simplest hypothesis being that this evolutionary split took place in
Asia. If so, some of the more adventurous members of the anthropoid
lineage later migrated to Africa, where many subsequent events in anthropoid evolution apparently occurred. Eventually, anthropoids even
reached South America, although no one believes anthropoids originated
there. On the other hand, there is no evidence that tarsiers ever left their
Asian homeland. A major problem, then, for anyone who would argue
that anthropoids originated in Africa is the absence of any living or fossil tarsiers from that landmass.

Long before there was any substantial fossil record for early humans,
Charles Darwin used similar logic to conclude that Africa may have been
the ancestral homeland for our own lineage. In *The Descent of Man,* Darwin noted that:

> In each great region of the world the living mammals are closely related
> to the extinct species of the same region. It is therefore probable that
> Africa was formerly inhabited by extinct apes closely allied to the gorilla
> and chimpanzee; and as these two species are now man's nearest allies, it
> is somewhat more probable that our early progenitors lived on the African
> continent than elsewhere.[10]

Decades after the original publication of *The Descent of Man* in 1871,
discoveries of early hominid fossils in Africa convincingly upheld Darwin's prediction about the geography of human evolution.

Although Darwin's logic remains impeccable, and despite the fact that his views were subsequently vindicated, it is still something of an intellectual leap to apply Darwin's approach to an event that happened so much farther back in time. I suspect that the antiquity of anthropoid origins is almost an order of magnitude greater than the birth of the hominid lineage (about fifty-five million years ago for anthropoids, and five to seven million years ago for hominids). Relying too heavily on the geographic distribution of living tarsiers to reconstruct such an ancient chapter in our evolutionary history has obvious drawbacks. Fortunately, the fossil record, fragmentary and imperfect though it may be, provides crucial evidence that bolsters an Asian origin for the lineage we share with monkeys and apes. Critical fossils from Myanmar and China form the second and third lines of evidence favoring an Asian origin for anthropoids.

The first putative fossil anthropoids to be unearthed in Asia were discovered in Myanmar during the early part of the twentieth century. After a series of wars between the Burmese and the British during the late nineteenth century, Burma was annexed to India, then a British colony. As a result, the first significant paleontological exploration of Myanmar was conducted by British paleontologists and geologists employed by the Geological Survey of India. In 1913 a British paleontologist named G. D. P. Cotter, working in Eocene strata in the region of the Pondaung Hills in central Myanmar, found three fossilized fragments of upper and lower jaws, all of which appeared to belong to a single individual. The specimens were so incomplete and so poorly preserved that they were not made known to science until fourteen years later, when they were formally described by Cotter's supervisor at the Geological Survey of India, Guy Pilgrim.

Pilgrim's analysis of these fossils, which he named *Pondaungia cotteri* in honor of his colleague, was meticulous, cautious, and surprisingly prescient. Pilgrim acknowledged that the scrappy nature of the specimens left open the possibility that *Pondaungia* might not be a primate at all. Nevertheless, he proceeded to point out anatomical details of the preserved cheek teeth that suggested, not only that *Pondaungia* was a primate, but that it was actually the most primitive anthropoid known at the time. In his own words, Pilgrim noted that:

> If my interpretation of the structure of the teeth in *Pondaungia* is correct, and if it really is a Primate, then it must represent an earlier Anthropoid stage than *Propliopithecus* [one of the few anthropoid fossils known at that time, from the Fayum region of Egypt]. . . . It seems, however, worthy

of consideration whether *Pondaungia* does not partially fill the gap between the definitely Anthropoid *Propliopithecus* and some Lower or Middle Eocene Tarsioid.[11]

By the time Pilgrim got around to publishing his description of *Pondaungia* in 1927, a second fossil primate had already been discovered in the same vicinity, this time by the famous American paleontologist Barnum Brown, primarily known for his expeditions to western North America, where he collected numerous dinosaurs for the American Museum of Natural History in New York. Brown and his wife, Lilian, traveled to the Pondaung Hills in early 1923, with a retinue of Burmese assistants and servants. Virtually impassable roads and primitive modes of local transportation hindered the expedition's work. The threat of malaria was constant, and it eventually claimed the life of one of Brown's Burmese servants. Brown himself contracted malaria later in the expedition, which prevented him from extending his paleontological exploration farther north, into China's Yunnan Province.[12] Despite these hardships, Brown's campaign succeeded in amassing an important collection of fossil mammals, some of which proved to be more nearly complete than those collected by the earlier Geological Survey of India expeditions. The vast majority of the specimens uncovered by Brown belonged to large mammals, including extinct rhinolike forms known as brontotheres and amynodonts and primitive hippolike animals called anthracotheres. When the collection was initially unpacked and curated at the American Museum, a single, rather innocuous-looking specimen was considered insufficiently important to warrant its own entry in the museum's permanent catalogue. Fourteen years later, it would finally be recognized as the second species of fossil primate from the Pondaung Hills.

The task of studying and describing the fossils collected by Barnum Brown's expedition to Myanmar fell to Edwin H. Colbert, who was then a young assistant curator of vertebrate paleontology at the museum. Like Brown, Colbert would eventually gain scientific celebrity for his work on dinosaurs. During the 1930s, however, the trajectory of Colbert's career was dictated by Brown's field expeditions in southern Asia, which aimed primarily to find and collect fossil mammals. As Colbert began his research on the Myanmar fossils, it became apparent that most of the specimens belonged to species that had already been described and named by Pilgrim and Cotter, whose teams had gotten there first. The most important exception was a fragment of a lower jaw preserving the crowns of three teeth and part of the region near the chin. This area, known as

Figure 3. Barnum Brown (on horseback), leading the American Museum of Natural History expedition to the Pondaung region of Myanmar (formerly Burma) that recovered the holotype lower jaw of *Amphipithecus mogaungensis*. Photograph courtesy of and copyright by American Museum of Natural History Library.

the symphysis, is the site where the two separate bones of the lower jaw meet to form a joint at the midline. Colbert's rapidly growing expertise on early mammals allowed him to recognize immediately that this broken bit of jawbone pertained to an early primate.

Most living and fossil species of mammals, including primates, can be distinguished from their closest relatives on the basis of their teeth alone. This may sound trivial, but for paleontologists, the evolutionary fingerprint stamped onto the anatomy of mammalian teeth is both critical and fortuitous. Early mammals owed their evolutionary success to the complicated structure of their teeth, which allowed them to chew their food prior to swallowing it. This ability, absent in birds and reptiles, lets mammals eat a wider variety of foods more efficiently than other vertebrates can. As mammals evolved, their diets often changed, and the anatomy of their teeth and jaws responded in kind. At the same time, mammalian teeth are the hardest, most durable parts of the mammalian body. How fortunate for paleontologists that the most diagnostic elements of the mammalian skeleton are precisely those that are most likely to be preserved as fossils.

The teeth of primates, like those of most mammals, can be segregated into four different classes. From front to back in the jaw, these basic tooth types include incisors, which in humans are roughly chisel-shaped; canines, which are simple and fairly conical in structure; premolars, which dentists call bicuspids because of their two main cusps; and molars, the relatively large teeth at the back of the jaw that do most of the actual chewing. Humans normally have two incisors, one canine, two premolars, and three molars (one of which is known as a "wisdom tooth" because it is the last tooth to erupt as teenagers reach adulthood) on each side of their upper and lower jaws.

In the jaw fragments of *Pondaungia cotteri* described by Pilgrim, only upper and lower molars were preserved. But the new specimen described by Colbert had two premolars and a single molar still intact. The rest of the teeth were broken away long ago, perhaps not long after the animal died. Thus, Colbert had the luxury of being able to analyze the anatomy of the premolars and the symphysis for the first time. These new pieces of the puzzle gave Colbert more confidence than Pilgrim had, although the two men reached virtually identical conclusions about the evolutionary position occupied by these Burmese fossil primates.

Colbert formally described the second Burmese primate, which he named *Amphipithecus mogaungensis,* in 1937.[13] Citing the peculiar anatomy of the premolars and the great depth and robusticity of the jaw, Colbert concluded that *Amphipithecus* represented an anthropoid rather than a relative of lemurs or tarsiers. A surprising feature shown by the lower jaw of *Amphipithecus* was that, in life, it would have possessed three premolars on each side. (Only two of these teeth remained intact in the fossil, but the presence of the other premolar could readily be inferred from its broken root.) Among living anthropoids, only the monkeys of Central and South America possess three premolars on each side of their lower jaws. All living anthropoids of the Old World resemble humans in having only two premolars. Rather than interpret *Amphipithecus* as a relative of South American monkeys that somehow happened to live in Myanmar, Colbert concluded that *Amphipithecus* was related to living and fossil anthropoids from the Old World, especially *Propliopithecus* from the Fayum region of Egypt. Possibly, the retention of an additional premolar that was lacking in other Old World anthropoids merely signified the primitive evolutionary status of *Amphipithecus.*

Pondaungia and *Amphipithecus,* from the Eocene of Myanmar, are roughly thirty-seven to thirty-eight million years old, which makes them about three to four million years older than *Propliopithecus* and its contemporaries from the early Oligocene of Egypt.[14] This fact alone caused

Figure 4. The holotype lower jaw of *Amphipithecus mogaungensis* from the Pondaung Formation of Myanmar, collected by Barnum Brown in 1923. Photograph courtesy of and copyright by American Museum of Natural History Library.

the Burmese fossils to play a central role in debates about anthropoid origins throughout the twentieth century. Yet, from the very beginning, these Burmese primates inspired controversy. For example, although Colbert's ideas about the evolutionary position of *Amphipithecus* converged neatly on those of Pilgrim regarding *Pondaungia*, Colbert himself doubted that the two Burmese primates were closely related. He even hinted that *Pondaungia* might not be a primate at all, referring to it derisively as a "supposed primate." In retrospect, it is clear that Colbert made too much of relatively minor anatomical differences between *Amphipithecus* and

Pondaungia. Indeed, the fragmentary specimens that were known at the time shared no parts in common. In a very real sense, Colbert was comparing apples and oranges.

Incomplete fossils, like all of the specimens of *Pondaungia* and *Amphipithecus* available to Pilgrim and Colbert, are almost inherently controversial. The problem is exacerbated in the case of fossils that lie near the origin of groups, like the anthropoids, that attract lots of scientific attention. From a purely practical perspective, the only way to resolve these sorts of disputes is by finding more—and preferably more complete—fossils. As the decades passed, however, only a few additional fragments of *Pondaungia* and *Amphipithecus* were collected and described, and these specimens added little new anatomical information.[15] During the second half of the twentieth century, Myanmar became politically isolated from much of the West because of its record of military dictatorship. Political isolation hindered scientific collaboration, and efforts to advance our understanding of *Pondaungia* and *Amphipithecus* effectively ceased. Over this same interval of time, the fossil record of early anthropoids in Africa grew by leaps and bounds. By the early 1990s, the disparity was so severe that most experts believed that anthropoids must have originated in Africa, and that *Pondaungia* and *Amphipithecus* might not be anthropoids after all.[16]

I remained agnostic about the geography of anthropoid origins until I began fieldwork in China in early 1992. That project, undertaken in collaboration with colleagues from the IVPP, focused on a newly discovered series of ancient fissure-fillings near the village of Shanghuang, not far west of Shanghai. Fissures form whenever limestone rock formations are exposed to the elements, because limestone dissolves in rainwater. Over time, as water percolates through structures that originated as tiny cracks, they enlarge. Forming low points on the local terrain, these limestone fissures naturally tend to fill up with mud and any other debris, such as animal bones and carcasses, that happen to wash into them. As luck would have it, the Shanghuang fissure-fillings formed during the middle Eocene, about forty-five million years ago. The abundant fossils that our team recovered there include small, primitive primates that are roughly seven or eight million years older than *Pondaungia* and *Amphipithecus.* For the first time, these fossils placed me squarely in the center of the debate over when, where, and how the common ancestors of monkeys, apes, and humans evolved.

Certain fossils require radical adjustments to theories of how various forms of life evolved. One of the small primates we found at Shanghuang rapidly became such a pivotal fossil. Like several other "missing links" in evolutionary biology, this new primate, which we later described as *Eosimias sinensis* ("dawn monkey from China"), possessed a unique combination of primitive and advanced anatomical features.[17] Eventually, its age and anatomy would force me to disagree with decades of earlier research on anthropoid origins. In retrospect, the poor quality of the fossil record of early anthropoids at the time meant that earlier theories were ripe for being overturned. As already noted, living anthropoids differ in numerous fundamental ways from living prosimians. Prior to our discoveries at Shanghuang, however, the fossil record did little to blur the distinction. The earliest fairly complete anthropoid fossils then known, from the Fayum region of Egypt, were obviously anthropoidlike in all major respects. Although the advanced anatomy of these Egyptian fossils rendered their anthropoid status uncontroversial, this also left a gaping hole in the fossil record that could only be filled by more primitive fossils. *Eosimias* clearly met this criterion. It wasn't immediately obvious to me (and it still isn't obvious to some of my colleagues) that, in stark contrast to the Fayum anthropoids, *Eosimias* is a primitive anthropoid. It resembled neither Eocene prosimians nor other anthropoids known at the time. Before I could fully comprehend its evolutionary significance, however, I had to undertake a thorough analysis of its anatomy.

Any anatomical study of a previously unknown fossil is constrained by the quality of the material that is recovered. Like Pilgrim and Colbert before me, at first I had only fragmentary jaws and teeth of *Eosimias,* and nothing more, to go by. The best specimen we unearthed from the Shanghuang fissure-fillings was a lower jaw with three teeth intact—the last premolar and the first two molars. Crucial features, like the anatomy of the incisors, the canine, and the front part of the jaw, remained ambiguous at best. To make matters worse, *Eosimias* was considerably more primitive than either *Pondaungia* or *Amphipithecus,* making it even more difficult to evaluate. Yet despite these problems, my examination of these first fragmentary specimens convinced me that *Eosimias* qualified fully as a primitive anthropoid. My confidence derived partly from the utter lack of evidence supporting a different position for *Eosimias* on the primate evolutionary tree. The anatomical details underpinning my views are discussed in chapter 7. The important point to make here is that, for most scientists, remarkable claims require remarkable evidence. By any standard, the first fossils of *Eosimias* we found at Shanghuang were unremarkable, at least in terms of their completeness. This led many experts to doubt the anthropoid status of *Eosimias*. As a result, our fateful expedition to the little ravine near the Yellow River was launched as a conscious effort to uncover anatomically superior specimens of *Eosimias*. Thanks to Wen's landmark discovery, we succeeded beyond our wildest expectations.

In fact, the discovery of this single specimen has catapulted *Eosimias* to an elite position among Eocene primates. Although many primates have been described from the Eocene, few of them are documented by reasonably complete remains. Fewer still are known from truly superior anatomical specimens—either skulls or complete or partial skeletons. Of those rare species that are represented by such extraordinary fossils, such as *Adapis parisiensis* from France and *Notharctus tenebrosus* and *Shoshonius cooperi* from Wyoming, all are clearly fossil prosimians. They are only distantly related to the lineage that ultimately gave rise to modern monkeys, apes, and humans.

Wen's specimen reveals that *Eosimias* differs dramatically from these Eocene prosimians. Like living anthropoids, *Eosimias* has deep, powerfully constructed lower jaws. Its front teeth or incisors resemble those of living anthropoids in both their vertical orientation and small size. Living and fossil prosimians almost always have jaws that are more gracile, especially up front near the symphysis. As a result, their incisors tend to protrude forward, rather than being erect like ours. Immediately behind the incisors, the large, daggerlike canine of *Eosimias* also looks distinctly like that of an anthropoid. The premolars of *Eosimias* are very primitive, but again they resemble those of other early anthropoids, including *Amphipithecus* from Myanmar, in being oriented obliquely in the jaw. In Eocene prosimians, the long axis of each premolar is oriented front to back. The molars of *Eosimias* differ from those of all other primates. They are primitive in the sense that an extra cusp called the paraconid is still present. This cusp was suppressed later in the evolutionary history of anthropoids. In other details of their anatomy, however, even the molars of *Eosimias* show anthropoid features. As in other early anthropoids, the rear part (or talonid) of the last molar is highly abbreviated in *Eosimias*. This region is often greatly enlarged in Eocene prosimians.

Although it's too early to speculate about what *Eosimias* might have looked like in the flesh, a few important details are already clear. For example, we have a good idea of how big *Eosimias* was, because the size of the lower molars correlates closely with body size in living primates. *Eosimias sinensis* from Shanghuang probably weighed about three and a half ounces (100 grams). Wen's *Eosimias,* which appears to document a new species, would have weighed slightly more (about four and a half ounces, or 130 grams). The smallest living monkeys, the pygmy marmosets of South America *(Cebuella pygmaea)*, overlap *Eosimias* in body size, but most living anthropoids are substantially larger, typically by an order of magnitude or more. Indeed, even most tarsiers would tip the scales at a heavier weight than *Eosimias*. Small body size alone would have forced *Eosimias* to consume a diet rich in calories. *Eosimias* therefore probably ate a variety of insects, small vertebrates, and fruits. The relatively foreshortened lower jaw of *Eosimias* indicates that its muzzle must also have been abbreviated, like that of most monkeys. All modern primates the size of *Eosimias* live in trees, not on the ground. It therefore seems likely that *Eosimias* was a denizen of the forest as well. Beyond this, it is premature to predict much about the biology of *Eosimias*. Its intermediate evolutionary position between modern prosimians and anthropoids means that *Eosimias* may have been either prosimianlike or anthropoidlike in most of its biological attributes. Such a transitional spot on the evolutionary tree hinders attempts to reconstruct the habits and appearance of *Eosimias,* at least until more complete specimens are found. Yet at the same time, this makes *Eosimias* crucial in the search for anthropoid origins.

Exceptional fossils serve as critical guideposts for deciphering evolutionary history. Fossils often demonstrate that real animals once possessed combinations of features that are never found together in their living relatives. The famous "feathered dinosaurs" from northeastern China provide a classic example of this phenomenon, because they show that animals with skeletons that are undeniably dinosaurian in overall form were also covered with an external coat of feathers like that of modern birds.[18] Such genuine chimeras from deep time can be pivotal when it comes to reconstructing the family tree of a group of organisms. In the example given above, new and spectacular specimens have dramatically illuminated the family tree encompassing birds and theropod dinosaurs. Exceptional fossils can also show the sequence in which certain anatomical features, and their associated functions, evolved. Again, in the case of the feathered dinosaurs, it now seems clear that feathers evolved long before other features that are characteristic of modern birds, like their toothless, horny beak. The relatively primitive forelimbs and breasts of the feathered dinosaurs demonstrate that these animals could not fly. Feathers must therefore have originally evolved to serve some other function, like courtship display or the conservation of body heat. At the same time, exceptional fossils testify that such transitional animals lived in a specific place at a certain time. This information can be crucial in determining when and where major lineages first evolved.

By any of these criteria, *Eosimias* qualifies as an exceptional fossil. For me, *Eosimias* functions as a Rosetta Stone for reconstructing the ancestry of monkeys, apes, and humans, in much the same way that feathered dinosaurs have fundamentally resolved the origin of birds. But not all scientists agree that *Eosimias* is so critical for understanding anthropoid origins. Indeed, not all scientists agree on the importance of feathered dinosaurs for reconstructing the origin of birds. Consensus rarely emerges along the cutting edge of any scientific issue. Yet the following two points seem beyond dispute. First, *Eosimias* is far more primitive than any other fossil thought to be related to the origin of anthropoids. It is so primitive, in fact, that some experts continue to deny that *Eosimias* has any relevance for solving the mystery of anthropoid origins. Second, *Eosimias* is millions of years older than any other fairly complete fossil thought to belong to the anthropoid lineage. It is so old, in fact, that its age alone conflicts with widely accepted theories about when the anthropoid lineage was born. At the core of these disagreements regarding *Eosimias* lie two very different paradigms for reconstructing the evolutionary history of primates.

I refer to these two evolutionary paradigms as the ladder and the tree. The older ladder paradigm has largely withstood the test of time, a major criterion bolstering the scientific impact of any theory or model. In order to convey the underlying philosophy, methods, and goals of these competing evolutionary paradigms, let's make an analogy between the large-scale evolution of life on Earth (known as phylogeny) and the much smaller-scale family trees that are more familiar to most of us (known as genealogy). The ladder paradigm attempts to establish the phylogenetic line of descent from a remote ancestor to whatever descendant species is of interest. In genealogy, a similar goal would be to chart your direct ancestors (great-great-grandparents and such), with little regard for determining your distant aunts, uncles, and cousins.

Within the field of paleoanthropology, the ladder paradigm owes much to the influence of Sir Wilfrid E. Le Gros Clark, a British anatomist and primatologist whose publications dominated the study of primate evolution for much of the mid twentieth century. Although one might easily oversimplify the complex views of such an important scientific figure, it is fair to say that Le Gros Clark perceived the entire span of primate and human evolution as a steady progression from primitive to advanced. In Le Gros Clark's view, the original gamble made by the earliest primates—to invade the trees and take on a highly arboreal lifestyle— led almost inexorably to a series of evolutionary trends that reached its climax with the advent of *Homo sapiens*. Le Gros Clark summed it all up rather nicely in his seminal book, *The Antecedents of Man:*

> Among the Primates of today, the series tree shrew–lemur–tarsier– monkey–ape–man suggests progressive levels of organization in an actual evolutionary sequence. And that such a sequence did occur is demonstrated by the fossil series beginning with the early plesiadapids [so-called "archaic primates" from the Paleocene] and extending through the Palaeocene and Eocene prosimians, and through the cercopithecoid [Old World monkeys] and pongid [apes] Primates of the Oligocene, Mio- cene, and Pliocene, to the hominids of the Pleistocene. Thus the founda- tions of evolutionary development which finally culminated in our own species, *Homo sapiens*, were laid when the first little tree shrew–like crea- tures advanced beyond the level of the lowly insectivores which lived dur- ing the Cretaceous period and embarked on an arboreal career without the restrictions and limitations imposed by . . . a terrestrial mode of life.[19]

According to Le Gros Clark's ladder paradigm of primate evolution, the origin of anthropoids was simply one of several important steps along the path from tree shrew to human. This particular step corresponds to a significant evolutionary transition, from more primitive prosimians to more advanced anthropoids, marked by such novel anatomical features as a bigger brain, more forward-facing eyes enclosed in bony eye sock- ets, and a reduction of the snout. Needless to say, because anthropoids evolved from prosimians, they must have originated later in time.

Later students of the primate fossil record eventually abandoned Le Gros Clark's concept that the evolution of this group entailed a steady progression toward humans. But Le Gros Clark's ladder continues to influence studies of primate evolution to this day. In terms of interpret- ing the primate fossil record, the ladder paradigm sustains modern at- tempts to link undoubted anthropoids with earlier fossil prosimians in a simple ancestor-descendant fashion.[20] Given this mind-set, the earliest anthropoids must have evolved from a group of anatomically advanced prosimians. Because most of these advanced prosimians lived toward the end of the Eocene, the idea that anthropoids originated relatively recently, near the Eocene-Oligocene boundary (about thirty-four million years ago), follows logically from Le Gros Clark's ladder. Indeed, this notion of a relatively recent origin for anthropoids is intimately related to the ladder's expectation that a sequence of fossils traversing the "prosimian- anthropoid boundary" will ultimately be uncovered.[21]

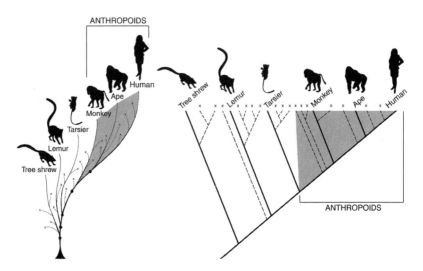

Figure 5. Divergent evolutionary paradigms lead to very different notions of how anthropoids (denoted by stippling) should be defined. According to the ladder paradigm (shown on the left), the earliest anthropoids not only follow prosimians in the fossil record but also differ from them in possessing most of the diagnostic features found in living anthropoids. In contrast, the tree paradigm (shown on the right) posits that the anthropoid lineage originated whenever the lineage leading to living tarsiers bifurcated away from it. In this case, the earliest anthropoids may have been quite ancient, and they may have lacked many, if not most, of the diagnostic features that characterize living anthropoids. Original art by Mark Klingler, copyright Carnegie Museum of Natural History.

The alternative tree paradigm flows from the work of the German entomologist Willi Hennig, whose methodology for reconstructing evolutionary relationships is known as cladistics. Returning once again to the analogy between phylogeny and genealogy, Hennig's approach makes no attempt to identify direct ancestors. Instead, the tree paradigm seeks to determine which species are closer evolutionary cousins. Identifying these closely related species hinges on documenting their shared biological features, especially those features that have arisen relatively recently in evolutionary history. Assuming that all of life on Earth ultimately derives from a single ancestral source, all species must eventually converge at some level on the tree of life. The goal of cladistics is to identify which limbs of this tree sprout nearest one another from a larger, common trunk.

The ladder and tree paradigms differ in several fundamental ways. The tree paradigm views the product of evolution as a constantly branching sequence of lineages, while the ladder paradigm envisions a simpler, ladderlike progression from primitive to advanced. Thus, the tree paradigm recognizes that the ancient bifurcation between tree shrews and humans established two independent lineages, each of which subsequently experienced its own unique evolutionary history. There is no reason to presume that tree shrews have been frozen in time since they split away from the human lineage, nor is it necessary to postulate that humans underwent a "tree shrew stage" at some early phase in their evolutionary history.

Paleontologists who follow the tree and ladder paradigms often interpret fossils in very different ways. Both sides agree that the quality of the fossil record varies dramatically across space and time. In a few special places, like the Bighorn Basin of Wyoming or the White River Badlands of South Dakota, several million years of evolutionary history are reasonably documented by abundant fossils. These rich sequences of fossil-bearing strata provide a great deal of information about the kinds of animals that inhabited these particular regions during a finite interval of time. Taking the exceptional fossil records from these areas as a kind of gold standard, it is clear that, even in such best-case scenarios, certain animals are well represented as fossils while others are not. In the latter case, there may be major gaps in our knowledge of their anatomy and evolutionary significance. Even if we disregard any distinction between well-known and poorly documented fossils, we must admit that both classes of fossils combined document only a tiny fraction of the Earth's ancient biological diversity. Once we acknowledge these inherent limitations of the fossil record, the slim chance that any fossil is the direct ancestor of another (or of a living species) becomes immediately apparent.[22] Accordingly, the tree paradigm treats fossil species in much the same way that it deals with living ones. They are assumed to be evolutionary cousins, not direct ancestors. The ladder paradigm, on the other hand, is fixated on the issue of direct ancestry. As such, followers of the ladder paradigm are far more likely to propose that a given fossil is directly ancestral to, or near the ancestry of, some later group of organisms.

How do these different evolutionary paradigms bear on the search for anthropoid origins? I believe the paradigms have exerted an enormous influence, because they have affected the way in which different scientists frame the entire debate. Under the tree paradigm, the anthropoid lineage was established, by definition, when the tarsier lineage bifurcated from that leading to anthropoids. It is at least conceivable under the tree paradigm that the origin of anthropoids was very ancient (corresponding to whenever the tarsier and anthropoid lineages split) and that the earliest members of the anthropoid lineage were extremely primitive in their anatomy and other biological attributes. In contrast, the ladder paradigm equates the origin of anthropoids with achieving an important evolutionary stage. In this case, anthropoids should differ from prosimians by their acquisition of some key set of anatomical features, or by having crossed a biological threshold separating primitive prosimians from more advanced anthropoids. Hence, the ladder paradigm predicts that the origin of anthropoids was relatively recent (at least compared to their prosimian ancestors) and that the earliest anthropoids must have been anatomically quite advanced.

Given its great antiquity and primitive anatomy, I suspect that *Eosimias* is the key to resolving this dispute about which evolutionary paradigm—the ladder or the tree—best illuminates the deep recesses of our distant past. Looking up from the rock that contains Wen's small treasure, the oblique rays of light now striking the walls of the ravine remind me once again of the massive timescales that are at play here. The limestone block in my hand containing the world's oldest fairly complete fossil anthropoid dates to the latter part of the middle Eocene, some forty thousand millennia before our time. The loess walls of the ravine, ancient themselves by human standards, began to be deposited some two thousand millennia ago. The evolutionary history of the human lineage corresponds, in large measure, to the loess. The origins of the anthropoid lineage are at least as ancient as the limestone. The stratigraphic unconformity separating the limestone from the overlying loess equals roughly thirty-eight thousand millennia.

Two conclusions emerge from the mathematics of the ravine's stratigraphic column. First, the common evolutionary history that we share with other anthropoids far outweighs the unique evolutionary history that is ours alone. Second, if we want to get a better picture of how *Eosimias* fits into the grander scheme of primate and human evolution, we've got to venture well beyond this ravine. Our itinerary begins on the other side of the vast Eurasian landmass, where the first chapter in this saga was written some two hundred years ago.

2
Toward Egypt's Sacred Bull

n the northern suburbs of Paris lies an artistic district known as Montmartre ("mount of the martyrs"). The area takes its name from events that transpired in the third century A.D., when a small cadre of Christian missionaries was dispatched to the Gallo-Roman city of Lutece, as Paris was known at the time. The prominence of Roman Catholicism throughout subsequent French history testifies to the effectiveness of these early evangelists in converting the local population. Yet, as might have been expected, the group's missionary zeal also contributed to their own downfall. Local pagan priests conspired with the ruling Roman authorities to have the Christian missionaries arrested and tortured. Eventually, some were executed on a hillside outside of town. Legend has it that Saint Denis, who was among those beheaded on the slopes of Montmartre, rose from the dead, picked up his decapitated head, and climbed to its summit, aided by an angel.

By the latter part of the eighteenth century, the hill that Saint Denis is said to have ascended so miraculously had been dramatically transformed. Paris had expanded far beyond its roots as a Gallo-Roman frontier town, but urban sprawl was only indirectly responsible for the changes around Montmartre. The city itself still lay to the south. Rather, the rapid expansion of Paris fueled a growing demand for building materials, especially the plaster that still bears the city's name. Gypsum is the most important ingredient in the production of plaster of Paris, a substance prized for its utility in finishing interior ceilings and walls. The rock strata that outcrop in the vicinity of Montmartre contain vast quantities of gypsum suitable for industrial exploitation. As a result, where wooded slopes had once witnessed the martyrdom of Saint Denis, commercial rock quarries now gouged the sides of Montmartre.

Occasionally, the laborers who toiled in the Montmartre gypsum quarries uncovered the fossilized remains of animals embedded directly in the rock. While the fossil bones themselves had no apparent commercial value, they did attract the attention of a young, brilliant, and highly ambitious scientist by the name of Georges Cuvier (1769–1832). Cuvier's meticulous research on the fossil bones unearthed near Montmartre would eventually launch paleontology as a respectable scientific discipline. Fittingly, efforts to decipher our own deep evolutionary history can also be traced directly to Cuvier, although Cuvier himself would never have admitted such a role. Indeed, Cuvier's part in our story highlights two of his most egregious scientific mistakes. The first of these was his decision to stand out as one of the leading antievolutionists of his day. The second was his failure to recognize one of the many fossils he described from Montmartre—a badly crushed skull and lower jaws for which Cuvier coined the name *Adapis*—for the lemurlike primate that it is. Yet, as we shall see, Cuvier's errors must be judged in historical context. To imagine the intellectual setting of the time, you must first forget everything you've ever heard about Charles Darwin and his scientific theories.

I find it counterintuitive that the science of paleontology antedates Darwin's theory of evolution by means of natural selection. Yet upon further reflection, what first seems like a quirk of history actually makes sense. The logical connections between paleontology and biological evolution that are so obvious today were much less apparent then. In fact, when Cuvier initially founded paleontology, his primary goal was to illuminate the history of Earth itself, not that of its inhabitants. Like any good scientist, Cuvier believed that all comprehensive theories needed to be grounded in direct, empirical observations. Some of the more outlandish theories in vogue during Cuvier's lifetime attempted to explain Earth history, often in terms that closely paralleled the biblical account in the Book of Genesis. Cuvier scorned such grandiose, yet weakly supported arm-waving. Instead, Cuvier thought that fossils, which he correctly interpreted as relics from Earth's antiquity, might provide precisely

Figure 6. Georges Cuvier, the father of paleontology and the scientist who described *Adapis parisiensis*, the first Eocene primate ever unearthed.

the sort of factual basis that all-encompassing geological theories required. In much the same way that archeological artifacts and the ruins of ancient cities illuminate prehistoric human civilizations, Cuvier thought that fossils might cast light on Earth's deepest antiquity. In reaching this conclusion, Cuvier relied on his considerable skills as a comparative anatomist to demonstrate that the vast majority of fossils known to him belonged to animals that were extinct. At the end of the eighteenth century, this was a radical idea.

Although the notion seems absurd today, most of Cuvier's contemporaries believed that extinction was a logical impossibility. After all, extinction would imply that God's creation was somehow imperfect. Even such an erudite product of the Enlightenment as Thomas Jefferson, who was hardly a lackey of any religious orthodoxy, doubted that fossils could pertain to animals that no longer roamed the Earth. Jefferson emphasized this widespread view with respect to fossils of *Mammut americanum* (the American mastodon), an animal that he mistakenly referred to as a "mammoth":

> It may be asked, why I insert the mammoth [in a list of animals common to Europe and North America], as if it still existed? I ask in return, why I should omit it, as if it did not exist? Such is the economy of nature, that no instance can be produced, of her having permitted any one race of her animals to become extinct; of her having formed any link in her great work so weak as to be broken. To add to this, the traditionary testimony of the Indians, that this animal still exists in the northern and western parts of America, would be adding the light of a taper to that of the meridian sun.[1]

Cuvier's own anatomical studies of fossil elephants, including both Siberian mammoths (in this case, "true" mammoths) and American mastodons, showed that these fossils differed in many ways from living African and Indian elephants.[2] Cuvier therefore concluded that the fossils pertained to extinct species. While this realization was controversial in its own right, Cuvier proceeded to draw inferences that went even farther. One of these related to the antiquity of humans. Cuvier knew of no instance in which fossil humans had been unearthed. The first human fossils—now known as Neanderthals—would not be discovered until several decades later, in 1856. Faced with this evident disparity, Cuvier decided that fossils such as those of the mammoth and mastodon dated to an earlier interval of life on Earth, before humans first appeared on the scene.

If mammoths and mastodons belonged to an ancient, prehuman world, Cuvier's growing knowledge of stratigraphy forced him to regard the fossils that were being recovered from the gypsum beds near Montmartre as relics of an era that was older still, one that antedated even the mammoth and the mastodon. The evidence supporting this view seemed straightforward to Cuvier, as the following excerpt makes clear.

> Some authors . . . have thought that the fossil bones of quadrupeds are always found in loose deposits, the most recent of all those that envelop the core of the earth. This is not generally so. Often they are embedded in true stone . . . in the natural beds of these rocks, and sometimes of very ancient rocks. In this way those around Paris are in the middle of enormous beds of plaster, covered in turn by beds of oysters and other marine shells. I even believe I have noticed a fact still more important . . . namely that the older the beds in which these bones are found, the more they differ from those of animals that we know today.[3]

The modern subdiscipline of paleontology known as biostratigraphy is founded upon Cuvier's observation that certain assemblages of fossils characterize specific rock strata. Biostratigraphy allows geologists to estimate the age of rock formations based on the fossils they contain, in the same way that a photograph of all the members of a college football team enables any fan to decipher the year the photo was taken. With a bit of detective work, virtually any team photo can be accurately dated because the team's composition is constantly changing. College football players graduate, turn pro, or eventually lose their eligibility, while new players are recruited each year to replace them. What mechanism might cause a similar pattern of episodic change among the ancient forms of life that once inhabited Earth? Cuvier's expertise as an anatomist led him to view all animals as being highly adapted to their different ways of life. Accordingly, Cuvier felt that only a major environmental perturbation or catastrophe could cause the extinction of a species. Because different types of fossils occur in successive rock formations, each corresponding to a different interval of time, Cuvier came to regard the history of life on Earth as consisting of numerous separate phases. Each of these sequential phases of prehistoric life was obliterated, in turn, by some type of catastrophe.

The delicate relationship between an animal's form and its mode of existence had a second important implication for Cuvier. Evolutionary change through time in an animal's anatomy would likely upset the complex balance between form and function, resulting in extinction. Moreover, because Cuvier believed that all parts of an animal worked together seamlessly as an integrated whole, minor evolutionary changes in any single part of an organism would likely cascade throughout the animal, again with decidedly unhappy consequences. For all of these reasons, Cuvier strongly opposed the pre-Darwinian evolutionary theories that floated among scientific circles of the time. The most prominent of these was expounded by Cuvier's colleague at the Muséum National d'Histoire Naturelle in Paris, Jean-Baptiste de Lamarck.

Lamarck's personal views on evolution emphasized the inheritance of features acquired during an organism's lifetime. A classic example of Lamarckian evolution depicts generations of giraffes developing progressively longer necks because of each generation's constant straining to reach higher and higher branches. We now know that, no matter how much effort a giraffe might expend stretching its neck during its lifetime, the fruits of that labor are not bestowed directly upon its offspring. Likewise, a human bodybuilder might pass along the genetic basis for a muscular physique to his or her offspring, but no one is born with the body of Charles Atlas. Darwin and Mendel showed that evolution results from the natural selection of certain genetically inherited traits over others, not from the simple accumulation of changes that take place during an animal's lifetime. In hindsight, Cuvier correctly rejected Lamarck's version of evolution, but his stubborn resistance to the entire concept stands as one of the great scientific miscalculations of all time.

As a result, Cuvier's conception of the historical pattern of life on Earth, although a significant advance over earlier ideas, never came full circle. Cuvier's most important scientific contribution was the demonstration that extinction was a real phenomenon—one that had evidently impacted ancient life repeatedly. However, Cuvier himself never explored the rather obvious need for a mechanism of generating new species to compensate for the losses caused by the catastrophes he postulated. Apparently, he regarded the entire issue of how new species originate as beyond the realm of science.[4]

While Cuvier's unique scientific views prevented him from adopting or improving upon the evolutionary theories of Lamarck, they also provided the basis for his most noteworthy intellectual triumphs. Perhaps the most famous of these resulted from Cuvier's analysis of a small and rather innocuous-looking mammal skeleton that was exhumed from the Montmartre gypsum beds. Based on his examination of its teeth, Cuvier readily identified the fossil skeleton as that of a marsupial. This alone was a startling verdict, given that marsupials do not occur naturally in Europe today, and fossil marsupials had never before been found there. Yet Cuvier was unwilling to settle for this single conclusion, because he realized that the specimen at hand could be used to showcase the predictive power of paleontology.

Exhibiting a natural flair for the dramatic, Cuvier confidently predicted that further preparation of the skeleton from the rock in which it was entombed would reveal the presence of "marsupial bones." These small but distinctive bones, located in front of the pelvic region, support the pouch in which newborn marsupials continue their development outside the womb. Of all living mammals, only marsupials shelter their offspring in this manner. And, as the name suggests, marsupial bones are only found in marsupials. Imagine how Cuvier's reputation as a scientist was enhanced as he publicly prepared the fossil skeleton from the rock, eventually revealing marsupial bones in exactly the spot where he had predicted they would occur![5] Not only did Cuvier verify his identification of the skeleton as that of a marsupial; he also demonstrated once and for all that his anatomical methods really worked. These techniques allowed Cuvier to predict the nature of unknown anatomical features in much the same way that chemists and physicists forecast the outcomes of their experiments. Cuvier's self-proclaimed ability to identify extinct animals on the basis of very limited fossil material, and even to reconstruct their overall appearance, had endured a very public and substantial test.

Despite the undeniable success of Cuvier's methods of identifying fossils, his procedures were hardly infallible. Then, as now, embarrassing misidentifications or misinterpretations of fragmentary fossils remained a distinct possibility. One of the more glaring mistakes made by Cuvier concerns a second fossil from the same gypsum beds that yielded his famous opossum skeleton. This specimen, a crushed and incomplete skull and lower jaws, would eventually be recognized as the first Eocene primate ever discovered. Sadly, Cuvier himself never realized its significance. The animal was formally described in 1822, in the second edition of Cuvier's magnum opus, *Recherches sur les ossemens fossiles (Researches on Fossil Bones)*. Cuvier gave his mysterious new fossil the name *Adapis*, an allusion to Apis, the sacred bull god of ancient Egypt (the Latin prefix *ad-* means "toward"). Cuvier considered *Adapis* to be a peculiar member of a group of mammals he called "Pachydermes." We now know that Cuvier's conception of the Pachydermes, with or without *Adapis*, is a hodge-podge of mammals lacking any special evolutionary affinity. The group included elephants, tapirs, rhinoceroses, hippopotamuses, pigs, hyraxes, and horses. Cuvier's addition of the primate *Adapis* to this motley assemblage shows that he really had no idea what type of animal *Adapis* was. But it would be unfair to criticize Cuvier too severely for this faux pas. After all, exhuming the remains of a lemurlike primate in the suburbs of Paris would have been at least as unexpected in Cuvier's day as his discovery of a fossil opossum there. In Cuvier's defense, such diagnostic features as the postorbital bar, which forms the outer margin of the eye socket in all primates, were not preserved in the specimen of *Adapis* available to him. The true identity of *Adapis* would only be revealed when better fossils were unearthed—and these would not be found for another fifty years.

Far to the south of the Parisian gypsum beds that yielded Cuvier's original material of *Adapis* lie the limestone plateaus of the Quercy region of south-central France. Anyone who visits this area today is struck by its warm, dry Mediterranean climate, a dramatic contrast to the cooler and more humid conditions that prevail near the French capital. These modern climatic differences must have been muted during the greenhouse conditions of the Eocene. Certainly the ancient environments in both northern and southern France suited *Adapis* and other Eocene primates, for their fossils occur in reasonable abundance in both regions. Despite Cuvier's precocious discoveries on the outskirts of Paris, most of what we now know about *Adapis* comes from the limestone-dominated terrain of southern France.

As is the case at Shanghuang in eastern China, the limestone plateaus of the Quercy region are riddled with fissures. Many of them contain fossils. These French fissure-fillings differ from their Chinese counterparts in being rich in calcium phosphate, which is economically valuable as a type of fertilizer. Soon after their discovery in 1865, commercial exploitation of the phosphatic fissure-fillings commenced on a massive industrial scale. A few years later, fossil vertebrates began to be recovered from some of the fissure-fillings, as a by-product of the search for fertilizer. In due course, the miners who labored in the Quercy phosphate pits would unearth a veritable treasure trove of fossils, including the world's first complete skulls of Eocene primates. The exquisite preservation of many of these fossils revealed even the most intricate details of their anatomy. Careful comparisons between the nearly pristine Quercy skulls and the crushed and distorted specimen of *Adapis* described five decades earlier by Cuvier demonstrated once and for all that Cuvier's small, peculiar "pachyderm" was actually a lemurlike primate.[6]

Ultimately, several types of lemurlike primates were exhumed from the Quercy fissure-fillings. Related species would later be found as far away as Wyoming, Egypt, and China. All of these species resemble *Adapis* in important aspects of their anatomy, and together they form a major branch of the primate evolutionary tree. Because *Adapis* was the first member of this diverse group to be discovered, the entire assemblage eventually took its name from the fossil that Cuvier christened in homage to the bull god of ancient Egypt. Today, we refer to this important group of extinct primates as adapiforms.

In many ways, adapiforms stand out as the most primitive fossil primates uncovered to date. The anatomy of their teeth and jaws, in particular, is more generalized than that of any other primate group. For example, the earliest adapiforms retain four premolars on each side of their upper and lower jaws, whereas most other primates have no more than three. Furthermore, the two front premolars in the lower jaws of the earliest adapiforms (technically referred to as P_1 and P_2, respectively) remain decidedly primitive in their size and the configuration of their roots. Even among the few other primates that retain all four premolars, this is not the case. The P_1 in early adapiforms is relatively large, whereas this tooth

Figure 7. Abundant and well-preserved fossils of *Adapis* and other primates were unearthed as a by-product of the commercial exploitation of phosphatic fissure-fillings in the Quercy region of southern France in the nineteenth century. Photograph courtesy of and copyright by Phosphatières du Quercy, provided by Mr. Thierry Pélissié.

is invariably reduced or absent in other Eocene primates. Similarly, P_2 retains two distinct roots in most adapiforms, while this tooth possesses a single root in other major primate groups. Both of these features suggest that basal members of the adapiform family tree differed from other Eocene primates in having relatively longer muzzles. As other primates reduced the length of their snouts to achieve more monkeylike or tarsierlike faces, their lower dentitions were forced to become more compacted from front to back. This compaction was accomplished by reducing and subsequently losing P_1 and by compressing the dual roots of P_2 to form a single root.

The archaic nature of most adapiforms has generated a great deal of controversy about their precise role in primate evolution. Anatomically primitive groups like adapiforms make attractive potential ancestors for their more advanced relatives. In fact, two very different groups of living primates, lemurs and lorises on the one hand and anthropoids on the other, have been cited as possible descendants of adapiforms. Living lemurs and lorises form the most basal branch of the primate evolutionary tree alive today. Accordingly, lemurs and lorises lie far from anthropoids in an evolutionary sense. The possibility that both the lemur/loris and anthropoid lineages evolved from adapiforms therefore seems remote. Nevertheless, an impressive variety of anatomical evidence has been marshaled in support of a possible evolutionary relationship between adapiforms and anthropoids. This subject is fully reviewed in chapter 5. For now, it is sufficient to note that most of the features that have been used to support an evolutionary link between adapiforms and anthropoids are simply primitive. These features were present in the common ancestors of all primates. Such traits do not provide compelling evidence for an exclusive grouping between adapiforms and anthropoids alone.

In contrast, the possibility of a close evolutionary connection between adapiforms and living lemurs and lorises appears to be genuine. Despite their generally archaic anatomy, the overall body plan of most adapiforms strongly resembles that of living lemurs. At least some of these detailed similarities appear not to have been present in all early primates. For example, the upper ankle joints of adapiforms and living lemurs match each other precisely. Both groups bear grooves for muscle tendons in exactly the same position on the back of the upper ankle joint. Likewise, the joint between the fibula (the smaller of the two bones of the lower leg, below the knee) and the talus (the uppermost ankle bone) has the same unusual shape in both lemurs and adapiforms. In both cases, all other primates differ from adapiforms and lemurs in these features.[7] These unusual traits suggest that adapiforms and lemurs share a unique common ancestry that excludes other primates. In this sense, adapiforms can be envisioned as extremely primitive lemurlike primates. Living lemurs and lorises evolved either from an adapiform or an unknown primate that would have been adapiformlike in most respects.

Adapiforms enjoyed their greatest evolutionary success during the Eocene (from fifty-five to thirty-four million years ago), when they lived on most of the world's continents (all except South America, Antarctica, and Australia). Europe served as a particularly important arena of adapiform evolution and diversification. Approximately fifty species of adapiforms have been described from that continent alone.[8] North America also hosted a great variety of Eocene adapiforms, more than twenty species having been described to date. So far, fewer adapiforms have been found in Africa and Asia, although this may reflect poor sampling of the fossil record on those continents rather than any actual biological pattern.

The nearly global diversity of adapiforms segregates roughly into four major groups—Adapidae, Notharctidae, Sivaladapidae, and Cercamoniinae. Three of these groups—those whose names end with the suffix –idae—are usually distinguished at the family level in taxonomy, the science of naming and systematically arranging the diversity of life. To understand the significance of this taxonomic distinction, recall that all living and fossil humans are ranked in the single family Hominidae. Indeed, many paleoanthropologists insist that living great apes (and all of their fossil relatives) must be included alongside humans in this family.[9] The general consensus that adapiforms belong in at least three families therefore reflects the large variation in anatomy and ecology across this extinct group of primates. In order to come to grips with their sheer diversity and to assess whether any of these animals might be related to anthropoids, we need to look at each of the four adapiform groups in turn. First, let's consider the Adapidae, the family of adapiforms that includes *Adapis* and its nearest relatives.

Like many of their closest relatives, adapids retain such primitive traits as having four premolars on either side of their upper and lower jaws. Yet it would be wrong to view these animals as typical Eocene primates. All undoubted adapids share diagnostic features that other adapiforms lack. Chief among these are their sharply crested cheek teeth, which were specialized for eating leaves, a dietary adaptation known as folivory. To the extent that their postcranial skeletons are known, the limbs of adapids were built for slow and powerful climbing and grasping rather than leaping. These distinctive adaptations show that adapids form a cohesive branch of the primate family tree, one that shares a single common trunk. This adapid branch of the family tree also happens to be moderately bushy. Roughly sixteen species have been discovered and described to date.[10] By far, the most thoroughly documented of these is *Adapis parisiensis*.

Today, *Adapis* ranks among the best-known of all fossil primates. Numerous complete skulls of *Adapis* have been found, and most of its major limb bones are also represented by multiple specimens. This unusual abundance of anatomical information paints a vivid portrait of what *Adapis* must have been like in life. Nonetheless, as is often the case in paleontology, lingering disagreements remain about certain aspects of its paleobiology.

In life, *Adapis* weighed about two and a half pounds (1.1 kilograms), making it roughly the size of a large loris or a small lemur. Although it resembled lemurs in many anatomical details, the differences would be immediately apparent were it possible to compare them side by side in a zoo. Among the more obvious differences between *Adapis* and modern lemurs (especially members of the living families Lemuridae and Indriidae) would be their overall body proportions. Most lemurs are lanky, agile animals. Relative to their torsos, lemurs have long, gangly limbs. Their hindlimbs are particularly well developed, being substantially longer than their forelimbs. As a result of these body proportions, when lemurs walk on all fours they look like "rear-wheel drive" animals, with their rear-ends higher up than their shoulders. Most lemurs spend relatively little time on the ground, however. Like the vast majority of other primates, they strongly prefer to move and feed in trees. There, their seemingly skewed limb proportions aid them immensely during their acrobatic leaps from tree to tree. Lemurs use their powerful and elongated hindlimbs to propel themselves forward at the beginning of each leap, which explains why their limb proportions are so unusual compared with more typical quadrupedal mammals.

In contrast to most lemurs, *Adapis* had relatively short, stubby limbs, and its forelimbs and hindlimbs were similar in length. The limb proportions of *Adapis* more closely approximate those of a small monkey than a lemur. These proportions are not what one would expect in a primate that leaps a lot. Other key aspects of hindlimb anatomy in *Adapis* support the view that it was not a proficient leaper. For example, *Adapis* lacked the specialized knee joint typical of primates that frequently leap. The extremely short ankle region of *Adapis* also contrasts with the condition in many leaping prosimians. In *Adapis* the entire ankle is reduced in length, mainly because of the abbreviated nature of the calcaneus— the bone that lies in the heel of the foot. In small prosimian primates that are accomplished leapers, such as tarsiers and bushbabies, the calcaneus (and thus the ankle as a whole) is greatly elongated. These acrobatic animals use their elongated feet to spring from a crouched resting posture into a full-fledged leap. *Adapis,* having none of these features, must have leapt seldom, if at all. Instead, its powerful forelimbs and very mobile ankle joints suggest that *Adapis* climbed in a slow, deliberate fashion as it moved through the trees.[11] Among living primates, lorises, which are renowned for their slow, powerful style of climbing, may provide the best available analogues for how *Adapis* moved.

However, these sorts of comparisons between living prosimians and

Eocene primates can only be pushed so far. *Adapis* was not a loris, and it hardly matches modern lorises in every detail of its skeletal anatomy. Indeed, one of the ongoing debates about the paleobiology of *Adapis* concerns exactly how similar its locomotion would have been to that of living lorises. For example, *Adapis* and lorises differ in the anatomy and functional capabilities of their hands. The hands of lorises are uniquely adapted for powerful grasping. In order to increase the power of the grip between the thumb and the remaining fingers, the index finger of lorises has been reduced to a tiny nubbin. Functionally, this increases both the angle and the gape between the thumb and the other fingers, giving lorises a highly effective "pincer" grasp that is otherwise unknown among primates. *Adapis* lacked the extreme reduction of the index finger characteristic of lorises.[12] Instead, the anatomy of its wrist and hand implies that *Adapis* walked quadrupedally on the tops of large branches, or even on the ground. If accurately reconstructed, this type of locomotion recalls that of small monkeys more than that of lorises.

Further aspects of the paleobiology of *Adapis* can be discerned from its skull, jaws, and teeth. The teeth and jaws of all primates reflect the broad range of diets these animals consume. By analyzing the structure of fossil primate teeth and jaws, we can infer a great deal about their ancient diets. This is particularly true in the case of *Adapis,* because its jaws and teeth are so specialized anatomically. Primates that consume relatively soft food, like fruits and the exudates of trees (sap and gum), have cheek teeth dominated by large, relatively flat cusps and basins. In contrast, primates that eat food requiring a great deal of cutting and slicing need teeth with sufficient shearing crests to get the job done. Like other adapids, *Adapis* possesses sharply crested cheek teeth. Particularly high and continuous crests line the outer sides of the upper and lower molars of *Adapis.* When these teeth occlude with one another during chewing, their outer crests shear against each other like the opposing blades of a pair of scissors, cutting apart any food that lies in their path. The lower molars of *Adapis* also bear transverse crests that augment this shearing function.

Given its sharply crested cheek teeth, we know that *Adapis* must have eaten food that had to be chewed up and sliced apart before being digested. Primates as a whole consume two basic classes of food meeting this description—insects and leaves. Which of these foods could have met the dietary requirements of *Adapis*? As it turns out, the dietary adaptations of primates are strongly constrained by their body size.[13] Insects are rich in calories, but they are little and difficult to catch in abundance.

As a result, only fairly small primates—those weighing less than about one pound—can meet their dietary needs by preying on insects. Leaves, on the other hand, yield fewer calories, but they occur prodigiously and are relatively easy to harvest. Many large species of primates survive on a diet of leaves alone. *Adapis* was roughly twice as large as any modern primate whose diet consists mainly of insects. We can therefore be confident that leaves, rather than insects, formed the bulk of its diet.

Other aspects of the anatomy of *Adapis* reinforce this conclusion. For example, *Adapis* differs from more primitive adapiforms in having the two halves of its lower jaw solidly fused together at the chin. This feature, known as a fused mandibular symphysis, occurs in several groups of living and fossil primates, including all living anthropoids. Although there is no simple relationship between an animal's diet and fusion of its mandibular symphysis, many of the living primates with this feature eat mainly leaves. Functional studies indicate that fusion of the mandibular symphysis is an anatomical response to cyclical repetitive chewing of tough or fibrous foods.[14] Anyone who eats lots of uncooked, leafy green vegetables knows that such a diet requires forceful, repetitive chewing. Hence, the fused mandibular symphysis of *Adapis* also points toward its having a specialized diet of leaves.

A folivorous dietary adaptation can also be reflected in the external anatomy of the skull, far beyond the teeth and jaws per se. Primates that specialize in eating leaves need massive jaw muscles to power the prolonged chewing mandated by this dietary regime. Among the most important muscles controlling jaw movements in general and chewing in particular are the temporalis muscles, which originate on both sides of the head and insert onto the top of the lower jaws. In many modern leaf-eating primates, the temporalis muscles are greatly enlarged, which means that the areas on the sides of the skull where these muscles originate must also increase in size. A frequent anatomical solution to this problem is to add crests to the top and back of the skull. These bony ridges, technically known as the sagittal (top) and nuchal (back) crests, vaguely recall the crests that once adorned the helmets of Roman centurions. *Adapis* possesses remarkably large sagittal and nuchal crests. Similar structures grace the skulls of gorillas, a modern primate folivore. Obviously, the temporalis muscles of *Adapis* were massive and powerful, a further indication of its leaf-eating habits.

Our survey has shown that *Adapis* had all of the anatomical equipment—sharply crested cheek teeth, a fused mandibular symphysis, and hypertrophied temporalis muscles anchored to well-developed sagittal and nuchal crests—needed by a dedicated primate folivore. In hindsight, I suspect that the specialized dietary adaptations of *Adapis* are what misled Cuvier when he originally tried to determine the kind of animal *Adapis* might be. Recall that Cuvier's interpretation of *Adapis* rested primarily on the anatomy of its cheek teeth, since the only skull available to him was so badly crushed and distorted. The general pattern of molar cresting in *Adapis* does not differ radically from that of rhinos and hyraxes, two members of Cuvier's "pachyderm" group of mammals. Hyraxes, which vaguely resemble medium-sized rodents like the common groundhog *(Marmota)*, feed extensively on leaves. So do their "pachyderm" brethren the rhinos. Although Cuvier had no way of knowing this, the similarities in molar anatomy shared by *Adapis,* rhinos, and hyraxes faithfully reflect their diets, but tell us little about their affinities. In other words, each of these animals solved the anatomical problems that eating leaves posed by independently evolving similar patterns of molar cresting. Cuvier erred only in assuming that these similarities indicated a close biological relationship. Placed in context, we must forgive Cuvier for his embarrassing determination that *Adapis* was a small new type of pachyderm. Cuvier reached the simplest interpretation of the data that were available at the time, a standard practice in modern science. Similar problems continue to baffle evolutionary biologists, because it can often be difficult or impossible to determine whether individual features evolved convergently rather than being inherited from a common ancestor.

One of the greatest obstacles to reconstructing the general biology and external appearance of an Eocene primate like *Adapis* lies in the nature of the fossil record itself. Hard anatomical tissues like teeth and bones readily become preserved as fossils. In contrast, we typically must infer or reconstruct the anatomy of soft structures like hair, skin, muscles, and internal organs, since these features generally can't be observed directly. In a few instances, fossilized teeth and bones do provide reliable clues as to the nature of soft structures nearby. For example, the upper incisors of *Adapis* disclose the soft anatomy of its nose. Living lemurs and lorises resemble dogs and many other mammals in having a hairless, moist region of skin surrounding the nostrils. This whole area, called the rhinarium, ultimately links up with a structure known as Jacobson's organ, which is located inside the snout above the roof of the mouth. In order to establish a physical connection between the moist external rhinarium and Jacobson's organ, lemurs possess a broad midline gap between the roots of their upper central incisors. Although we still don't understand the exact function of this anatomical arrangement, it probably serves a fundamental biological purpose, because it occurs across a broad spectrum of mammals, ranging from opossums to hedgehogs and lemurs. In most mammals in which its role is known, Jacobson's organ detects pheromones and other chemical stimulants related to reproductive behavior. Jacobson's organ remains large and prominent in lemurs, but this structure is reduced or absent in tarsiers and anthropoids. Therefore, the wet external nose of lemurs may somehow be involved in collecting and transporting pheromones to Jacobson's organ.[15] Like those of living lemurs, the upper central incisor roots are broadly spaced in *Adapis*, although their crowns contact each other in the midline. This anatomical configuration is consistent with a direct connection between the rhinarium and Jacobson's organ, suggesting that the nose of *Adapis* was wet and hairless like that of a lemur. If so, the sex life of *Adapis* may have relied more on its sense of smell than its sense of vision. While such a predilection is common among lemurs, the dominance of visual pornographic media, rather than scratch-and-sniff cards, demonstrates that the same pattern does not hold among at least some anthropoids.

At the most basic level, the behavior of any animal is constrained by its sensory capabilities. The anatomy of the brain, in particular, yields useful information about mental abilities and sensory development in primates. Like virtually all soft anatomical structures, brains rarely become fossilized. However, natural or man-made molds of the inside of the skull, known as endocasts, can reveal aspects of the external anatomy of the brain in certain fossil primates, including *Adapis*.[16] Endocasts of *Adapis* show that its brain sported exceptionally large olfactory lobes, confirming that its sense of smell remained acute. This emphasis on the sense of smell was carried over from the earlier mammalian ancestors of all primates and is an important indication of the primitive nature of the brain in *Adapis*. This conclusion is underscored by the diminutive size of its brain as a whole, which checks in at a paltry 8.8 cubic centimeters—less than the volume of a standard walnut. Relative to its body weight, the brain of *Adapis* was smaller than those of most, if not all, living lemurs. Of course, anthropoids have brains that are larger yet.

As we noted in chapter 1, the vast majority of living anthropoids are diurnal—that is, they are mainly active during daytime. Prosimians differ in that many species, including tarsiers, lorises, and some lemurs, are mainly active at night—a nocturnal activity pattern. Among living primates, a correlation exists between the size of the eye sockets and their daily activity pattern. Nocturnal species have relatively larger orbits—and hence larger eyeballs—than diurnal species. *Adapis* has exceptionally small orbits for its body size. Assuming that the correlation between orbit size and activity pattern among living primates applies equally to their Eocene relatives, *Adapis* must have resembled some lemurs and most anthropoids in being active during daytime.

In contrast to our thorough understanding of *Adapis,* we know relatively little about its nearest relatives in the family Adapidae. Like *Adapis,* all adapids possess sharply crested cheek teeth, suggesting that they too had folivorous diets. However, jaws and teeth are all that document most species of adapids. As a result, aside from their penchant for eating leaves, most aspects of their paleobiology remain obscure. The major exception to this rule is *Leptadapis magnus,* whose fossils have been discovered alongside those of *Adapis* in the Eocene fissure-fillings of southern France. *Leptadapis* was larger and considerably more robust than *Adapis,* but otherwise these two primates appear to have been very similar. *Leptadapis* weighed between ten and twenty pounds (4.5–9 kilograms), making it larger than most, if not all, living lemurs. Skulls of *Leptadapis* bear strong sagittal and nuchal crests like those of *Adapis,* providing similarly enlarged areas for the origin of the temporalis muscles. The few known limb bones of *Leptadapis* reveal it to have been much more muscular than *Adapis. Leptadapis* perhaps compensated for its large body size with increased brute strength. It probably moved slowly through the trees by climbing and clambering from branch to branch, much like its smaller evolutionary cousin.

As the rich fossil record of the French Quercy fissure-fillings attests, adapids reached the pinnacle of their evolutionary success during the second half of the Eocene. Although adapids obviously diversified in Europe, these animals did not originate there. Instead, they show up suddenly in the European fossil record about forty-four million years ago. Despite the presence of a variety of adapiform primates in Europe prior to the invasion of true adapids, none of these animals is a plausible ancestor for *Adapis* and its nearest relatives. Until recently, the geographic source for the adapid invasion of Europe remained a mystery. The ancient geographic setting of Europe, which was effectively an island continent at the time, only compounded this puzzle. During most of the Eocene, a shallow seaway bisected the Eurasian landmass east of the modern Ural Mountains, isolating Europe and Asia from each other. At the same time, an eastward extension of the modern Mediterranean known as the Tethys Sea severed all direct land connections between Europe and Africa through the Near East. Given this geographic context, adapids must have arrived on European shores with wet feet. But from which

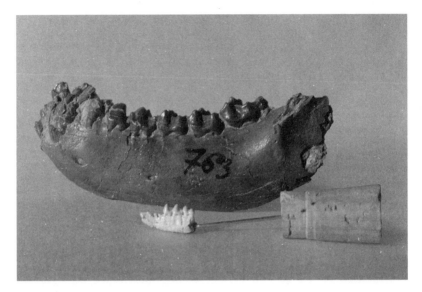

Figure 8. Lower jaws of two European adapiforms illustrate the vast differences in size and paleobiology within this group. The small specimen in the foreground is *Donrussellia,* one of the earliest and smallest known adapiforms. Like other primates in this size range, *Donrussellia* probably ate mainly insects and fruits. The large specimen in the background is *Leptadapis,* an advanced adapiform of large body size. The tooth and jaw structure of *Leptadapis* show that it fed mainly on leaves. Photograph courtesy of and copyright by Marc Godinot.

direction—Africa to the south or Asia to the east—did the invading adapids hail?

Serendipity impacts many scientific pursuits. Its role in paleontological fieldwork is genuine, albeit impossible to quantify. Often, those who make significant new discoveries in paleontology simply happen to be in the right place at the right time. My team's contribution to understanding the early history of adapids underscores this point. When my Chinese colleagues and I initiated our cooperative project on the fossil mammals of the Shanghuang fissure-fillings, we had no intention of illuminating the problem of adapid origins. We were motivated instead to determine the origin of anthropoids, and our discovery of *Eosimias sinensis* at Shanghuang marked our first significant advance in this field. Yet we also recovered fossil primates there that had no bearing on anthropoid origins. One of these turned out to be an early adapid. We named our new primate *Adapoides troglodytes* ("*Adapis*-like inhabitant of caves"), in recognition of its close affinities with *Adapis* and its discovery in a limestone fissure-filling.[17]

My team recovered numerous teeth and jaws of *Adapoides* at Shang-huang, as well as a few of its ankle bones. Yet even these fairly fragmentary specimens shed light on the origins and early evolution of adapids. They show, for example, that *Adapoides* was a small, relatively primitive member of the Adapidae. In life, *Adapoides* would have weighed only about ten to twelve ounces (300–350 grams). Most other adapid species are considerably larger than *Adapoides,* and some—like *Leptadapis magnus*—exceed it in size by well over an order of magnitude. Only one other adapid, *Microadapis sciureus,* approaches the small body size of *Adapoides.* Significantly, *Microadapis* is the oldest European adapid, and its cheek teeth are more primitive than those of any other member of the family. The development of molar crests in *Adapoides* is intermediate between that of *Microadapis* on the one hand and *Adapis* and its close relatives on the other. This precocious emphasis on molar shearing suggests that *Adapoides* had already committed to a folivorous diet, despite its small body size.

The discovery of *Adapoides* settles two long-standing questions surrounding the evolutionary history of adapid primates. The first of these concerns the evolution of body size among these animals. Most groups of mammals get larger through time, a tendency known as Cope's Rule (in honor of Edward Drinker Cope, who, as we shall see, was one of the most important American paleontologists of the nineteenth century). Our own lineage certainly conforms to this pattern, as demonstrated by the diminutive stature of the early hominid "Lucy," who stood only about three and a half feet tall in life. In fact, only a few instances of lineages becoming smaller through time have been cited among primates. Adapids have the distinction of belonging to this minority. Yet, as is so often the case in paleontology, the strength of this view depends on how one reconstructs the evolutionary history of adapids. Some scientists, such as Philip Gingerich of the University of Michigan, regard *Adapis* as a direct descendant of the much larger *Leptadapis*.[18] If this were the case, the *Leptadapis-Adapis* series would qualify as a dwarf lineage—a violation of Cope's Rule. However, other accounts of adapid phylogeny do not recognize *Adapis* as a direct descendant of *Leptadapis*. These conflicting versions of adapid evolution recall our earlier discussion of the ladder and tree evolutionary paradigms. Positing that *Adapis* evolved directly from *Leptadapis* falls squarely within the ladder paradigm, since it implies a simple ancestor-descendant relationship between these two fossils. For those who favor the competing tree paradigm, a more plausible evolutionary hypothesis views *Leptadapis* and *Adapis* as close cousins, not ancestor and descendant. Recent work on Quercy adapids, based on new excavations that have focused on establishing the age of fossils from individual fissures, supports a treelike pattern of adapid phylogeny. This research shows that *Leptadapis* and *Adapis* largely overlap one another in time, precluding the possibility that one evolved directly from the other.[19] Rather, the two most basal twigs of the adapid family tree consist of *Microadapis* and *Adapoides,* both of which are substantially smaller than either *Adapis* or *Leptadapis.* Hence it appears that adapids, like humans, conform to Cope's Rule—both lineages generally became larger through time.

A second outstanding issue that *Adapoides* helps to resolve is the geographic origins of adapids. Given their sudden appearance in the fossil record of Europe, where no suitable ancestors existed earlier in the Eocene, adapids must have migrated to Europe from elsewhere. Prior to our work at Shanghuang, most authorities believed Africa to be the ancestral homeland of adapids, reflecting a persistent bias that viewed Africa as the source of most major primate groups.[20] Our discovery of *Adapoides* in China shows that Asia is a more likely birthplace of the adapid lineage. Indeed, *Adapoides* constitutes the only hard evidence that true adapids ever lived anywhere other than Europe. Of course, more distantly related adapiforms inhabited Africa and North America. Let's turn our attention now to the North American cousins of Eurasian adapids, which form an entirely separate branch of the adapiform family tree.

If Cuvier had a hard time envisioning lemurlike primates on the outskirts of Paris, it should come as no surprise that the first discoveries of comparable fossils in the New World were equally unexpected—and similarly misinterpreted. Even today, most people who visit or reside in the Rocky Mountain region of the western United States have difficulty imagining that this area was once home to a thriving subtropical ecosystem, replete with such rain-forest icons as primates. Once again, the fossil record reveals that the truth can be stranger than fiction. We now know that a second major group of adapiform primates, the family Notharctidae, evolved mainly in North America. Their fossils were first discovered in Wyoming Territory in 1870, as an expanding nation turned its attention toward the largely unexplored West.

The first adapiform fossil ever discovered in North America was a fossilized right lower jaw bearing seven worn teeth. Members of a scientific survey team underwritten by the federal government found the specimen in badlands near Black's Fork of the Green River, in southwestern Wyoming's Bridger Basin. At the time, wide-ranging scientific surveys routinely preceded or accompanied the westward expansion of the United States. The Bridger Basin survey team of 1870 worked under the supervision of Ferdinand V. Hayden, whose additional claims to fame include exploration of the region set aside today as Yellowstone National Park. Hayden sent the Bridger Basin fossils back east to Philadelphia, where they were studied by Joseph Leidy, a local physician who had developed an abiding interest in natural history and paleontology.[21] Leidy felt compelled to describe the fossils without delay.

Despite their official government positions, by the summer of 1870 Hayden's men had no monopoly on the Bridger Basin fossils. A "gold rush" mentality prevailed among North American fossil collectors of the time. As soon as a new discovery of fossil vertebrates was announced from some remote corner of the American West, multiple competing teams converged on the scene to secure specimens for their eastern patrons. In the case of the Bridger Basin, Leidy announced the first discovery of fossils there in 1869.[22] Predictably, Othniel Charles Marsh, a prominent professor of paleontology at Yale University, followed up on Leidy's announcement the next summer, traveling with his own team to the Bridger Basin. Upon their arrival, Marsh's team had the audacity to demand that Hayden's men stand aside to give them sole access to the region's fossil beds.[23] In the Bridger Basin and elsewhere, fierce competition among such rival teams of fossil collectors regularly took place, sometimes escalating to include fisticuffs and even gunplay. By no means did the competition end when one team made an exciting discovery in the field. Rather, scientific credit flowed to whichever team published its findings first. As a result, the earliest descriptions of most fossil vertebrates from western North America consist of terse summaries of their anatomy, serving little purpose beyond allowing the fossil in question to be named. In this highly charged atmosphere, the same extinct animal commonly received multiple names from competing paleontologists. At the same time, those who coined names for these new fossils rarely appreciated their evolutionary significance. The fossil primate from Wyoming that Leidy would name *Notharctus tenebrosus* illustrates the prevailing conditions well.

The shipment of Bridger Basin fossils that Hayden sent to Philadelphia in 1870 contained mostly turtles and crocodiles, but Leidy identified the remains of three new mammals among the lot. Leidy hastily described and named all three in a paper that took up little more than a page in his local scientific journal, the *Proceedings of the Academy of Natural Sciences of Philadelphia*. In his original description of *Notharctus*, Leidy failed to compare his new fossil with primates. "The specimen apparently indicates an animal allied to the raccoon, than which it was nearly a third smaller in size," he wrote.[24] Given the concise nature of Leidy's verbal description (which was not supplemented by photographs or drawings of the specimen), we can understand the cascade of names that were proposed for the same animal over the following two years. Marsh alone coined three additional names for *Notharctus* during this interval, *Limnotherium*, *Thinolestes*, and *Telmatolestes*, which are all now regarded as scientifically redundant names for *Notharctus*. By 1872, Edward Drinker Cope, like Leidy a resident of Philadelphia, joined the melee when he christened a slightly more complete jaw of *Notharctus* from the Bridger Basin *Tomitherium*. Making sure to cover all his bases, Leidy himself added to the growing confusion by proposing *Hipposyus*. Less than two years after it was formally described, *Notharctus* was known by no fewer than six different aliases!

Like Leidy before them, neither Marsh nor Cope initially had any inkling of what type of animal *Notharctus* (a.k.a. *Limnotherium*, *Thinolestes*, *Telmatolestes*, *Tomitherium*, and *Hipposyus*) might be. Following in the intellectual footsteps of Cuvier, Marsh first compared *Limnotherium* with "pachyderms," while asserting that *Thinolestes* showed affinities with carnivorous mammals. A breakthrough occurred in November 1872, when Marsh published the following paragraph in Yale's in-house scientific journal, the *American Journal of Science and Arts*:

> An examination of more complete specimens of some of the extinct Mammals already described by the writer from the Eocene deposits of the Rocky Mountain region, clearly indicate [*sic*] that among them are several representatives of the lower Quadrumana [at that time, nonhuman primates were usually classified in an order separate from humans called Quadrumana, or literally "four-handed"]. Although these remains differ widely from all known forms of that group, their more important characters show that they should be placed with them. The genera *Limnotherium*, *Thinolestes*, and *Telmatolestes*, especially, have the principal parts of the skeleton much as in some of the Lemurs, the correspondence in many of the larger bones being very close. The anterior part of the lower jaws is similar to that of the Marmosets [a group of living New World monkeys], but the angle is more produced downward, and much inflected. The teeth are more numerous than in any known Quadrumana.[25]

The following year, Leidy began to stake out a similar position regarding the primate affinities of *Notharctus*, although his views were based on the anatomy of its lower jaw and teeth, rather than its limb bones. Leidy apparently reached this conclusion independently from

Marsh, because it was published in a much longer monograph that was largely completed prior to the publication of Marsh's note. Curiously, Leidy also supported the view that *Notharctus* was somehow related to "pachyderms," thus abandoning his earlier notion that it was an extinct relative of raccoons. At the beginning of Leidy's account of *Notharctus,* he referred to it as a "small extinct pachyderm . . . probably as carnivorous in habit as the raccoon and bear."[26] Three pages later, Leidy embraced an entirely different, if not schizophrenic, view regarding its evolutionary significance:

> In many respects the lower jaw of *Notharctus* resembles that of some of the existing American monkeys quite as much as it does that of any living pachyderms. *Notharctus* agrees with most of the American monkeys in the union of the rami of the jaw at the symphysis, in the small size of the condyle, in the crowded condition of the teeth, and in the number of incisors, canines, and true molars, which are also nearly alike in constitution. . . . The resemblance is so close that but little change would be necessary to evolve from the jaw and teeth of *Notharctus* that of a modern monkey.[27]

As we shall see in chapter 5, Leidy's discussion of how the jaw of *Notharctus* might easily be transformed into that of a monkey anticipated much of the ensuing debate over anthropoid origins for the next 125 years.

Over the next several decades, continued exploration of the rich fossil beds of the Bridger Basin uncovered several partial skeletons of *Notharctus.* Today, *Notharctus* and its close relative and contemporary *Smilodectes* are among the best-known fossil primates of any age (both date to the early part of the middle Eocene, about forty-eight million years ago). These North American notharctids differ from European adapids like *Adapis,* especially in the structure of their limbs and teeth. They also differ from *Adapis* in that their evolutionary precursors are less mysterious. Both *Notharctus* and *Smilodectes* evolved from more primitive notharctids that inhabited North America during the early Eocene (from fifty-five to fifty million years ago).

The skeleton of notharctids diverges from that of adapids in several important ways, all of which relate to a greater propensity for vertical clinging and leaping.[28] Notharctids display lemurlike limb proportions, with hindlimbs that are appreciably longer than the forelimbs. Recall that adapids have forelimbs and hindlimbs of roughly similar length, yielding limb proportions that are more monkeylike than lemurlike. The knee joint of notharctids resembles that of living lemurs in being tall and narrow. In contrast, *Adapis* possesses a shallower, wider knee, recalling that of living lorises. Because of the increased length of their calcaneus, the ankles of notharctids are more elongated, and hence lemurlike, than those of adapids. Considering that *Notharctus* and living lemurs are separated in time by nearly fifty million years, the close similarity shown by their skeletons—first noted by Marsh in 1872—is truly remarkable.

The structure of its skeleton indicates that *Notharctus* was an active and agile primate that had no difficulty leaping from tree to tree. It seems likely that *Notharctus* frequently used fairly vertical postures while clinging to trunks and larger branches. In order to execute these kinds of acrobatic maneuvers, notharctids needed muscular limbs and grasping hands and feet. Not surprisingly, these attributes too are reflected in the skeletal anatomy of these animals. Notharctids had long, flexible fingers. As is typical of primates, their fingers and toes bore nails rather than claws. The big toe of notharctids was capable of powerful grasping. Notharctid forelimbs bear extremely well-developed bony crests and ridges marking the areas where the major muscles of the shoulder and forearm originate and insert. Indeed, one of the major distinctions between the skeletons of notharctids and living lemurs lies in the greater robustness of notharctid forelimbs. To get a reasonable approximation of what notharctids must have looked like in life, think of lemurs on steroids.

Like *Adapis* and its close European relatives, *Notharctus* and *Smilodectes* appear to have been adapted for eating leaves. These animals weighed from four and a half to nine pounds (two to four kilograms), making them too large to have specialized on a diet of insects. Their cheek teeth bear strong crests, especially compared to their earlier North American relatives. Nevertheless, the anatomical pattern of molar crest development in notharctids differs considerably from that of adapids, implying that these different adapiform groups evolved their folivorous diets independently. The dental anatomy of the earliest North American notharctids supports this view, because their molar crest development is subdued compared to that of adapids and later notharctids. The diet of these early North American notharctids likely consisted largely of fruits rather than leaves.[29]

Beyond yielding insights about their ancient diet, the teeth of Eocene primates potentially illuminate more interesting aspects of their behavior. In particular, patterns of social organization have been inferred from the anatomy of the canines in a few Eocene primates, including *Notharctus* and *Adapis*. Males and females of some species of living primates differ in the size and shape of their canines, a distinction known as sexual dimorphism. Strong canine sexual dimorphism occurs in many primate species (such as baboons) in which individuals live in complex social groups characterized by intense competition among males for access to mates. Weak to nonexistent canine dimorphism exemplifies species that are solitary (like tarsiers), relatively monogamous (like gibbons), or characterized by female social dominance (like most lemurs). Hence, if their canines could be shown to be highly dimorphic, this might indicate that adapiforms lived in relatively large social groups in which males competed with each other to form harems.

Unfortunately, the level of canine dimorphism that may have been present in *Adapis* remains unclear. Certainly, the large sample of skulls of *Adapis* reveals significant variation in their overall size, as well as in the size and shape of their canines. Some leading experts on adapids, such as Philip Gingerich, interpret this variation as sexual dimorphism.[30] If so, *Adapis* probably lived in large, polygynous social groups. Others, such as Marc Godinot of the Institut de Paléontologie at the Muséum National d'Histoire Naturelle in Paris, interpret the same data as evidence for multiple species of *Adapis* instead.[31] Detailed analyses of limb bones of *Adapis* from the Quercy fissure-fillings support the view that several species of *Adapis* occur there. According to Godinot, *Adapis* limb bones fall into different groups, each of which he believes is a distinct species with its own unique pattern of locomotion.[32] If several species of *Adapis* are indeed present in the Quercy fissure-fillings, each may have shown little, if any, sexual dimorphism. In this case, *Adapis* may have lived a fairly solitary existence, may have formed monogamous pairs, or may have lived in larger groups dominated by females, like many modern lemurs.

In contrast to the ambiguous case for sexual dimorphism in *Adapis*, the evidence for *Notharctus* is clear-cut. To find out why, we must detour briefly to yet another intermontane basin in Wyoming. Soon after I joined the scientific staff of the Carnegie Museum of Natural History, I signed on to the museum's long-standing field project in the Wind River Basin of central Wyoming. There, an especially rich set of localities known as the Buck Spring Quarries samples an ancient pond environment of about fifty million years ago. Our excavations at the Buck Spring Quarries uncovered two varieties of *Notharctus*. Some specimens have long, saberlike canines, while others have short, stubby canines that are more rounded in cross-section. Unlike the Quercy sample of *Adapis*, we know that *Notharctus* individuals from the Buck Spring Quarries inhabited a restricted area at the same time. Aside from the obvious differences relating to their canines, the *Notharctus* sample from the Buck Spring Quarries is anatomically monotonous. Moreover, these specimens vary little in size. As a result, they all appear to represent the single species *Notharctus venticolus*. These specimens embody the earliest well-documented evidence for sexual dimorphism in primates.[33] Presumably, male individuals of *Notharctus* competed intensely to form harems and to achieve social dominance over their rivals.

We have already seen how adapids attained their broadest evolutionary success after invading Europe from elsewhere, probably Asia. Likewise, the ecological importance and diversity of North American notharctids was predicated upon their successful colonization of the New World. The fossil record shows that notharctids arrived suddenly in North America at the beginning of the Eocene, about fifty-five million years ago. In earlier Paleocene fossil sites throughout North America, not only are possible notharctid ancestors conspicuously lacking, but no primates of any sort have ever been found. Although notharctids lack a "smoking gun" fossil like *Adapoides* that points toward Asia as their likely birthplace, we have good reason to believe that this was indeed the case. Perhaps the most compelling evidence supporting this view comes from the fact that notharctids first appear in North America alongside a host of other immigrant mammals, all of which seem to hail from Asia.[34]

Once notharctids arrived in North America, they rapidly rose to a position of ecological prominence. Notharctids are almost always the most abundant primates in sites of early and middle Eocene age in North America. The earliest notharctids, classified in the genus *Cantius*, weighed about 1.5 pounds (700 grams). Early species of *Cantius* had very primitive teeth, quite unlike those of *Notharctus*. Fruits, rather than leaves, seem to have been its chief food. Despite these important differences in body size and diet, the skeletal anatomy of *Cantius* is remarkably similar to that of *Notharctus*. Rich sequences of fossil-bearing rock strata, especially in the Bighorn Basin of northwestern Wyoming, allow us to trace the evolution of *Cantius* lineages through millions of years of early Eocene time. During this interval, *Cantius* became progressively larger and its teeth evolved new structures, including two extra cusps on its upper molars.[35] These evolutionary modifications suggest that an advanced species of *Cantius* may have been ancestral to *Notharctus*.

While both adapids and notharctids probably originated in Asia, neither of these groups achieved ecological prominence there. Instead, a third family of adapiforms—the sivaladapids—filled the ecological niches in Asia that were occupied by adapids in Europe and by notharctids in North

America. The history of Asian sivaladapids lasted for more than thirty million years (from the late middle Eocene, about forty million years ago, to the late Miocene, about seven or eight million years ago). Despite their long duration, little is known about sivaladapid anatomy. The only traces of their former existence that have been unearthed so far consist of teeth, jaws, and a single bone fragment showing that sivaladapids possessed a big toe that was capable of grasping.[36]

Given the sparse nature of their fossil record, we can draw only the barest of inferences as to what sivaladapids might have been like in life. The smallest known sivaladapid, *Hoanghonius stehlini*, weighed about 1.5 pounds (700 grams), making it about the same size as a modern bamboo lemur (genus *Hapalemur*). At the large end of their size range, *Sivaladapis palaeindicus* tipped the scales at roughly 10 pounds (4.5 kilograms). Like *Notharctus* and *Adapis,* all sivaladapids have sharply crested cheek teeth that were specialized for eating leaves. Until their skulls or partial skeletons are exhumed, we can infer no more about the biology of these intriguing primates.

The fourth major group of adapiforms, known as the Cercamoniinae, lived primarily in Europe prior to the invasion of true adapids. Nearly thirty species of cercamoniines have been reported from the European fossil record alone. Additional species have been discovered as far away as Egypt and Texas.[37] Although the anatomy of cercamoniines is not as thoroughly documented as that of *Adapis* and *Notharctus,* these animals are pivotal to our story, because cercamoniines are the most anthropoidlike of all adapiforms. If anthropoids evolved from adapiforms, cercamoniines are the prime suspects. The evidence that potentially links anthropoids with cercamoniines is examined more fully in chapter 5. Here, we need to achieve a basic understanding of the biology of these animals, emphasizing the similarities and differences between cercamoniines and other major adapiform groups.

Cercamoniines evolved a wide range of body sizes. Among the smallest cercamoniines are species of *Anchomomys*, some of which weighed no more than 3.5 ounces (100 grams). The largest cercamoniine, *Cercamonius brachyrhynchus,* weighed almost nine pounds (four kilograms). Given this broad range of body sizes, it should come as no surprise that different species of cercamoniines ate different types of food. The smaller forms, such as *Anchomomys*, probably preyed on insects and other small animals. Many medium-sized cercamoniines likely consumed mainly fruits, a diet that would have been supplemented with insects and small vertebrates. Like most other adapiforms, the larger cercamoniines probably relied heavily on a diet of leaves. None of the cercamoniines show the extreme specializations for leaf-eating that are found in true adapids, however. For example, cercamoniines lack the large sagittal and nuchal crests on their skulls that are characteristic of adapids. This suggests that cercamoniines possessed smaller temporalis muscles, making them less efficient chewing machines than adapids must have been.

Partial skeletons or disarticulated limb bones document a few cercamoniines. These specimens demonstrate that cercamoniines resemble notharctids rather than adapids in having hindlimbs that were appreciably longer than their forelimbs. Accordingly, cercamoniines probably moved acrobatically through the Eocene forests of Europe mainly by climbing and leaping. It is difficult to determine whether cercamoniines were primarily active during daylight or at night, because partial skulls in which the size of the eye sockets can be determined are so rare. Apparently, both diurnal and nocturnal species were present. For example, *Europolemur koenigswaldi* from the famous middle Eocene fossil site of Messel in Germany possesses relatively small eye sockets, indicating that it was mainly active during daytime. *Pronycticebus gaudryi* has much larger eye sockets, however, suggesting that this species was probably active at night.

Perhaps because there were so many species of cercamoniines, their basic biological adaptations seem to have been more heterogeneous than those of other adapiform groups. Often, this type of "shotgun approach" yields dividends in terms of evolutionary success—the fossil record is littered with extinct animals that made the mistake of putting all their biological eggs in one basket by becoming too specialized. Yet for reasons that remain largely unknown, this strategy failed to pay off for cercamoniines. Almost all of the European cercamoniines became extinct once true adapids invaded that continent about forty-four million years ago. The only cercamoniine that survived this onslaught was tiny *Anchomomys,* whose diminutive size and insectivorous diet placed it beyond competition with the Asian adapid immigrants.

What emerges immediately from this survey of the four major groups of adapiforms is the strong geographical component to their evolutionary history. All four groups likely originated in Asia, but each evolved more or less in isolation, in a region apart from its close evolutionary cousins. Thus, with minor exceptions, notharctids evolved in North America, adapids radiated in Europe, and sivaladapids endured for tens of millions of years in Asia. Likewise, the cercamoniines were primarily a European group, although their European preeminence would ultimately be overwhelmed by the invading adapids. What does this geographical element contribute to our understanding of adapiforms—or, for that matter, to our knowledge of evolution in general?

Popular accounts of the history of life on Earth often depict evolution as an endless struggle for existence among competing species of plants and animals. If so, it stands to reason that those species that "win" the evolutionary game must be biologically superior to those that "lose." However, species that never interact, because they inhabit different regions, cannot engage in any meaningful struggle for existence. Accordingly, geography has the potential to throw a wrench into the strictly Darwinian machinery of evolution, inserting an element of chance into what might otherwise be a constantly escalating arms race between evolutionary competitors. Making matters even more unpredictable, geography itself evolves through the long span of Earth history due to plate tectonics, continental drift, and the rise and fall of sea levels. Superimposed on the shifting sands of geography is the added specter of climate change, with the potential to transform what was once an Eocene rainforest in Wyoming into terrain that today supports sagebrush, tumbleweed, and little else.

Climatic and geographic factors impacted the evolution of adapiform primates right from the start. It is hardly a coincidence that adapiforms first appear in the fossil record of Europe and North America at the very beginning of the Eocene. At this exact time, Earth experienced a brief, but dramatic, episode of global warming, apparently caused by a rapid release of methane into the atmosphere from sediments on the ocean floor.[38] Since methane is a highly effective greenhouse gas, this influx led to dramatic warming, especially near the polar regions. These high-latitude areas were precisely where ancient land bridges connected the three continents of the Northern Hemisphere. For example, where Siberia and Alaska are currently separated by the narrow expanse of water known as the Bering Strait, a dry spit of land allowed animals to move freely between Asia and North America at the beginning of the Eocene. A second land bridge, long since submerged by the tectonic spreading of the North Atlantic, likewise connected North America with Scotland and the rest of Europe. Once these high-latitude regions became sufficiently warm, early adapiform primates (and other tropical plants and animals) were free to spread from their Asian homeland to North America and on to Europe. Needless to say, upon reaching these vast tracts of virgin territory, migrating adapiforms were highly successful. They evolved into the great diversity of forms that we now know as notharctids and cercamoniines.

So long as the greenhouse conditions of the Eocene persisted, adapiform primates thrived on the northern continents. Yet we know that something went horribly awry, because these animals no longer reside in the leafy suburbs of Paris, nor do they inhabit the windswept plains along Black's Fork of the Green River in southwestern Wyoming. What transpired was an unlucky coincidence of changing geography and changing climate, which would forever erase adapiforms from the biological inventory of Europe and North America.

Toward the end of the Eocene, while the effects of continental drift in the Northern Hemisphere were fairly inconsequential, major alterations occurred in the south. There, rifting between the two most stubborn remnants of the former supercontinent known as Gondwana finally separated Australia from Antarctica about forty million years ago. This led to new patterns of oceanic circulation and the development of glaciers on Antarctica for the first time in the Cenozoic Era. The cooler waters around Antarctica eventually formed a conveyor belt to the rest of the world, supplying cold ocean currents to regions that had previously basked in the greenhouse conditions of the Eocene. In North America, subtropical forests gave way to open woodlands, and then to drier grassland conditions. In Europe, too, the climate became significantly cooler, causing the flora to shift from subtropical forest to more seasonal and temperate woodland.

For tropically adapted animals like adapiform primates, the deterioration in global climate across the Eocene-Oligocene boundary posed dire consequences, but these varied greatly depending on geography. Equatorial regions continued to harbor tropical plants and animals, while areas like the Rocky Mountain West were no longer suitable for them. Accordingly, the likelihood that any particular adapiform group might weather the storm depended heavily on where that group happened to live. To gain a better understanding of this process, let's take a closer look at the geographic pattern of adapiform extinction as climate deteriorated.

North American notharctids apparently bowed out early, because their fossils have never been found in strata younger than about forty million years ago. The cercamoniines were also decimated early on, although their decline probably resulted from the invasion of adapids into western Europe rather than climate change per se. European adapids persisted longer, right up to the Eocene-Oligocene boundary about thirty-four million years ago. But by far the most resilient of all adapiform groups were the sivaladapids, who survived in southern Asia until roughly seven or eight million years ago, more than twenty-five million years after the demise of their brethren on other continents! What accounts for this massive temporal disparity? Were sivaladapids biologically superior to other adapiforms? Or were they merely in the right place at the right time?

I strongly suspect the latter to be the case. Neither Europe nor North America provided access to tropical refuges for the unlucky adapiforms that happened to live on those continents as global climate deteriorated. Unlike today, at the end of the Eocene, North America was not directly connected to the South American tropics by the isthmus of Panama, which is a much more recent geological feature. As such, North American adapiforms literally had no place to hide as the entire continent became cooler and drier. Similarly, European adapiforms were unable to cross the marine barrier separating them from the African tropics. But the situation in Asia was quite different. There, many of the tropical islands comprising what is now Indonesia would have been directly connected to the Asian mainland, depending on the rise and fall of sea level. These areas, as well as adjacent parts of the Malay Peninsula, would have remained warm and humid even as Earth's climate cooled, since they all bordered the equator. Sivaladapids were thus the only adapiform group that was fortunate enough to inhabit a landmass with an emergency escape route to the tropics as the bottom fell out of the Eocene greenhouse climate.

Despite their good fortune of living in Asia, it is conceivable that the resilience of sivaladapids owed more to their biology than to geographic circumstance. We know so little about sivaladapid anatomy that future discoveries may yet reveal that they possessed certain biological attributes that ensured their longevity. However, from what we already know about sivaladapids and their close evolutionary cousins, this possibility seems remote. Sivaladapids, like most other fairly large adapiforms, apparently made their living as specialized leaf-eaters. Their teeth bore effective shearing crests for slicing through leafy vegetation, but I would argue that sivaladapid teeth were no better at chewing leaves than those of notharctids and adapids. All three adapiform groups independently fused the symphyses of their lower jaws, although notharctids and adapids accomplished this feat millions of years before sivaladapids did so. Finally, although the cranial anatomy of sivaladapids remains unknown, I doubt that the skulls of these creatures could have been better adapted for leaf-eating than that of *Adapis,* for example. Recall that

Adapis possesses strong sagittal and nuchal crests on its skull, providing expanded areas to accommodate the enlarged temporalis muscles that are so important for powerful and prolonged chewing. Clearly, all large adapiforms were highly effective leaf-eaters. Adapids in particular were veritable leaf-eating machines. It's hard to see how sivaladapids were able to outlive their closest relatives by more than twenty-five million years on the strength of their superior biology alone.

Whether because of biology or geography, the Asian reign of sivaladapids lasted for most of the Cenozoic. Possibly, sivaladapids still inhabited the forests of southern Asia when our own lineage bifurcated from that leading to chimpanzees in Africa. Given their long duration, I can't help but feel disappointed that sivaladapids are no longer part of Earth's living biota. But they are not. Walking through the rain forest in northern Borneo a few years back, I recall looking up to catch a glimpse of a loud group of primates making their way through their arboreal domain. It required little imagination to picture a troop of *Sivaladapis* up there, munching contently on the lush tropical foliage that flourished all around me. Yet the animals I saw in their place were leaf monkeys of the genus *Presbytis*—higher primates like you and me. Whether by coincidence or not, the earliest known fossils of leaf-eating monkeys in southern Asia date to roughly seven million years ago—precisely the time that sivaladapids finally disappear from the fossil record.

The extinction of adapiforms reminds me of the well-known example in chaos theory of a hurricane in the Caribbean being caused by the beat of a butterfly's wings in China. Over the long span of time on which evolution operates, events half a world away can have exactly this effect. Who could have guessed that the final fragmentation of Gondwana into Australia and Antarctica would refrigerate the Eocene global greenhouse, rendering France and Wyoming unsuitable for primates? From the point of view of *Adapis* and *Notharctus,* this was a catastrophe of monumental proportions. Whatever distant events allowed leaf-eating monkeys to migrate from their African homeland to southern Asia proved to be equally catastrophic for sivaladapids. Somewhere out there, I imagine, Cuvier must feel vindicated.

3

A Gem from the Willwood

n 1880, most of Wyoming Territory lay, at least proverbially, at the ends
of the earth. This was particularly true of Wyoming's northern two-
thirds, a remote and sparsely populated region that had been deliber-
ately bypassed by the Union Pacific Railroad, whose completion about
a decade earlier had brought both economic stimulus and rapid and reliable transportation to Wyoming's southern tier. The railroad's decision
not to challenge the rugged terrain of northern Wyoming matched that
of most early visitors to the area. For decades, white settlers had consistently forsaken the region, which they trudged across in covered wag-
ons in their quest to reach more promising lands in Oregon. Only four
years previously, Sioux and Cheyenne warriors led by Crazy Horse and
Sitting Bull had massacred all U.S. Cavalry troops under the command
of General George Armstrong Custer at the Battle of the Little Bighorn,
just across the territorial boundary in southern Montana.

Jacob Wortman was still a young man in 1880, barely twenty-four
years old. He himself was very much a product of America's westward
expansion. Wortman's parents had traversed the Oregon Trail in the summer of 1852, eventually settling on a farm near Oregon City. Jacob was
born four years later. Growing up in the American West, Wortman's interests were naturally inclined toward the world around him. The ma-

jestic landforms and colorful rock formations of his native Oregon, as well as its abundance of wild plants and animals, intrigued young Wortman. The dramatic scientific developments that took place as he grew up only exacerbated Wortman's predilection toward all things natural. Chief among these was Darwin's theory of evolution and its radical influence on how humanity viewed itself in relation to the rest of nature. The political upheavals of the Civil War notwithstanding, Darwin captured the public's imagination and Jacob Wortman's fancy. By the time Wortman enrolled in college, he knew he wanted to pursue a scientific career that would somehow advance Darwin's cause. But Wortman fell into despair whenever he pondered how he might make a living for himself in this field. Fittingly, the same siren song that lured Wortman toward the natural sciences in the first place eventually provided him with exactly the break he needed. Oregon's rich fossil record caught the attention of Edward Drinker Cope just as Wortman was wrapping up his college studies. Within a few short years, their collaboration would highlight the clues offered by the American West regarding Darwin's theory and our own deep evolutionary history.

Despite his youthful age, Wortman already boasted substantial experience as a field paleontologist by the summer of 1880. His budding career had received a major boost three years earlier, when Charles Sternberg—who collected fossils for Edward Drinker Cope—hired Wortman to help him search for fossil mammals in the John Day beds of eastern Oregon. In due course, Sternberg trusted Wortman enough to leave him in charge of the expedition for extended periods. Wortman rapidly developed the skills of a field paleontologist, and he soon graduated to work directly for Cope himself. Cope's first instructions directed Wortman to explore the extensive Eocene badlands of the Wind River drainage in central Wyoming, with the goal of securing the first fossil vertebrates from that region. Wortman had already met with phenomenal success. Now he needed to decide whether to gamble on finding an entirely new fossil field to the north, or to play it safe by staying where he was and finding more of the same.

Wortman's base of operations in 1880 was Fort Washakie, an isolated military outpost named in honor of the Shoshone chief who was both a friend to white settlers and a scourge to the Crow. Located near the modern town of Lander, Wyoming, the fort offered reasonable access to the variegated but mostly buff-colored Eocene strata known as the Wind River Formation. Wortman had already collected about forty-five species of fossil vertebrates from the Wind River beds, many of which appeared to be new to science. After dutifully crating his cache of Wind River fossils for shipment back east to Cope in Philadelphia, Wortman mulled over the leads he had obtained from drifting miners and ranchers he met at the fort. These men told Wortman about the vast expanses of badlands that lay across the mountains to the north, in the Bighorn Basin. Their accounts suggested that the potential for finding fossils there might exceed that of continued exploration in the vicinity of Fort Washakie. Were their stories true or merely the tall tales of men who relished the rugged individualism of the western frontier? Wortman decided that he had to find out for himself. He outfitted his team of packhorses and struck out to cross the stark mountains that separated him from whatever might lie to the north.[1]

Wortman apparently undertook the expedition with genuine trepidation. Years later, he recalled his experience as follows:

> At the time of my first trip into the Big Horn Basin in 1880 the country was a wild, uninhabited region, save for the occasional visits of roving bands of hostile Indians, and any explorations there by a small party were attended by no small amount of risk to one's personal safety. In fact, I was advised by the commander of Fort Washakie, at that time the base of our operations, that the trip was a hazardous one, and that he would not undertake to answer for our safe conduct.[2]

Regardless of how dangerous it might have been, Wortman's journey to the Bighorn Basin ranks among the most momentous excursions ever undertaken by a North American paleontologist. Riding from the mountain front into the basin's interior, Wortman traversed one of the planet's most continuous stratigraphic columns, extending from the Cambrian to the Eocene. More important for his immediate purposes, Wortman was able to verify the stories he had heard at Fort Washakie about the Bighorn Basin's extensive Eocene badlands. He succeeded in finding fossils of three early Eocene mammals, all of which Cope described as new species later that same year. At the time, Wortman had no way of knowing just how significant the fossil record of the Bighorn Basin would become, but his trek opened wide one of the world's most extraordinary portals onto the Eocene. Soon enough, the Bighorn Basin would yield an unprecedented bounty of paleontological treasures. One of these would create brand new possibilities for reconstructing the deep evolutionary history of humans and other higher primates.

The following year, Cope asked Wortman to concentrate all of his collecting efforts on the expansive Eocene outcrops in the Bighorn Basin, strata now known as the Willwood Formation. Once again, Wortman exceeded his employer's expectations, collecting some sixty-five species of fossil vertebrates from the Willwood badlands, almost half of which Cope soon described as new to science. Most of the Bighorn Basin fossils discovered by Wortman during the summer of 1881 were fragmentary. Many were encased in an extremely hard matrix known as hematite, rendering them difficult to study. Yet one of Wortman's fossils stood out from all the rest because of its excellent state of preservation and its obvious scientific potential. Wortman's prize specimen—the nearly complete skull of a tiny primate—measured little more than an inch in length. Its large eye sockets, ample cranial capacity and short muzzle distinguished it from anything Cope had ever seen.

Ever mindful of being scooped by his competitors, Cope wasted little time before proposing the new species *Anaptomorphus homunculus* for Wortman's primate skull.[3] By naming the new species, Cope established scientific priority over this creature, ensuring that his latest contribution to science would be forever acknowledged in some way. Nevertheless, Cope's referral of Wortman's skull to the genus *Anaptomorphus* was uncharacteristically conservative. Cope had coined the name *Anaptomorphus* almost a decade earlier, when he described the lower jaw of a small primate from southwestern Wyoming's Bridger Basin as *Anaptomorphus aemulus*. Because Wortman's skull from the Bighorn Basin lacked its lower jaws, it was impossible to determine how similar the two species might be—the two fossils simply shared no parts in common. Cope could have easily made the case that Wortman's skull deserved a new genus of its own, especially given the prevailing tendency to name every fossil. Decades later, when lower jaws from the Bighorn Basin were finally discovered, they differed appreciably from *Anaptomorphus aemulus*. As a result, Wortman's skull is now universally regarded as belonging to a distinct genus, *Tetonius*.

Right from the start, Cope emphasized that *Tetonius* represents a separate branch of primate evolution, one that diverged substantially from the other major group of North American Eocene primates, the lemurlike notharctids. The diminutive size of *Tetonius* implied an animal whose biology must have differed fundamentally from that of the much larger notharctids. In life, *Tetonius homunculus* weighed only about three ounces (ninety grams), making it roughly twenty times smaller than *Notharctus tenebrosus*. While its tiny size immediately set *Tetonius* apart from most other Eocene primates known to Cope, the anatomy of its skull and teeth hinted that *Tetonius* might play a key role in reconstructing

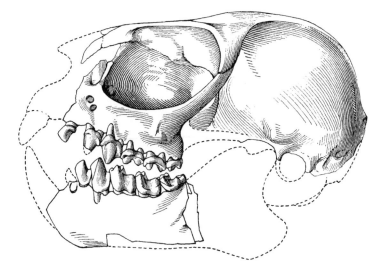

Figure 9. Jacob Wortman's skull of *Tetonius homunculus,* the first omomyid primate skull ever unearthed in North America. Reproduced from Matthew and Granger 1915.

primate and human evolution. Cope noted that *Tetonius* resembles living tarsiers in a number of ways, including its enlarged orbits (although not so large as those of tarsiers) and details of its ear region. At the same time, Cope regarded the upper premolar teeth of *Tetonius* as being similar to those of anthropoids in having complete lingual lobes known as protocones. This unusual combination of features led Cope to conclude rather breathlessly that *Tetonius* occupied a pivotal spot on the evolutionary tree of primates:

> In conclusion, there is no doubt, but that the genus *Anaptomorphus*
> [i.e., *Tetonius*] is the most simian lemur yet discovered, and probably
> represents the family from which the true monkeys and men were derived.
> Its discovery is an important addition to our knowledge of the phylogeny
> of man.[4]

As we shall see in chapter 5, Cope's assertion regarding *Tetonius* anticipated one of the main schools of thought that would dominate the study of anthropoid origins for most of the twentieth century.

As someone who has spent numerous field seasons following in the footsteps of Jacob Wortman, I have developed a deep sense of admiration for his skill, his perseverance, and his luck (which logically follows from the previous two factors). Wortman was an exceptionally gifted fossil collector, as his subsequent career in paleontology clearly attests. In the Bighorn Basin in 1881, he also had the advantage of collecting in virgin terrain, where no one had previously searched for the fossilized remains of prehistoric life. But all of these factors fail to explain Wortman's good fortune in finding what is still the oldest primate skull known from North America. More than 120 years of subsequent exploration in the Bighorn Basin—using the latest camping gear and off-road vehicles, global positioning systems and detailed topographic maps—has failed to locate a second specimen of *Tetonius* showing such superior preservation. By this criterion, the skull of *Tetonius* found by Jacob Wortman in 1881 is as extraordinary—and in some ways as priceless—as the Hope Diamond. Knowing this, it is easy to appreciate why some of the paleontologists who have spent a large portion of their professional lives looking for fossils in the Bighorn Basin habitually refer to *Tetonius* as "the gem of the Willwood."[5]

Just as *Adapis* and *Notharctus* exemplify their own diverse branches of the primate family tree, *Tetonius* pertains to a much larger group of extinct primates known as the Omomyidae. The group takes its name from the first fossil primate ever discovered in North America, a Bridger Basin species that Joseph Leidy described as *Omomys carteri* in 1869. To date, roughly sixty-nine species of omomyids have been reported from North America alone.[6] This extraordinary level of biological diversity ranks North American omomyids among the most successful primate radiations known from any place or time. Omomyids have also been found in Europe and Asia, but fossils that can be unambiguously assigned to this group have never been unearthed in Africa. Most of the omomyid-like primates from Europe belong to a different, but closely related, group known as the Microchoeridae. In order to evaluate the possible role of omomyids in the origin of anthropoids, we first need to understand something about their basic biology and mode of evolution. Unfortunately, the quality of the omomyid fossil record pales in comparison to that of their distant cousin, *Notharctus*. Fairly complete omomyid skeletons have not yet been discovered. To get a general picture of what omomyids must have been like, we must therefore consider the range of anatomy exhibited by this diverse group.

Most species of omomyids are documented only by their teeth and jaws. Accordingly, we can infer a great deal about the types of food these animals consumed. The vast majority of omomyids weighed less than a pound (450 grams), making it highly unlikely that any of these animals ate substantial quantities of leaves.[7] This finding alone renders the evolutionary radiation of omomyids distinct from that of the lemurlike adapiforms, most of which were specialized folivores. Broadly speaking, omomyids must have eaten the same sorts of foods preferred by the smallest primates alive today. Living mouse lemurs, bushbabies, lorises, tarsiers, and small monkeys like the pygmy marmoset eat such things as insects, small vertebrates, fruits, gums, and sap.[8] Omomyids as a group show such a broad range of jaw and tooth structure that they probably ate all of these types of food. At the same time, individual species undoubtedly specialized in one or two of these basic food groups, while forsaking others. Let's briefly consider some of the evidence indicating dietary specialization among particular species of omomyids.

The general relationship between dental anatomy and diet is discussed in chapter 2 in the context of reconstructing the dietary habits of the lemurlike adapiforms. The same basic principles apply to omomyids. That is, animals that eat foods requiring lots of cutting and slicing prior to digestion need cheek teeth with sufficient shearing ability, so folivorous adapiforms (like *Notharctus*) have sharper cheek teeth than those of species (like *Cantius*) that specialized in softer food, such as fruits. Given that few, if any, omomyids were large enough to have been committed folivores, those species with sharply crested cheek teeth must mainly have consumed insects and small vertebrates. In contrast, omomyids with rounded and blunt molars likely concentrated on softer foods that are more easily masticated. Most North American omomyids lack the strong molar shearing crests characteristic of living primates (like *Tarsius*) that mainly consume insects and small vertebrates.[9] One of the most significant exceptions to this rule is a lineage of omomyids known as the Washakiini. This relatively insectivorous or carnivorous branch of the omomyid family tree includes *Shoshonius cooperi,* an animal that will figure prominently later in our story.

While *Shoshonius* and a few other omomyids apparently preferred live animal prey, the majority leaned strongly toward vegetarianism. Most living primates consume fruits of various types, a pattern that almost certainly held for omomyids as well. Many omomyids have molars with weakly developed shearing crests and wide, blunt basins useful for chewing soft foods like ripe fruits. The species from the Bridger Basin that Cope originally described as *Anaptomorphus aemulus* is an excellent example of such an animal. *Anaptomorphus* and its close relatives, such as *Absarokius*, stand out as small primate frugivores in the conventional sense. *Tetonius* and several other omomyids have equally blunt and rounded cheek teeth, but their specialized incisors suggest that they were adapted to a different dietary regime, one that focused on tree exudates.

Certain species of primates are among the few living vertebrates that routinely exploit the gums and saps of flowering trees, also known as angiosperms. Gums and saps can be plentiful and nutritious food sources in many forests, yet their location on the trunks of trees puts them out of reach of most animals. In order to feed extensively on gums and saps, not only must an animal be able to climb, but it must also cling to vertical tree trunks for prolonged periods of time. Modifications of the jaw and front teeth facilitate this specialized dietary niche. The most committed living gum-feeders have lower incisors that are robust, pointed, and procumbent, which they use to gouge the bark of flowering trees to stimulate the flow of exudates. Modern humans employ the same basic strategy to collect the sugary exudates of maple trees in New England. Living mammals that engage in this sort of bark-gouging behavior include certain species of marmosets (small South American monkeys) and the Australian marsupial sugar glider *(Petaurus)*. Many other living primates (including mouse lemurs and bushbabies) consume gums and saps, but they lack the robust incisors necessary for active gouging.

The earliest and most primitive primates possessed small, delicate lower incisors that were not built for gouging bark. Several omomyids, including *Shoshonius, Anaptomorphus,* and *Absarokius,* retain this primitive incisor arrangement, which they inherited from earlier primate ancestors. Other omomyids, including *Tetonius, Anemorhysis,* and *Trogolemur,* evolved greatly enlarged and pointed lower incisors that protruded forward out the front of their mouths—exactly the sort of incisors required for active bark-gouging behavior. These particular omomyids also reinforced the bony part of their lower jaws in the region of the chin, another common adaptation to bark-gouging behavior among living gumfeeders. A further clue to the ancient diet of these specialized omomyids lies farther back in their mouths. The molars of *Tetonius, Anemorhysis,* and *Trogolemur* exhibit low relief and poor development of shearing crests, indicating that these omomyids did not rely heavily on insects and small vertebrates for food. Taken as a whole, the jaws and teeth of *Tetonius, Anemorhysis,* and *Trogolemur* suggest that these particular omomyids were specialized gum-feeders, although they probably ate other types of food whenever the opportunity arose.

From the preceding discussion of omomyid diets, we can infer that some omomyids exploited narrow feeding strategies, while others remained more conservative and versatile. Whether their locomotion was equally

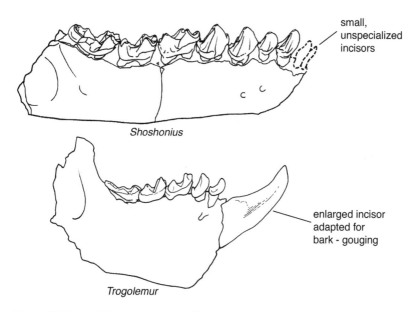

small, unspecialized incisors

Shoshonius

enlarged incisor adapted for bark - gouging

Trogolemur

Figure 10. Divergent incisor morphology reflects the range of dietary specialization among North American omomyid primates. Enlarged and procumbent incisors like those found in *Trogolemur* suggest that certain omomyids were specialized gum-feeders who used their incisors to gouge the bark of trees to stimulate the flow of sap and gum. Based on the development of shearing crests on its cheek teeth, *Shoshonius* ate mainly insects and small vertebrates. Its incisors were small and relatively unspecialized. Original art by Mark Klingler, copyright Carnegie Museum of Natural History.

varied remains an open question, however. Limb bones have been found for only a small fraction of the known species of omomyids. Even in these cases, only a few elements are known for each species, making any reconstruction of posture and style of movement tentative. At the same time, the differences in skeletal anatomy that have been documented so far among omomyids are all fairly minor. As a result, a reasonably detailed picture of what might be considered the "typical" omomyid pattern of movement has emerged. As new discoveries are made, it may become feasible to assess the range of omomyid locomotor adaptations more precisely, but this is beyond our current power of resolution.

In general, omomyids can be depicted as active tree-dwelling primates with a clear propensity for leaping.[10] Their overall pattern of locomotion probably falls somewhere between those of living mouse lemurs and bushbabies. The forelimbs of omomyids were surprisingly muscular, and the structure of their shoulders and elbows allowed wide ranges of movement at these joints. These characteristics suggest that omomyids were particularly adept at climbing. Like all primates, omomyids possessed nails rather than claws on their fingers and toes, and their big toes were capable of strong grasping. Their hips were built to accommodate many different hindlimb postures. In particular, the thighs of omomyids could be widely splayed apart—a necessary posture for clinging to large, vertical supports. Their knees and ankles indicate that omomyids were accomplished leapers, even compared with their larger cousins, the lemurlike adapiforms. Omomyids differ conspicuously from adapiforms in having more elongated ankle regions. The longer ankles of omomyids gave them greater mechanical leverage, making them more powerful leapers than *Notharctus* and other adapiforms.

Despite the poor sampling of omomyid skeletal elements noted earlier, the fossil record is sufficient to reveal important differences among omomyids in a few key features. These anatomical distinctions indicate that various omomyid species must have moved in different ways. One example comes from the anatomy of the omomyid tibia and fibula, the two bones that form the lower part of the leg between the knee and ankle. In tarsiers, rabbits, kangaroos, and many other mammals that habitually leap, the tibia and fibula become fused at their lower end near the ankle joint. Fusion between the tibia and fibula strengthens the lower leg during strictly fore and aft motion, but at the expense of decreased flexibility. Most primates, including humans, apparently need this extra mobility between the tibia and fibula, and the two bones remain separate to accomplish this. An intermediate condition exists whereby the tibia and fibula do not actually fuse, but they become so closely approximated that each bone leaves an obvious scar on its mate. From a purely functional perspective, this anatomically intermediate condition mimics that in which the two bones are fully fused—in both cases movement between the tibia and fibula is effectively halted. Unlike in the case of tarsiers, the tibia and fibula remain separate in all omomyids for which the relevant anatomy is known. However, the intermediate condition occurs in some omomyids, particularly *Shoshonius* and *Absarokius*. Based on their tightly bound lower leg bones, both *Shoshonius* and *Absarokius* must have been powerful and agile leapers. Other omomyids, including the relatively large middle Eocene species *Hemiacodon gracilis*, retain far greater flexibility at this joint, suggesting that these animals practiced a more eclectic pattern of locomotion that was not so focused on leaping.[11]

From what we've learned so far, omomyids can be characterized as small, fairly acrobatic primates that ate a wide variety of foods (almost everything except leaves). They apparently moved through their arboreal domain by climbing, leaping, and walking quadrupedally on the tops of branches. To understand more about their behavior and way of life, we need to look more closely at the anatomy of Wortman's *Tetonius* skull. This unique specimen provides many important clues regarding the daily activity pattern and sensory development of omomyids.

The most obvious feature shown by the skull of *Tetonius* is its large eye sockets, as Cope himself observed as soon as he examined the specimen.[12] Relative to the length of its skull, the eye sockets of *Tetonius* are larger than those of living primates that are active during daytime. *Tetonius* resembles small nocturnal primates such as bushbabies in this respect, suggesting that it too was mainly active at night. Upper jaw fragments of additional species of omomyids preserve the lower margin of the eye socket, giving some indication of orbit size across a broader spectrum of omomyids. Although the evidence is less compelling than in the case of *Tetonius*, enlarged eye sockets appear to have characterized these species as well. Most, if not all, North American omomyids therefore seem to have been active primarily at night.[13] A similar activity pattern is characteristic of tarsiers, lorises, bushbabies, and many lemurs, but only the South American owl monkey *(Aotus)* shows a similar preference among anthropoid primates.

We can reconstruct the sensory adaptations of *Tetonius* on the basis of the natural endocast of its brain preserved in Wortman's skull.[14] Dating from the early part of the Eocene (roughly fifty-four million years ago), *Tetonius* provides the earliest evidence regarding brain anatomy in any primate. Even at this early date, the brain of *Tetonius* shows some advanced features that set it apart from those of contemporary mammals. In *Tetonius*, the cerebral cortex is expanded in the temporal and occipital regions of the brain, areas that are related to vision and hearing. At the same time, the sense of smell was being downplayed in *Tetonius*, as evidenced by the reduced size of the brain's olfactory lobes compared with those of contemporary mammals. Nevertheless, the olfactory lobes of *Tetonius* remain large by the standards of living primates, especially compared to the tiny olfactory lobes of living tarsiers. The absolute size of the brain in *Tetonius* is truly unimpressive—it took up a meager volume of only 1.5 cubic centimeters. Even when we recall that *Tetonius* weighed little more than three ounces (ninety grams), its diminutive cranial capacity ranks it below living primates in the important ratio of brain size to body size.

Tetonius and other omomyids had much smaller canines than most other living and fossil primates. Indeed, the lower canines are nothing more than tiny vestigial teeth in several omomyid species, including *Pseudotetonius ambiguus* and *Trogolemur myodes*. Sexual dimorphism in the size and shape of omomyid canines is nonexistent. This important anatomical distinction between omomyids and at least some adapiforms (especially notharctids) implies substantial differences in the social organization of the two groups. In particular, we can dismiss any reconstruction of omomyid behavior that depicts them as gregarious species in which males competed intensively with one another to form harems. Rather, it seems more likely that omomyids resembled living tarsiers in their social organization. Tarsiers live a relatively solitary existence, although adult males and females form pair bonds of variable strength and endurance. Adult tarsiers of either sex exclude other adults of the same sex from their territories.[15] However, comparing the social behavior of omomyids with that of living tarsiers is risky, and anatomical differences between omomyids and tarsiers suggest that they must have behaved differently as well. For example, territorial fighting between living tarsiers typically entails biting, frequently resulting in scars and broken bones. The teeth of living tarsiers are well suited to this role—their canines and upper central incisors are large and sharply pointed. Because omomyids lacked such formidable weapons, they must have been more tolerant of each other, or else they fought in an entirely different manner.

Our overview of omomyid anatomy has demonstrated that these Eocene primates differed from contemporary adapiforms in many important ways. Omomyids were almost uniformly smaller than adapiforms, although the two groups did overlap to a minor extent. As a result of their small size, omomyids failed to exploit the potentially lucrative leaf-eating ecological niche, which was the commonest dietary strategy among adapiforms. Their smaller size also led omomyids to adopt an acrobatic style of locomotion emphasizing leaping. Although many adapiforms were also highly capable leapers, at least some of them (such as *Adapis*) downplayed this ability, focusing instead on a more deliberate mode of clambering through the trees. Most if not all omomyids seem to have been nocturnal, whereas the majority of adapiforms appear to have been diurnal. Finally, the social organization of omomyids apparently differed from that of at least some adapiforms, because there is no evidence for sexual dimorphism in canine size and shape among omomyids. In North America, omomyids also differed from adapiforms in being far more diverse—roughly sixty-nine species of North American omomyids have been described to date, compared with about twenty-one species of

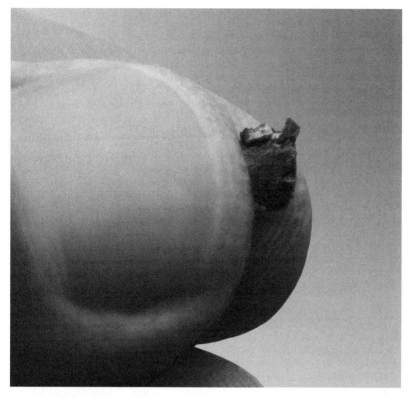

Figure 11. The bulk of the fossil record of North American omomyids consists of relatively fragmentary specimens such as this lower jaw of *Teilhardina* from the Bighorn Basin. Photograph courtesy of and copyright by Marc Godinot.

notharctids. How can we account for this discrepancy, especially given that omomyid fossils, being much smaller, are more difficult to find than those of notharctids?

I suspect that small body size itself was an important factor contributing to the high biological diversity of omomyids. The most diverse groups of living mammals are all small—think especially of rodents and bats—whereas mammals of larger body size, like rhinos and elephants, tend to be represented by only a few species. While the disparity in body size between omomyids and notharctids was considerably less than that between a mouse and an elephant, the same causal factors may well be at play. One aspect of biology that is particularly crucial in this regard is the ability of a species to disperse across long distances or geographic barriers such as mountain chains and rivers. In general, larger primates range over comparatively vast territories, while smaller species inhabit smaller domains. Living tarsiers, for example, have been reported to inhabit territories of from two to four hectares, while living lemurs that are similar in size to notharctids occupy home ranges ranging from twenty to three hundred hectares.[16] Speciation—the evolutionary splitting of one parent species into two daughter species—results from interrupting the flow of genes among different populations of the parent species. The most widely accepted means of achieving such reproductive isolation is to segregate different populations of the parent species geographically. Accordingly, if small species like omomyids are more prone to this sort of geographic isolation because of their inability to disperse across large distances or geographic barriers, we should expect them to show higher levels of diversity as a result.

The intermontane basins of the Rocky Mountain West, where the vast majority of North American omomyids have been found, provided an ideal setting for isolating individual populations of omomyids from one another. Mountainous uplifts bounded each basin, creating formidable barriers to the free dispersal of omomyids across the American West. During the Eocene, most of these mountain ranges were geologically young, and their topography would have been more rugged as a result. Against this geographic context, it is not surprising that different species of omomyids often inhabited adjacent basins during the Eocene. For example, three species of omomyids—*Tetonoides pearcei, Arapahovius gazini,* and *Anemorhysis savagei*—have been described from early Eocene strata in the Washakie Basin of southern Wyoming.[17] None of these species has ever been found in rocks of the same age in northern Wyoming's Bighorn Basin, although many more omomyid specimens have been recovered there.

Appropriately, many North American omomyids are named for the types of geographic barriers—especially mountain ranges and rivers—that may have fomented their diversity. *Tetonius* itself illustrates this trend rather well, being named for the Teton Range that dominates Wyoming's skyline, far to the west of where Wortman found his celebrated skull.[18] Other omomyids are also named for mountain ranges or topographic highpoints in the Rocky Mountain West—*Absarokius* for the Absaroka Range, *Uintanius* for the Uinta Range, *Tatmanius* for Tatman Mountain, and *Jemezius* for the Jemez Mountains. In a similar vein, some omomyids are named for streams and rivers—*Chlororhysis* for the Green River, *Anemorhysis* for the Wind River, and *Strigorhysis* for Owl Creek. Several of these geographic features, including the Absaroka Range, Tatman

Mountain, and Owl Creek, are located in or near the Bighorn Basin. This concentration is anything but random. It reflects the unsurpassed role of the Bighorn Basin's vast fossil fields in generating our current understanding of omomyids and how they evolved. During the 120 years since Wortman first explored the Eocene badlands of the Bighorn Basin, this region has become the gold standard for paleontologists interested in the plant and animal life of the early Eocene.[19] The thick sequence of sedimentary rocks that make up the Willwood Formation yields a nearly continuous record of life over the course of several million years. A few other Eocene sites—such as Messel in Germany—have produced individual specimens that are more dramatic, but no place on Earth documents change through time better than the Bighorn Basin. This natural laboratory provides us with a unique vantage point to assess how omomyids evolved during the early Eocene. In particular, we can examine the Bighorn Basin record of omomyids to see which evolutionary paradigm—the ladder or the tree—better reflects their evolutionary history.

Even in the abundantly fossiliferous Bighorn Basin, answering such an ambitious question requires years of effort in the field, coupled with painstakingly diligent studies in the lab. Assembling the fossil database is the obvious prerequisite for anything else, yet this is the most difficult step of all. I know because, as a graduate student, I spent five summers prospecting for fossils in the Bighorn Basin with my mentors Ken Rose and Tom Bown, who were then working on precisely this problem.[20] Day after day, I dutifully scoured what seemed like endless Willwood outcrops, only to find that I had failed to discover a single omomyid among the ample fossils I had collected. As it turns out, omomyids were never very abundant, judging by their representation in the large samples of early Eocene mammals known from the Willwood Formation. They typically amount to only 1.5 percent of all the fossil mammals recovered at a given site. In order to examine how omomyids evolved, we collected detailed stratigraphic data for each specimen we uncovered. This allowed hundreds of Bighorn Basin omomyids to be arranged in stratigraphic order, a useful proxy for time. Only after all of this preliminary work was completed could Bown and Rose attempt to decipher how Bighorn Basin omomyids evolved. The patterns they uncovered show that, at least in an ideal setting like the Bighorn Basin, both the ladder and the tree paradigms find some basis in reality.

The evolutionary pattern shown by *Tetonius* in the Bighorn Basin is particularly interesting, because it provides compelling evidence for gradual change in an evolving lineage of primates. With the exception of Wortman's skull, all *Tetonius* specimens from the Bighorn Basin consist of upper and lower jaws. Accordingly, the evolutionary changes among these animals that were documented by Bown and Rose are restricted to this anatomical region. Functionally, I suspect that the modifications that transpired during the evolutionary lifetime of the *Tetonius* lineage indicate increasingly efficient gum-feeding in these small primates. The latest members of the lineage developed features suggesting that they were more effective at gouging bark than their ancestors would have been. Other aspects of the anatomy of these animals—such as their limbs and their brains—may also have evolved during the early Eocene, but we have no way of evaluating this at present.

The earliest and most primitive specimens of *Tetonius* from the Bighorn Basin belong to a species called *Tetonius matthewi*. These small primates have elongated lower jaws in which the front teeth (from front to back, two incisors [abbreviated as I_1 and I_2], a canine, and three premolars [abbreviated as P_2, P_3, and P_4, respectively]) retain very primitive proportions with respect to one another. In particular, the I_1 in *Tetonius matthewi* remains fairly small; the canine is larger than I_2; P_2 is present, and two distinct roots support P_3. Over the course of the next two million years or so, this primitive pattern of tooth and jaw structure underwent several successive modifications. The entire lower jaw became shorter from front to back and deeper in the region of the chin. This severe shortening of the jaw required a great deal of compaction in the lower teeth. One of the premolars (P_2) was lost entirely. The widely splayed roots of a second premolar (P_3) were first compressed and then fused into a single root that supported a much smaller tooth crown. The canine also became diminished, so that it eventually resembled one of the incisors (I_2) in size and shape. At the same time, the other lower incisor (I_1) became hypertrophied and progressively more chisel-shaped. The youngest specimens of the *Tetonius* lineage in the Bighorn Basin differ so radically from *Tetonius matthewi* that they are referred to an entirely separate genus and species, *Pseudotetonius ambiguus*.

The Bighorn Basin record of *Tetonius* has played an important role in scientific debates about the mechanics of evolution in general. Notably, it conflicts with the popular model of evolution known as punctuated equilibrium, whereby gradual evolutionary change within lineages is regarded as insignificant. Instead, the mode of evolution shown by the *Tetonius-Pseudotetonius* lineage conforms to a more classical interpretation of Darwin's theory. It also recalls Le Gros Clark's depiction of primate evolu-

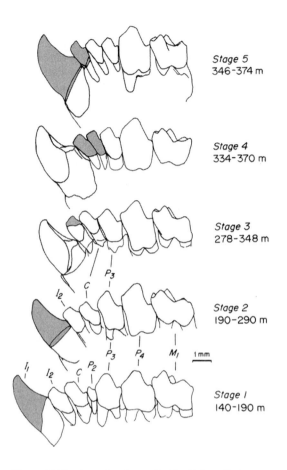

Figure 12. The dense fossil record in the Bighorn Basin of Wyoming reveals gradual evolution in certain lineages of omomyid primates there. The lineage connecting primitive *Tetonius matthewi* (stage 1, at the bottom) with its more advanced descendant *Pseudotetonius ambiguus* (stage 5, at the top) is particularly well documented. Anatomical change over time produced severe compaction of the front part of the lower dentition, possibly reflecting a progressive adaptation to gum-feeding. Reproduced from Bown and Rose 1987, copyright The Paleontological Society, Inc. Reproduced by permission.

tion as a simple ladderlike progression from tree shrew to human. Accordingly, in the Bighorn Basin at least, the ladder paradigm appears to have some relevance for understanding how omomyids evolved. But is the ladder paradigm alone sufficient to reconstruct the evolutionary history of Bighorn Basin omomyids?

The answer is unequivocally no. *Tetonius* itself appears abruptly in the Bighorn Basin fossil record, apparently as an immigrant from somewhere else. In order to reconstruct how *Tetonius* fits on the family tree of all omomyids, we cannot merely connect the dots to some earlier omomyid ancestor. Instead, we must utilize the tenets of cladistics—the intellectual basis for the competing tree paradigm. The same is true for the large number of other Bighorn Basin omomyids that appear suddenly in the fossil record there. One of the most important examples is that of *Steinius vespertinus,* the most primitive North American omomyid yet discovered.[21] Despite its very primitive anatomy, *Steinius* does not occur in the oldest rock strata of the Bighorn Basin. By the time *Steinius* finally appears—more than halfway through the thick layer cake of rock strata forming the Willwood Formation—the *Tetonius-Pseudotetonius* lineage was already extinct. Yet in most of its known features, *Steinius* is more primitive than the earliest *Tetonius matthewi!* Once again, if we want to understand how *Steinius* relates to other omomyids, we are forced to use the evolutionary tree paradigm. The ladder simply isn't applicable, because no sequence of fossils documents the previous ancestry of *Steinius.*

Even in the ideal setting of the Bighorn Basin, the example of *Steinius* shows that we cannot simply read evolutionary history as if it were successive pages in a book. Instead, we must regard the fossil record as highly variable. In the spectacular badlands of the Bighorn Basin, the fossil record can be remarkably rich, as in the case of the *Tetonius-Pseudotetonius* lineage. In these rare instances, it often makes sense to employ the ladder paradigm to interpret how certain long-lived and densely sampled lineages evolved. Unfortunately for paleontology, once we leave the confines of the Bighorn Basin the fossil record typically becomes more diffuse. Instead of sampling entire chapters of the evolutionary saga in one place, we are usually lucky to unearth a page here and a paragraph there. The only way that we can make sense of such a far-flung and incomplete fossil record is by working under the assumptions of the tree paradigm. Beneath the shade of the tree, there are no ancestors—only the phylogenetic equivalent of aunts, uncles, cousins, and siblings.

If we expand our scope to consider continents other than North America, we find that omomyids closely resembling those from the American

West also lived in Europe and Asia during the Eocene. A separate group of omomyidlike primates known as the Microchoeridae evolved in Europe. The relationship between European microchoerids and North American omomyids is analogous to that between European adapids and North American notharctids. In each case, we know that the European and North American groups of Eocene primates are closely related, even if we don't yet know precisely how. In an evolutionary sense, the Eurasian omomyids stand even closer to their North American brethren. Let's briefly consider this Eurasian record in order to reach a broader understanding of their possible role in our own deep evolutionary history.

To date, the Asian record of omomyids is meager, especially given the amazing diversity of these creatures in North America. By far the most significant discoveries have been unearthed in China, where several different omomyids from the Eocene show remarkably close affinities to North American species.[22] One of these Chinese omomyids, called *Asiomomys changbaicus,* is the nearest known relative of *Stockia powayensis* from southern California. My team discovered a second Chinese omomyid, which we named *Macrotarsius macrorhysis,* during our fieldwork in the Shanghuang fissure-fillings near Shanghai. These Chinese fossils of *Macrotarsius* establish an even closer link with North America, because all other species of *Macrotarsius* hail from that continent. Given that two different groups of middle Eocene omomyids—*Macrotarsius* and the *Asiomomys-Stockia* group—have now turned up on both sides of the Pacific, we are forced to conclude that omomyids successfully crossed the Bering land bridge about forty-five million years ago. This primate dispersal event took place roughly ten million years after the initial migration of primates across the Bering land bridge at the beginning of the Eocene (see chapter 2). I suspect that additional waves of migration flowed between these two continents during this time, although direct fossil evidence for this is lacking. For example, is it possible that the sudden appearance of *Steinius* in the Bighorn Basin reflects yet another episode of dispersal from Asia?

Against the backdrop of the Earth's modern geography, it seems counterintuitive that the omomyids of North America and Asia were so similar, while an entirely different group—the microchoerids—lived in Europe. However, once the North Atlantic land bridge became submerged during the early Eocene, Europe was permanently segregated from North America. At the same time, an epicontinental seaway east of the modern Ural Mountains sequestered Europe from Asia during most of the Eocene. These geographic barriers prevented North American and Asian omomyids from interacting with their European relatives, allowing European microchoerids to evolve under conditions that George Gaylord Simpson famously referred to as "splendid isolation." The microchoerids that resulted were omomyidlike primates with a few distinguishing features of their own.

Microchoerids achieved a moderate level of diversity in Europe, where they are represented by about eleven species.[23] Although microchoerids occur at many Eocene localities in Germany, Switzerland, England, and Spain, the best anatomical material has been recovered from the Quercy fissure-fillings of southern France. There, a microchoerid known as *Necrolemur antiquus* lived alongside *Adapis, Leptadapis,* and a variety of other mammals. *Necrolemur* differs from omomyids chiefly in the specialized anatomy of its hindlimbs.[24] As is the case with tarsiers, but in contrast to omomyids, in *Necrolemur* the tibia and fibula are fused and the calcaneus is extremely elongated. Both of these features indicate that *Necrolemur* engaged in a great deal of leaping. Paradoxically, the hip of *Necrolemur* resembles those of anthropoids and differs appreciably from those of tarsiers and omomyids. The femoral head (the "ball" part of the ball-and-socket hip joint) in *Necrolemur* is shaped like a sphere, whereas this structure is more cylindrical in tarsiers and omomyids. Most small primates that habitually leap (especially tarsiers and bushbabies) possess cylindrical femoral heads, reflecting their preference for flexed and widely splayed hip postures. *Necrolemur* would have been capable of a wide range of hip postures, but it is difficult to reconcile its very tarsierlike ankles and lower legs with its monkeylike hips.

Like tarsiers and omomyids, microchoerids had large eye sockets for their size, implying that they too were nocturnal. The teeth and jaws of all microchoerids share certain advanced characteristics, indicating that they evolved from a single common ancestor. For example, microchoerids differ from most omomyids in possessing only two lower premolars (having lost the small premolar known as P_2). Despite their uniformity in terms of tooth number, microchoerids evolved a wide variety of cheek tooth patterns. These differences in molar morphology indicate that microchoerids specialized in eating a wide range of food. For example, *Pseudoloris parvulus* possesses sharply crested cheek teeth like those of tarsiers. *Pseudoloris* probably fed mainly on insects and other small animals. On the other hand, the cheek teeth of *Necrolemur* emphasize broad basins and blunt cusps, suggesting a very frugivorous diet. Microscopic wear patterns on the enlarged and slightly procumbent lower central incisors of *Necrolemur* indicate that its front teeth were sometimes used for grooming fur, a behavior that is more reminiscent of lemurs than tarsiers.[25]

Precisely how microchoerids fit on the primate family tree remains something of a mystery. They share many features with omomyids, and most paleontologists believe that the two groups are closely related. An alternative view holds that microchoerids and tarsiers are descendants of a common ancestor they share to the exclusion of omomyids. The second interpretation rests mainly on the similarities in hindlimb anatomy between tarsiers and *Necrolemur*—the fused tibia and fibula and the similarly elongated calcaneus. However, these hindlimb features may have evolved independently in tarsiers and *Necrolemur*, as similar adaptations for leaping. The fact that the hips of tarsiers and *Necrolemur* differ so dramatically supports this view. If so, a unique evolutionary connection between tarsiers and *Necrolemur* (excluding omomyids) would be as spurious as that proposed by Cuvier for *Adapis* and "pachyderms."

If microchoerids are not the closest evolutionary cousins of tarsiers, they probably represent a lineage that diverged early on from North American omomyids. At the very beginning of the Eocene, an anatomically primitive omomyid known as *Teilhardina* lived on both sides of the Atlantic Ocean. Its fossils have been unearthed at a site called Dormaal in Belgium and in the Bighorn Basin.[26] Possibly, *Teilhardina* gave rise to both the European microchoerids and a large fraction of the later omomyids of North America. If so, the two groups must have gone their separate ways once the land bridge spanning the North Atlantic became submerged, as it did shortly after the time when *Teilhardina* lived.

In many respects the evolutionary history of omomyids and microchoerids closely tracked that of their larger relatives, the lemurlike adapiforms. Both groups achieved broad geographic distributions across North America and Eurasia, although adapiforms inhabited Africa as well. The severe climatic deterioration of the Eocene-Oligocene boundary decimated all major groups of Eocene primates, although sivaladapids weathered the storm in tropical Southeast Asia for a while. From a humanistic perspective, our understanding of these Eocene primates is founded upon the efforts of pioneers in the field of vertebrate paleontology—scientists like Cuvier in the case of *Adapis*, and Wortman and Cope in the case of *Tetonius*. Cuvier's role has already been discussed in considerable detail. Let's turn our attention now to Wortman and Cope, whose collaboration says a great deal about the status of American paleontology toward the end of the nineteenth century.

Edward Drinker Cope and Jacob L. Wortman shared similar interests and ambitions, but they hailed from very different worlds. Cope was born into a family of wealthy Quakers in Philadelphia. His personal fortune allowed him to pursue his passion for natural history with little regard for holding down a salaried position. Wortman, on the other hand, grew up under far humbler circumstances in Oregon. He had to rely on his profession for his livelihood, not vice versa. Both men were brilliant scholars, and their scientific achievements reflect their innate intelligence, combined with a high degree of personal motivation. Yet their different backgrounds meant that their relationship was hierarchical rather than balanced. Cope personified most American paleontologists of the time. Wortman embodied a new breed of scientists, for whom success depended on raw talent, hard work, and no small measure of luck. Like mammals at the Cretaceous-Tertiary boundary, scientists like Wortman would soon inherit the earth. But Wortman himself was ahead of his time, the academic equivalent of a Mesozoic mammal lurking among the shadows of the dinosaurs.

Despite his early role as a hired-hand fossil collector for Cope, Wortman always aspired to become a scientist in his own right. Early in 1882, at the tender age of twenty-five, Wortman published an original account of Bighorn Basin geology as a preface to Cope's much longer treatise describing the fossils Wortman collected for him there. Reading between Wortman's lines, you can detect the beginnings of some class-based tension even then:

> During the summer of the present year [1881] the writer has been engaged in further exploration of this interesting region, which resulted in the collection of a large number of extinct vertebrates, obtained exclusively from the lower Eocene horizon of the Big-Horn, and which have all been submitted to Prof. Cope, at whose instance the party was organized and equipped.[27]

By modern standards, Wortman would certainly be justified in feeling that his scientific achievements were being shortchanged by Cope. Wortman discovered the spectacular fossil fields of the Bighorn Basin entirely on his own initiative. Yet he had no choice but to hand over to Cope all of the paleontological riches that flowed from his discovery, including the precious skull of *Tetonius homunculus*. After completing his stint as Cope's field collector, Wortman was divorced from paleontology for a time, although apparently not by choice. He worked at a number of odd jobs and even completed a medical degree at Georgetown. But when the opportunity arose to return to paleontology, Wortman abandoned any plans he may have had for a career in medicine to pursue his passion for fossils. Cope offered Wortman new employment, this time to undertake the delicate task of cleaning and preparing the fossils he had collected from the hard matrix that encased them. While this new arrangement kept Wortman near his beloved fossils, Cope continued to block him from publishing on them, despite Wortman's growing reputation as a first-rate comparative anatomist.[28]

Wortman finally gained a measure of scientific autonomy by moving out from the shadow of Cope and into that of another well-heeled patron of nineteenth-century science, Henry Fairfield Osborn. Like Cope, Osborn came from a fabulously wealthy family—he was the nephew of J. P. Morgan, the famous railroad and banking tycoon. Unlike Cope, Osborn leveraged his social skills and family connections to advance the scientific agenda of an entire institution. Beginning in 1891, Osborn presided over the paleontology program at New York's American Museum of Natural History, eventually rising through the ranks to serve as its president. Osborn fully appreciated the public's fascination with dinosaurs and other prehistoric beasts, and he decided to hire Wortman to bring display-quality specimens back from the Rocky Mountain West. Wortman could not refuse the carrot that Osborn dangled in front of him. It was the opportunity, at long last, to publish on the fossils he had dedicated so much of his life to collecting and preparing for others to study.[29]

For most of the 1890s, Wortman excelled in his curatorial position in the department of vertebrate paleontology at the American Museum. He led multiple expeditions to his old haunts in the Bighorn Basin and other parts of the American West. Back in New York, he published extensively on geology and paleontology, focusing on the anatomy and evolution of large mammals like camels, carnivores, and the slothlike taeniodonts (for whom he coined the name Ganodonta, a term that has since been abandoned). Wortman's newfound academic freedom even allowed him to criticize his former employer Cope in print, something he obviously relished.[30] Yet Wortman remained beholden to Osborn, who had the authority to order him to pursue scientific objectives that sometimes conflicted with Wortman's own agenda.

Wortman's brief but tumultuous career as a dinosaur paleontologist illustrates the dilemma he faced. By the late 1890s, Osborn became committed to satisfying the public's growing demand for dinosaurs, and he directed Wortman to focus his field efforts on Jurassic-age sites, where he might find and collect dinosaur skeletons. Wortman complied, but his own scientific interests remained in the area of fossil mammals. At the same time, Osborn hired Barnum Brown (whom we encountered in chapter 1, as the discoverer of the Burmese fossil primate *Amphipithe-cus mogaungensis*), who would soon earn a reputation as one of the greatest dinosaur paleontologists of all time. Wortman and Brown failed to hit it off, possibly because of professional rivalry.[31] In any case, Wortman resigned his position at the American Museum in 1899 to accept a curatorial post in paleontology at the brand-new Carnegie Museum in Pittsburgh.

On paper, the move looked like a promotion. At the Carnegie Museum, Wortman was the sole curator of vertebrate paleontology, which put him in charge of that department. Within the narrow confines of his academic unit, he didn't have to answer to anyone like Osborn. But in reality, Wortman faced exactly the same predicament in Pittsburgh that had caused him to flee New York—the Carnegie Museum's authoritarian director, W. J. Holland, called all the shots. And what Holland demanded of Wortman must have sounded pretty familiar. The museum's founding patron, Andrew Carnegie, had also become infatuated with dinosaurs. He desperately wanted one to form the centerpiece of his museum. Ever conscious of living up to Carnegie's expectations, Holland decided to mount a serious expedition to satisfy the museum's most important benefactor. As soon as Wortman arrived in Pittsburgh, Holland sent him off to Wyoming with explicit orders to find an exhibit-quality dinosaur for the Carnegie Museum.[32]

Wortman's years of experience searching for fossils in Wyoming's badlands paid off handsomely and in short order. His team located the nearly complete skeleton of a sauropod—the group that includes the largest dinosaurs, notable for their long, giraffelike necks—at a remote site in southeastern Wyoming known as Sheep Creek. Detailed study of the dinosaur's anatomy revealed it to be a new species of the genus *Diplodocus*, and the fossil was tactfully named *Diplodocus carnegii*. *Diplodocus* had an immediate impact on paleontology in general and the Carnegie Museum in particular. Holland referred to it as "the animal which made paleontology popular," decades before dinosaurs starred in blockbuster Hollywood films.[33] Wortman's fabulous specimen still presides over the Carnegie Museum's Dinosaur Hall, only a few yards from my office. Today, many would consider the discovery of *Diplodocus carnegii* Jacob Wortman's crowning achievement in paleontology. With the exception of his discovery of the skull of *Tetonius* eighteen years earlier, I would

Figure 13. Jacob Wortman (standing, second from right) and an all-star cast of American paleontologists in the field near Sheep Creek, Wyoming, in 1899. Wortman had just resigned his curatorial position at the American Museum of Natural History to lead the Carnegie Museum's search for Jurassic dinosaurs. Standing to the left of Wortman is William J. Holland, the director of the Carnegie Museum, who is holding a rifle. Standing to the left of Holland is Henry Fairfield Osborn, future president of the American Museum of Natural History and Wortman's former boss. Standing at the far left is William Diller Matthew. Kneeling in front of Matthew is Walter Granger, who participated in numerous fossil campaigns in the Rocky Mountain West, the Fayum region of Egypt, and the Central Asiatic Expeditions to Mongolia. Negative 1086 in the archives of the Section of Vertebrate Paleontology, Carnegie Museum of Natural History, copyright Carnegie Museum of Natural History. Reproduced by permission.

have to agree. Wortman himself apparently felt the same way. Less than a year after accepting his job at the Carnegie Museum, Wortman resigned in favor of a financially unstable position at Yale's Peabody Museum. There, he had the chance to continue studying and publishing on fossil mammals.

Wortman's last substantial contributions to paleontology brought his career full circle. In 1903 and 1904, he published a series of articles in Yale's *American Journal of Science* that, among other topics, explored the significance of *Tetonius* for understanding primate evolution. Looking back at Wortman's published views today, they appear surprisingly modern, especially considering the quantum advances we've made in fleshing out the primate fossil record during the intervening century. Wortman supported the view that tarsiers and living monkeys, apes, and humans shared a common ancestry after the evolutionary divergence of lemurs. He also believed that fossil primates such as *Tetonius* and *Necrolemur* were closely related to tarsiers. Indeed, Wortman went so far as to classify tarsiers and their fossil relatives (including *Tetonius* and *Necrolemur*) as one of the three main branches of Anthropoidea, for which he coined the name Paleopithecini. Only in a few instances do Wortman's views now seem dated, but we have the benefit of a much fuller fossil record than did he.

In the end, Wortman never succeeded in escaping the powerful grip that old money had on American vertebrate paleontology during his lifetime. After a few years at Yale, Wortman was forced to abort his plans to study all of the fossil mammals in the great collection amassed by O. C. Marsh at the Peabody Museum. Like his bitter rival Cope, Marsh himself was able to pursue a career in paleontology only because of the deep pockets of his relatives. Those funding streams evaporated when Marsh died. Unfortunately for Wortman, the infrastructure of American science remained mired in the nineteenth century. Yale apparently lacked the financial resources to continue the type of highly visible research in vertebrate paleontology that had brought fame and notoriety to Marsh— and by extension to Yale itself. Wortman, being a man of only modest means, finally quit the field in disgust. He spent the rest of his life running a drugstore in Brownsville, Texas. It would take six more decades for Yale to reappear as a player in the search for our deepest evolutionary roots.

4

The Forest in the Sahara

On the fringe of Egypt's immense Western Desert, about sixty miles southwest of Cairo, a series of escarpments rises above a brackish lake known as Birket Qarun. In antiquity, the lake provided early Egyptian farmers with the rare opportunity to cultivate crops beyond the narrow strip of arable land lining the Nile Valley. Successive Egyptian dynasties controlled the level of the lake by regulating the flow of water through a canal linking it with the Nile, an indication of their technological prowess. Ancient roads, temples, and other archeological features abound in the surrounding region, a topographic basin known as the Fayum Depression. To the west, Saharan dune fields stretch farther than the eye can see, more or less continuously to the Atlantic coast of Morocco.

For more than a century, archeologists interested in the origins of human civilization have studied the temples and monuments in the vicinity of Birket Qarun. During the same interval, paleontologists interested in reconstructing the common ancestry of monkeys, apes, and humans have gravitated to the series of escarpments to the north. The rock strata forming these cliffs range in age from the end of the Eocene to the beginning of the Oligocene (about thirty-three to thirty-six million years ago).[1] Just as nearby antiquities testify to the life and times of Egypt during the reign of the pharaohs, the sequence of rocks known as the Jebel Qatrani Formation illuminates a much earlier chapter of our evolutionary history. Until recently, virtually all that was known about the earliest anthropoid primates came from this single rock formation in northern Egypt. The vast majority of the roughly five hundred primate fossils unearthed there have been discovered by a series of expeditions led by Elwyn L. Simons.

For several decades, Simons has spearheaded the quest for anthropoid origins. No one has spent more time searching for fossil primates in the field—or in a wider variety of places. During his long tenure as a professor, Simons has also trained an astounding number of students, many of whom have gone on to chart distinguished careers of their own. In some ways, I count myself among this group, although Simons might hesitate to claim me as his intellectual offspring. Nevertheless, I was first introduced to the world of fossil primates as an eager undergraduate in 1982. The course that changed my life was unpretentiously entitled "The Primate Fossil Record," and the professor was Dr. Elwyn Laverne Simons. I have yet to meet a more intriguing person.

For me, what sets Simons apart is his unique melding of personality traits that are rarely encountered together—qualities like offbeat eccentricity coupled with a peculiar form of charisma, stubborn bullheadedness matched by anxious insecurity, and simple absent-mindedness alongside scholarly brilliance. Whether by accident or design, Simons projects an image that harks back to the great explorers of a bygone era. He often sports vintage 1890s muttonchops for facial hair, and he prefers Greek fisherman's caps to more fashionable headgear. Yet Simons is equally at home schmoozing with multimillionaires who might be potential benefactors or crawling through a dank cave in the wilds of Madagascar, searching for the bones of recently extinct lemurs that were the size of an ape. Over the years, I have interacted with Simons in a variety of ways—we've camped together and collected fossils side by side in Wyoming; we've shared a microscope to compare fossil primates from China and Egypt, and we've debated the meaning of certain fossils at large professional meetings and small symposia.

From my earliest interactions with him as an undergraduate, I regarded Simons as a particularly intimidating professor. Later experiences showed me that this wasn't the case at all. Once he realized that I was truly interested in fossil primates, Simons and I developed a genuine sense of camaraderie. Like many other paleontologists, I always found Simons to be more relaxed and approachable in the field than in the lab or classroom. I fondly remember spending long summer evenings around the campfire in Wyoming listening to Simons spin yarns about his exploits in the field and his scientific battles with anyone who dared to question his views. These intellectual tirades, lubricated by whatever brand of bourbon happened to be in camp, often lasted long into the night. Yet the next morning Simons would reappear ready and eager to march once again into the badlands to search for clues bearing on our distant evolutionary history. Only rarely did the previous night's lecture prevent Simons from collecting his quota of fossils the following day. Even then, Simons often revealed himself to be uniquely at home in the desert. Once, nearly twenty years ago, I skirted an outcrop in the Bighorn Basin and spotted Simons lying spread-eagled face down on the Willwood Formation. Fearing the worst, I drew nearer, only to find that my former professor was merely taking a nap!

Many years have passed since Simons and I worked together in the field, and I can no longer claim that we are close. Because we now disagree entirely on the early evolution of anthropoids, I can easily imagine who serves as a current target of Simons's late-night rants by the campfire. Nonetheless, to understand how Fayum primates shed light on our deepest evolutionary roots, we must also become familiar with Elwyn Simons. Through his protracted series of Fayum expeditions, Simons maintained a virtual monopoly on the issue of anthropoid origins for several decades. In paleoanthropology, having the oldest fossils is often the surest way to establish yourself as the reigning expert. As we shall see, whether the point of contention was *Ramapithecus* or the origin of anthropoids, Simons has heeded this simple precept throughout his career. At times, this has fostered conflict between Simons and those whom he perceives as scientific rivals. Sometimes, men and ideas become inherently intermingled.

Simons first set foot in the Fayum Depression in the autumn of 1961, as a young tenure-track professor at Yale.[2] It had been nearly six decades since Jacob Wortman left New Haven to pursue a life devoid of paleontology. The intervening years had been good for American science. The creation of the National Science Foundation in 1950 meant that basic research in fields such as paleontology no longer depended solely on the deep pockets of wealthy patrons. Meanwhile, new discoveries of early hominids had stoked popular interest in the search for fossils bearing on humanity's remote ancestry. Only two years previously, Louis Leakey had announced with great fanfare that he and his wife Mary had found a startling new australopithecine skull at Olduvai Gorge in Tanzania.[3] The specimen, which was christened *Zinjanthropus boisei,* had been recovered in association with primitive stone tools and broken animal bones, possibly representing the leftovers of some prehistoric feast. Heralded by Leakey as "the oldest well-established stone toolmaker ever found," *Zinjanthropus* quickly captured the public's imagination.[4] *Zinjanthropus* had a relatively small braincase, adorned by sagittal and nuchal crests; an abbreviated lower face, in which the incisors and canines form a straight line; and massively enlarged molars adapted for prolonged and powerful chewing. After a long interval during which paleoanthropologists had focused on Europe and Asia as potential cradles of humanity, Africa once again took center stage. Having been trained as a vertebrate paleontologist, Simons wanted to push the pedigree of humans and our nearest primate relatives even farther back into the remote past. Egypt seemed like a good place to start.

As might be expected, however, Simons did not launch a major international expedition into the Egyptian desert on a lark. Previous researchers had already demonstrated that several types of early anthropoids had roamed the Fayum region during Oligocene time. Indeed, it was Henry Fairfield Osborn—Wortman's former boss at the American Museum of Natural History—who described the first of these in 1908.[5] Osborn's specimen consisted of a lower jaw preserving four low-crowned cheek teeth with remarkably blunt cusps. The fossil differed substantially from anything Osborn had ever seen, so he gave it the new name *Apidium phiomense.* By coincidence, Osborn followed the same etymological trajectory in constructing *Apidium* (from Apis, the sacred bull god of ancient Egypt, and the Latin diminutive suffix *-idium;* literally meaning "little Apis") that Cuvier had pioneered in naming the French Eocene primate *Adapis* nearly a century earlier. Like Cuvier before him, Osborn couldn't be sure what kind of animal his fossil represented. He noted that *Apidium* resembled primates in certain ways, but he also compared it with piglike artiodactyls. Rather than make an embarrassing mistake, Osborn decided to leave *Apidium* in taxonomic limbo. Thus, the first fossil of an early anthropoid ever found was not immediately recognized as such.

Within a few years of Osborn's description of *Apidium,* any doubts that might have lingered about the presence of early anthropoids in the Fayum were erased. Osborn had organized his expedition with the primary goal of recovering the fossils of small mammals. Earlier Fayum expeditions had located the remains of strange, elephantine beasts known as arsinoitheres, along with a menagerie of other exotic animals ranging

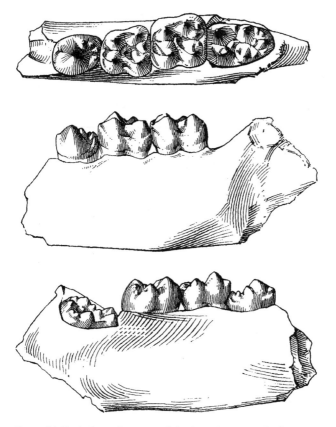

Figure 14. The holotype lower jaw of *Apidium phiomense*, the first of many fossil anthropoids to be described from the Fayum region of Egypt. Reproduced from Osborn 1908.

in size from medium to large. Osborn's experience collecting fossils in the American West led him to speculate that many additional types of mammals, most of them small, remained to be discovered in the Jebel Qatrani Formation. This turned out to be true. However, Osborn himself was only indirectly responsible for confirming the prediction. Instead, it fell to one of the local members of Osborn's field team—a man named Richard Markgraf—to demonstrate that small mammals could indeed be retrieved from the sandy fossil beds of the Fayum.

Markgraf was a German expatriate who lived in the nearby village of Sennuris. Even before Osborn mounted his expedition to the Fayum region, Markgraf had a history of collecting fossils there. Some of these he sold to a museum back home in Stuttgart.[6] Because Markgraf already possessed such practical experience, Osborn was pleased to offer him a position on his team once he arrived in Egypt. This decision paid off handsomely, because it was Markgraf who found the lower jaw Osborn would later name *Apidium phiomense.*[7] After the American Museum team returned to New York, Markgraf continued to collect fossils from the Jebel Qatrani Formation. As before, he sold some of the new fossils he found to the museum in Stuttgart, but he now offered fossils for sale to the American Museum as well. Whether by chance or design, Markgraf's best Fayum primate specimens all went to Stuttgart, however, no doubt much to Osborn's chagrin.

Once Markgraf's shipment of fossils arrived in Stuttgart, it was handed over to Max Schlosser, a German paleontologist based in Munich. Schlosser was duly impressed by the undeniable quality of Markgraf's collection. But the new Fayum primate fossils were truly remarkable. In contrast to the fragmentary specimen of *Apidium* described three years earlier by Osborn, Markgraf's new material included the virtually complete lower jaws of two new species of Fayum primates. A third primate fossil failed to measure up to such high standards, but even it appeared to document a new species. Taken together, this new collection of Fayum primates revealed an astonishing diversity of early anthropoids there. To communicate this newfound diversity to other scientists, Schlosser felt compelled to describe three new genera and species of early anthropoids—*Propliopithecus haeckeli, Parapithecus fraasi,* and *Moeripithecus markgrafi.*[8] It was now clear not only that early anthropoids occurred in the Fayum but also that the group had already diversified by early Oligocene time. For Schlosser, the three new Fayum anthropoids hailed from very different parts of the evolutionary tree—one came from the main trunk, while the other two belonged to peripheral, insignificant twigs.

As Schlosser saw it, the Fayum primate that was metaphorically climbing the trunk of the family tree was *Propliopithecus.* Its apelike cheek teeth and stoutly constructed lower canines made *Propliopithecus* a credible ancestor of living and fossil apes and humans alike. This idea was not so subtly conveyed in the construction of the new fossil's name—*Propliopithecus* literally means "before *Pliopithecus.*" In contrast to the completely novel Fayum primates, *Pliopithecus* was already familiar to Schlosser and other paleontologists of the day, because its fossils had turned up in a variety of sites across Europe. Despite its name, *Pliopithecus* antedates the Pliocene; it lived roughly fourteen million years ago, in the middle of the preceding Miocene epoch. Most of Schlosser's scientific peers regarded *Pliopithecus* as one of the oldest fossils directly related to modern apes. Schlosser therefore interpreted the similarity between European *Pliopithecus* and Egyptian *Propliopithecus* as indicating that the ape and human lineage could now be traced back as far as the Oligocene.

If *Propliopithecus* represented the main line of primate and human evolution, the other two Fayum primates described by Schlosser could only pertain to its outer margins. This was patently the case with *Parapithecus*, a primate that seemed so weird that Schlosser was forced to create a new family—the Parapithecidae—for it alone. Once again, Schlosser advertised his evolutionary conclusions regarding *Parapithecus* in the construction of its name—*Parapithecus* translates roughly as "alongside apes and monkeys." Even though all he had to work with was a lower jaw, Schlosser had ample reason to relegate *Parapithecus* to the sidelines of anthropoid evolution. Each half of its lower jaw supported only one incisor, where living apes and monkeys have two. At the same time, *Parapithecus* retained three lower premolars bilaterally, a resemblance to South American monkeys (all other anthropoids have only two lower premolars). Faced with this unique combination of features, Schlosser concluded that parapithecids were strange but primitive anthropoids that had reached an evolutionary stage roughly equivalent to that of South American monkeys. Because *Moeripithecus* was based on such incomplete material—a single lower jaw fragment bearing the first two molars—Schlosser offered no precise opinion about its position on the anthropoid family tree.

Despite the early successes of Osborn, Markgraf, and Schlosser, the Fayum's obvious potential remained largely untapped until Elwyn Simons burst on the scene in the late 1950s. Simons was young, ambitious, and completely immersed in the giddy intellectual ambience of the Ivy League. He boasted not one but two doctoral degrees—one from Princeton and a second from Oxford, where he had studied European fossil primates under the legendary British anatomist Sir Wilfrid E. Le Gros Clark. Soon after his return to the United States, Simons accepted a position at Yale's Peabody Museum of Natural History. There, he inherited the scientific mantle of Othniel Charles Marsh—one of the founding fathers of American paleontology.[9] The high-profile academic niche that Simons sought to establish would wed vertebrate paleontology with paleoanthropology, disciplines that were typically segregated under the earth sciences and social sciences, respectively. Already acknowledged as one of the preeminent scholars of fossil primates in the United States, Simons now hoped to unearth glamorous fossils of his own. With luck, these specimens might even compete in the popular arena with those being found by the Leakeys in East Africa.

Given our natural inclination to be interested in our own evolutionary history, two main criteria determine whether a fossil attracts the glare of the public spotlight. The first of these is its degree of kinship with us humans. The second is its age. The first criterion explains why the discovery of the earliest known hominid will always be more newsworthy than finding, say, the earliest marsupial. Early hominids are so close to us in an evolutionary sense that the issue of kinship completely trumps the greater antiquity of the early marsupial. But kinship alone isn't the only principle at work here. If this were the case, most paleoanthropologists would be searching for additional remains of Cro-Magnon people, the earliest representatives of our own species, *Homo sapiens*. But almost everyone admits that unearthing the world's oldest hominid is more exciting than finding more Cro-Magnon bones. Here, the factor of age dominates that of kinship, since all hominids bear uniquely on our evolutionary history. Simons knew that he couldn't compete head-to-head with the Leakeys on the issue of kinship, but he might easily surpass them when it came to age. The trick would be to make the connection between the much older fossils he hoped to find and the origins of the human lineage.

Simons's first Fayum expedition in late 1961 succeeded in adding to what was already a crowded field of early anthropoids from the Jebel Qatrani Formation. Two additional species were unearthed and described.[10] One of these turned out to be a smaller, older, and more primitive relative of *Apidium phiomense*. Simons named this new species *Apidium moustafai* in honor of one of his Egyptian collaborators. While *Apidium moustafai* hinted that—as in the case of the Bighorn Basin—it might eventually be possible to trace evolving lineages in the Fayum, the species differed only marginally from Osborn's *Apidium phiomense*. In contrast, the other new Fayum primate appeared to represent an entirely different branch of the anthropoid family tree. Simons named this species *Oligopithecus savagei* in honor of its discoverer, Donald E. Savage, a well-known paleontologist from the University of California at Berkeley.

Initially, a single lower jaw was all that documented *Oligopithecus*. The specimen showed an unusual combination of primitive and advanced features, suggesting that mosaic evolution—whereby different traits evolve sequentially, rather than all at once—played an important role in anthropoid origins. Advanced features in the lower jaw of *Oligopithecus* pointed toward a close affinity with living anthropoids, especially Old

World monkeys, apes, and humans. These included its relatively tall, stout canine, the presence of only two lower premolars (P_3 and P_4), the occurrence of a "honing facet" for the upper canine on P_3, and the overall depth of the lower jaw, which exceeds that of most Eocene prosimians. On the other hand, the lower molars of *Oligopithecus* retained enough primitive characteristics that they looked more like those of Eocene prosimians than living or fossil anthropoids. The front part of each lower molar—a structure known as the trigonid—projected high above the back part of each tooth, called the talonid. Most anthropoids differ from *Oligopithecus* in having molar crowns that are more uniform in height. Additionally, the first lower molar of *Oligopithecus* possessed an extra cusp known as the paraconid that was lost in other anthropoids. Given its unique combination of features, Simons equivocated on where to place *Oligopithecus* on the anthropoid family tree. He thought it might be related to Old World monkeys, but he couldn't rule out the possibility that *Oligopithecus* was a primitive relative of apes and humans instead. As we shall see, the question of where *Oligopithecus* fits on the family tree would eventually become even more ambiguous.

From a strictly scientific perspective, Simons's first expedition to the Fayum in 1961 met with phenomenal success. His team discovered new fossil localities, most of which were older than those established by earlier researchers. This alone demonstrated that the Fayum had not been exhausted by earlier exploration, as some contemporary scientists had apparently suspected.[11] More to the point, some of these new localities yielded the fossils of small mammals, including the new primates that were the expedition's main objective. But if Simons wanted to vie with the Leakeys for the public's attention, he would need to unearth more compelling fossils than *Apidium* and *Oligopithecus*. In due course, the Fayum would comply. Meanwhile, Simons launched a second paleoanthropological program—one that aimed to beat the Leakeys at their own game.

The story of how Elwyn Simons came to promote *Ramapithecus* as the world's oldest hominid is only incidentally related to my main thesis. Nevertheless, it bears repeating here, because the ensuing debate had a major impact on paleoanthropology in general and Simons's career in particular. Once Louis and Mary Leakey published their dramatic discovery of *Zinjanthropus* at Olduvai Gorge, it became apparent that even older hominids might soon be found. Documenting the ancestry of *Zinjanthropus* and its australopithecine kin became a focal point for paleoanthropological research. Simons happily waded into the fray, armed with some fragmentary fossils that had languished for decades in the collections of the Peabody Museum, just a few floors below the office where he worked.

In contrast to the exquisite skull of *Zinjanthropus,* the fossil evidence bearing on *Ramapithecus* was always meager. At first it consisted of nothing more than a scrappy upper jaw that had been collected from Miocene strata in the Siwalik Hills of northern India by a Yale graduate student in the 1930s.[12] Simons noted that the specimen possessed a relatively small—and therefore hominidlike—upper canine. Although the upper incisors were missing, their root sockets suggested that the face of *Ramapithecus* had been abbreviated like that of humans—not prognathic like those of apes. Potentially of even greater significance was Simons's reconstruction of the upper dental arcade (the shape of the complete upper jaw, as viewed from below) of *Ramapithecus.* This too resembled humans' in being parabolic or arcuate, rather than more U-shaped, as it is in apes. Despite the paucity of anatomical information available to him, Simons enthusiastically endorsed *Ramapithecus* as a primitive forerunner of australopithecines. If his view could be upheld, *Ramapithecus* would easily eclipse *Zinjanthropus* as the world's earliest hominid. Simons himself minced no words in pointing this out: "*Ramapithecus punjabicus* is almost certainly man's forerunner of 15 million years ago. This determination increases tenfold the approximate time period during which human origins can now be traced with some confidence."[13] As if this dramatic extension of the human lineage weren't enough, Simons and his graduate student David Pilbeam soon proceeded to argue that *Ramapithecus,* like *Zinjanthropus,* might have been an upright-walking biped able to manufacture and use stone tools.[14] Remarkably, neither limb bones of *Ramapithecus* nor its putative stone tools were available to support these claims.

The *Ramapithecus* challenge thrown down by Simons hardly went unnoticed. Almost immediately, it fueled a running feud with Louis Leakey, who had no intention of sitting idly by while his Olduvai hominids were left to choke on the dust of the Miocene. No sooner did Simons revive *Ramapithecus* than Leakey announced the discovery of his own hominidlike fossil from the Miocene.[15] Leakey's specimen—which he dubbed *Kenyapithecus wickeri*—had the implicit advantage of coming from East Africa, the same general region that was home to such undoubted hominids as *Zinjanthropus.* Simons responded by asserting that *Kenyapithecus* was merely an African variant of *Ramapithecus.*[16] As such, any hominidlike features present in *Kenyapithecus* simply confirmed his own findings regarding the hominid status of *Ramapithecus*. For his part, Leakey maintained that *Kenyapithecus* and *Ramapithecus* differed significantly from each other, a point that he emphasized by naming a second species of *Kenyapithecus* a few years later.[17]

The growing animosity between Simons and Leakey could not be concealed for long. Each began to publish papers that went out of their way to denigrate the other's views and interpretations. Simons and Pilbeam finally pushed Leakey too far when they proposed that a second Miocene ape that had been the subject of Leakey's work in Kenya was also unfounded. According to Simons and Pilbeam, the well-known genus *Proconsul*—which was among the most thoroughly documented of all Miocene apes, being represented by a nearly complete skull and a partial skeleton—was nothing more than an African version of a European ape called *Dryopithecus*.[18] When Pilbeam presented this interpretation to a small gathering of paleoanthropologists in Chicago, Leakey—who was also in attendance—jumped to his feet, shouting that Pilbeam was out of order and that "we don't have to listen to all this again."[19] Simons used the occasion to justify his ongoing critiques of Leakey's work: "It is because of things like that [Leakey's shouting match with Pilbeam] that people were prepared to be not very polite to him in print. It wasn't just the blunders he made that encouraged people to criticize him. It was this kind of blustering and arm-waving."[20] Around the same time, someone tacked a Louis Leakey dartboard onto a wall in the Peabody Museum at Yale. It was a fitting emblem of the intense rivalry that had developed between two of the world's leading students of fossil primates.

Despite their very public differences of opinion, Simons and Leakey agreed on one critical point—the human lineage must have diverged early on from all other living primates. Both men believed that *Ramapithecus* and *Kenyapithecus*—whether these were one and the same or not—were bona fide hominids from the Miocene. If so, the evolutionary split between apes and humans might conceivably extend back to the Oligocene. This idea appealed to both Leakey and Simons, if only because it left vast holes in the hominid fossil record that remained to be filled by scientists like themselves. Fortunately for Simons, even Leakey acknowledged that the Fayum Depression was the most promising place in Africa to find fossils that might validate such an early origin for the human lineage.[21]

Against the backdrop of his ongoing debates with Leakey, Simons continued to lead expeditions into the Fayum badlands. However, his second campaign there—which began in late 1962—yielded only modest results in terms of new fossil primates. A major reason seemed to

Figure 15. Elwyn Simons (right), who succeeded in bridging the disparate fields of vertebrate paleontology and physical anthropology, and his colleague and protégé Tom Bown, who solved the riddle of the Fayum's ancient environment, in Egypt's Western Desert. Photograph courtesy of and copyright by Richard F. Kay.

be that a dense pavement of desert gravel covered most of the surface outcrops of the Jebel Qatrani Formation. The Arabic term for this distinctive type of landscape is *serir*: literally, "pebble desert." The Fayum's *serir* conditions hampered the team's ability to prospect for fossils in two important ways. First, the ubiquitous rubble made it difficult to spot small fossils among the larger clasts and cobbles covering the desert floor. Most of the large fossil bones that protruded through the *serir* had been collected decades earlier by Osborn, Markgraf, and their predecessors. Second, the desert pavement hindered the natural erosion of the Fayum's fossil-bearing strata, in the same way that grass or any other plant cover prevents suburban yards from eroding. Simons needed to find some way of increasing the rate of erosion—or at least of penetrating the rocky debris that obscured all the small fossils—if his expeditions were to succeed.

Because of the low precipitation in Egypt's Western Desert, wind is the major agent of erosion there. Harnessing the awesome power of these winds altered the history of fieldwork in the Fayum. Yet like so many advances in paleontology, this innovation happened entirely by accident. In late 1963, a severe windstorm swept through the Fayum expedition's campsite. The following morning, when the team returned to the same locality where they had been working the previous day, they noticed that many new fossils were suddenly visible. The high winds had blown away so much sand and other debris that numerous fossils lay exposed for the first time. One of these freshly exhumed fossils was the nearly complete lower jaw of a primate. Smaller than *Propliopithecus,* but with remarkably large canines and small third molars, the new specimen appeared to document yet another new Fayum primate. Simons christened it *Aeolopithecus chirobates,* an allusion to Aeolus, the god of winds in Greek mythology.[22] The pagan gods of antiquity apparently appreciated this honor, because Aeolus has smiled upon Fayum expeditions ever since. From that day forward, Simons has systematically exploited the strong desert winds to uncover fossils.

The technique that Simons adopted was simple but effective. He instructed his crew to remove any large cobbles from the surface of sites that showed potential for yielding fossils. Smaller, pebble-sized debris was swept away with brooms. Each of the Fayum's productive fossil quarries was cleaned this way prior to the end of every field season. During the ten months or so between successive field seasons, the delicate task of excavating fossils from the sandy matrix in which they were entombed was left to the wind. The reward came the following field season, when the freshly exposed surface of each quarry held a bounty of new specimens.

As surely as the wind blows in the Egyptian desert, the pace of fossil primate discoveries in the Fayum began to quicken. The same locality that yielded *Aeolopithecus* after that fateful windstorm also produced lower jaws of a much larger anthropoid. Simons named this species *Aegyptopithecus zeuxis.*[23] Its imposing name underscored the new fossil's significance. Many paleontologists and paleoanthropologists become intimately associated with the most important fossils that they discover. Thus, Jacob Wortman will always be tied to *Tetonius* and *Diplodocus,* Louis and Mary Leakey are forever affiliated with *Zinjanthropus,* and Donald Johanson has a similar bond with "Lucy," the most famous representative of the species *Australopithecus afarensis*. More than any other fossil in the long list that he is responsible for having discovered and named, *Aegyptopithecus* is the beast that made Simons famous.

Like all other Fayum primates known at the time, *Aegyptopithecus* was documented at first by nothing more than its lower jaws. These revealed *Aegyptopithecus* to have been a gibbon-sized primate, with stout lower canines, two lower premolars on each side of its jaws, and molars that increased in size from front to back. Although *Aegyptopithecus* was of medium build by the standards of modern monkeys and apes, it remains the largest primate ever found in the Fayum. In its size and the enlargement of its molars from front to back, it resembled later primates from East Africa, especially Miocene apes, such as *Proconsul,* that Leakey and others had uncovered in Kenya. Simons interpreted this as evidence that the major lineages of living apes were already established by the Oligocene. *Aegyptopithecus,* he argued, was an early member of the group that included *Dryopithecus* and the living great apes (orangutans, gorillas, and chimps); *Aeolopithecus* was a primitive relative of living lesser apes (gibbons and siamangs); and—just possibly—*Propliopithecus* marked the beginning of the lineage that included *Ramapithecus, Australopithecus,* and humans.[24] Suddenly, the antiquity of the human lineage no longer seemed quite so far-fetched.

Despite its crucial position on the family tree, had better fossils of *Aegyptopithecus* not turned up, this animal would warrant no more space in introductory anthropology textbooks than its more arcane contemporaries. Its celebrity was guaranteed when Grant Meyer, who was responsible for the logistics of the early Fayum expeditions, spotted the top part of a primate skull protruding through the sand at one of the Fayum's most productive quarry sites. Like any experienced field collector, Meyer resisted the strong temptation to dig the specimen out from the surrounding rock right there on the spot. Instead, he impregnated the exposed part of the skull—and everything else in the immediate vicinity—with stabilizing glues, wrapped it all in a sturdy plaster jacket, and shipped the whole thing back to New Haven. It took several hundred man-hours of exacting and dexterous preparation to reveal just what Meyer had found. The specimen's completeness exceeded everyone's wildest expectations. Roughly a year after the fossil was found, Simons wrote, with obvious satisfaction: "It would be hard to say whether our surprise or our delight was the greater when the cleaning process revealed

Figure 16. The face of *Aegyptopithecus zeuxis,* as discovered in situ at Quarry M in the Fayum region of Egypt. Photograph courtesy of and copyright by Herbert Covert.

that Meyer had found not just a few skull fragments but a nearly whole skull of *Aegyptopithecus.*"[25]

Meyer's skull rapidly attained the stature of an icon in paleoanthropology. Its public unveiling likewise captured the attention of the popular press. In many ways, *Aegyptopithecus* conformed to what many people had in mind when they pondered the notion of a "missing link." Its apelike teeth and forward-facing eye sockets looked remarkably advanced—even human. At the same time, its long, doglike snout and small cranial capacity branded *Aegyptopithecus* as something distinctly primeval, a holdover from the Eocene. Simons leveraged the remarkable specimen to his considerable advantage. Taking a swipe at Louis Leakey's two most outstanding discoveries, Simons crowed: "Not only is the skull [of *Aegyptopithecus*] some eight to ten million years older than any other fossils related to man, but it is better preserved than any that are older than 300,000 years." Leakey's famous skulls of *Zinjanthropus* and *Proconsul* both fell well beyond Simons's arbitrary cutoff date of 300,000 years, implying that both were inferior to *Aegyptopithecus* in terms of their preservation. And just in case its relevance for deciphering human ancestry was lost on anyone, Simons now placed *Aegyptopithecus* at an even more critical juncture on the family tree: "*[Aegyptopithecus]* stands near the very base of the genealogical tree leading to later Great Apes and man," he declared. "It represents a major stage in the documentation of the fore-runners of man."[26]

By the time Simons announced the discovery of his prize skull to the world, the possibility of undertaking further exploration in the Fayum had been significantly diminished. Long-standing political tensions be-tween Israel and the various Arab states—especially Egypt—finally erupted into full-scale war. In early June 1967, in response to a massive buildup of Egyptian troops and armor along Israel's southern border, the Israeli Air Force preemptively attacked and destroyed most of Egypt's fleet of warplanes. Supported by overwhelming air superiority, Israeli tanks and infantry soon rumbled across the Sinai Peninsula, stopping only at the banks of the Suez Canal. For Egypt, the decisive military defeat spelled a political disaster. Predictably, the conflict also affected Fayum paleontology. For years, Egypt had been aligned politically with the So-viet Union, while Israel enjoyed the tacit support of the United States. Coming as it did at the height of the Cold War, the Arab-Israeli Six-Day War meant that American scientific expeditions to what remained of Egypt's sovereign territory were viewed with far greater suspicion than before. When a few young members of Simons's field crew made a wrong turn in the desert on their way back to Cairo, they stumbled right into an Egyptian military encampment. For the Egyptian government, this was the last straw. The Fayum region was officially closed to scientific research and would remain so for most of a decade.

While geopolitics disrupted Simons's series of expeditions to the Fayum, it did not prevent the study of fossils that were already out of the ground. Most paleontologists with active field programs accumulate many more fossils than they can ever analyze and describe alone. Mul-tiple field projects made Simons particularly rich in undescribed fossils, and many projects had to be farmed out to colleagues and graduate stu-dents. As a result, Simons became a magnet for the best and brightest paleontology students in America. Many of my colleagues refer to this interval—from the late 1960s to the early 1970s—as the "Golden Age" of paleoanthropology at Yale. Clearly, a large percentage of those who would become leading figures in American paleontology and paleoan-thropology passed through New Haven at this time. The list includes such distinguished personalities as Tom Bown, Glenn Conroy, John Fleagle, Philip Gingerich, Rich Kay, David Pilbeam, Ken Rose, and Ian Tatter-sall. Some of these young scientists pushed research on Fayum primates beyond anything that had been done previously. For example, while they were still students at Yale, Conroy and Fleagle pioneered the study of primate limb bones from the Fayum, providing the first glimpses of how the earliest anthropoids moved through their arboreal domain. Others contributed to complementary areas of research, ranging from studies of the earliest potential primates from the Paleocene to detailed analyses of how primates and other mammals chew their food. As a cohort, the crop of young scientists spawned by Yale during its "Golden Age" was impressive indeed. Their multiple talents would take Fayum research in brand-new directions once fieldwork there again became feasible.

By 1976, the dust of Middle Eastern politics had settled enough for Simons to negotiate a new bilateral agreement with the Egyptian authorities. Simons, who had recently left Yale to accept a new position at Duke University, could finally resume his Fayum fieldwork. After lying fallow for most of a decade, the Jebel Qatrani Formation again yielded a rich harvest of fossilized remains. But what was arguably more important than the new Fayum fossils was a changing emphasis on scientific research there. The major goal of Simons's early expeditions—finding and naming additional species of Fayum primates—became supplanted by a desire to understand more about the biology of these extinct animals. To some extent, this shift was dictated by the simple fact that so many Fayum primates had already been recovered and named. Any that remained to be discovered must have been extremely rare. At the same time, so little was known about how these animals moved, what they ate, and the environment in which they lived that these topics demanded greater attention. The interdisciplinary team assembled by Simons—dominated by his current and former students—made rapid progress on all of these fronts.

Reconstructing the Oligocene environment of the Fayum seemed like an obvious first priority. Aside from a few species that are highly adapted to life in the desert, the dune fields and *serir* landscapes surrounding the Fayum today are remarkably devoid of life. It would be difficult to imagine a place less likely to be inhabited by primates. Yet Simons and his predecessors had shown that a wide variety of anthropoids lived in the Fayum region during the remote past. Moreover, the fossil remains of some of these animals—especially *Apidium*—were so common that they must have been fairly abundant in life as well. Obviously, the modern Fayum ecosystem failed to provide whatever ancient habitats were required to sustain foraging troops of *Aegyptopithecus* and *Apidium*. Environmental conditions must have differed in the Oligocene, but to what degree? As is often the case in geology and paleontology, two competing versions of the truth vied for credibility by the time Simons kicked his new series of Fayum expeditions into gear. These radically different environmental reconstructions also implied markedly divergent lifestyles for the earliest African anthropoids.

The two alternative visions of what the Fayum was like in the Oligocene agreed on one point—it must have been wetter then than it is now. Beyond this common ground, the theories depicted environments that were poles apart. According to one theory, during the Oligocene, the Fayum would have been analogous to what one finds immediately south of the Sahara today. The semiarid ecosystem then prevailing in the Fayum would have resembled the terrain that typifies much of modern Mali, Niger, Chad, and Sudan. Geographers refer to this broad, relatively barren belt of north-central Africa as the Sahel. For convenience, then, let's call this first reconstruction of the ancient Fayum ecosystem the Sahelian model. Its roots go back to the earliest studies of Fayum geology, but its modern incarnation is the work of a Dutch scientist named Adriaan Kortlandt.[27]

According to the Sahelian model, the Fayum was only marginally wetter during the Oligocene than it is currently. The Mediterranean coast, which lies roughly 120 miles north of the region today, was much closer to the Fayum at that time. A large river—which Kortlandt referred to as the "Proto-Nile"—traversed the Fayum on its way to the nearby sea. Despite these adjacent bodies of water, the Sahelian model portrayed terrestrial Fayum environments as being prone to severe droughts. Mud cracks, salt crystals, and other geological indicators of dry conditions supported this interpretation. Hence, local rainfall must have been highly seasonal, resulting in rapid runoff through ephemeral drainages. Plant cover was minimal in some areas. Elsewhere, scrubby bush and dry forests alternated with grasslands. In any case, the large petrified trees that occur in great abundance in parts of the Fayum could not have grown locally. Instead, it was argued, they were driftwood that floated down the Proto-Nile from more humid and equatorial parts of Africa far to the south.

If the Sahelian model were true, it would severely limit the range of lifestyles that could have been adopted by the Fayum's early anthropoids. Any primates living in such dry, open environments would probably resemble baboons and vervet monkeys in their general behavior and ecology. Like these modern primate residents of African savannas, the early anthropoids of the Fayum would have been fully capable of climbing trees. They would probably have slept among the branches to defend themselves from ground-dwelling predators. But they would have spent most of their lives foraging on the ground and moving between occasional stands of small to medium-sized trees. Such a terrestrially oriented lifestyle departs dramatically from the strictly arboreal habits of the Eocene prosimians that are regarded as possible ancestors of the Fayum primates. At the same time, if the earliest anthropoids were already so committed to life on the ground, what were the ecological factors that instigated the birth of the hominid lineage? To Simons and his cadre of young scientific colleagues, none of this made much sense.

It also failed to withstand detailed geological scrutiny. Spurred into action by Kortlandt's resurrection of the Sahelian model, an interdisciplinary group of scientists decided to settle the controversy surrounding the Fayum's ancient environment once and for all.[28] The driving force behind this effort was Tom Bown, a brilliant young geologist employed by the U.S. Geological Survey. Like so many others, Bown had passed through Simons's lab at Yale during its "Golden Age." There, Simons sent Bown on several field campaigns to collect fossil primates and other vertebrates, including an epic overland adventure in 1969 that began in northern India and wound through Afghanistan, Iran, Turkey, and southeastern Europe before landing in Libya just as Muammar Qaddafi seized control of the government. Later, Bown organized his own series of expeditions to Wyoming's Bighorn Basin, where he and Ken Rose would greatly enrich our knowledge of omomyid evolution (see chapter 3). In the Bighorn Basin, Bown also earned a reputation as a leading researcher on ancient soils (also known as paleosols) and "trace fossils." Trace fossils are what extinct organisms leave behind in addition to their lithified body parts. These include tracks, nests, burrows, and the like. As it turns out, paleosols and trace fossils each provide detailed and highly reliable information about ancient environments. As a prominent member of Simons's new series of expeditions to the Fayum, Bown naturally developed an abiding interest in the paleosols and trace fossils of the Jebel Qatrani Formation. What he saw didn't jibe at all with the Sahelian model.

For one thing, the geological structures that Kortlandt interpreted as mud cracks turned out to result from ancient soil formation instead. Worse yet, the specific type of soil responsible for forming these pseudo–mud cracks develops under poorly drained—or even waterlogged—conditions, the opposite of what would be expected if the Sahelian model were correct. The abundance of fossilized tracks, nests, and burrows that Bown uncovered in the Fayum indicated that the region had supported a diverse

Figure 17. Fossil *Epipremnum* fruits from the Fayum, part of the abundant evidence amassed by Thomas Bown suggesting that the region once supported a lush coastal forest. Photograph courtesy of and copyright by Herbert Covert.

assemblage of invertebrates. Many of these organisms can only survive in environments where the substrate is relatively wet. Freshwater limestones bearing the microscopic fossils of algae and other organisms demonstrated that ponds and other bodies of sluggish or standing water dotted the ancient landscape. Cross-bedded sandstones marked the paths of ancient streams and rivers, but none of these were large enough to have been the "Proto-Nile." These small sandstones raised a serious problem for the Sahelian model. If an ancestral version of the Nile didn't flow through the Fayum, how could one account for its abundance of fossil trees? There were literally hundreds of them.

Bown and his team showed that the jumble of fossil logs known from the Jebel Qatrani Formation derives from an ancient forest that grew right there on the spot. Some of these fossil trees retain their roots and branches, delicate structures that would have been shorn off quickly had the trees been rafted for long distances. At certain sites in the Fayum, entire fossilized root systems mark the exact locations where large trees were once anchored. Tropical lianalike vines climbed the trunks of these trees, as shown by the characteristic fossil *Epipremnum* fruits that the team found in abundance. Even more so than the fossilized roots and branches, these delicate fruits were too dainty to survive a long float trip. But the final nail in the coffin of the Sahelian model came from the family trees of the trees themselves. Many of the Fayum's fossil trees are related to modern species that live in tropical parts of Southeastern Asia—nothing like them survives in more equatorial parts of Africa. How could an ancestral Nile with headwaters in equatorial Africa transport Southeast Asian trees to the Fayum? Burdened by so much extraneous baggage, the Sahelian model had to be jettisoned entirely. Clearly, the Fayum once supported a subtropical to tropical ecosystem that was sufficiently moist to sustain broad stands of tall trees. Such an idyllic scene agreed with the diversity of fossil primates that Simons and his predecessors had uncovered in the Fayum.

What was life like for these ancient denizens of a forest in the Sahara? If we could transport ourselves through time to visit the Fayum of about thirty-three million years ago, we would find that certain elements of the ecosystem were oddly familiar, while others would be utterly alien. Stepping out of that time machine, we would see that the environment differs markedly from that of the Fayum today, but it hardly qualifies as unearthly. The landscape offers little topographic relief, occupying a broad, lush, and tropical coastal plain on the southern margin of the Tethys Sea—a much more extensive forerunner of what will later become the Mediterranean. Mangrove forests predominate along the shoreline. Farther inland, swamps and ponds abound, supporting lily pads and other masses of floating vegetation. The wide variety of birds and reptiles exploiting these habitats offers few surprises to anyone familiar with their modern African counterparts. Similar assemblages of birds and reptiles occur today in parts of Uganda bordering Lake Victoria.[29] Big-footed jacanas appear to stride miraculously across open water, using adjacent lily pads as stepping-stones. Flamingos, storks, and herons flourish in the shallows. Cormorants dive for fish from the water's surface, while ospreys and fish eagles swoop down on their aquatic prey from high overhead. Crocodiles bask along the shores of sluggish rivers. Nearby, tortoises amble through the undergrowth. We even catch a glimpse of the odd side-necked turtle known as *Pelomedusa,* a frequent inhabitant of modern African water holes.

Just as we begin to convince ourselves that the ancient Fayum ecosystem faithfully mirrors more tropical parts of Africa today, however, we notice some mammals and realize that fundamental differences exist. For one thing, the most obvious members of modern African mammal faunas are nowhere to be found. There are no gazelles, no wildebeest, no impala, no Cape buffalo—not a single member of the currently diverse family of browsing and grazing artiodactyls known as the Bovidae. Piglike creatures such as warthogs are missing as well. Nor are there any rhinos or zebras—the modern African representatives of the odd-toed ungulates known as Perissodactyla. Similarly absent are the mammalian carnivores that prey upon this vast diversity of ungulates. Hence, we observe no lions, cheetahs, leopards, hyaenas, mongooses, civets, genets, jackals, or African hunting dogs.

In place of these missing groups of modern mammals, an exotic menagerie of unfamiliar beasts roams the Fayum landscape. The dominant terrestrial herbivores are related to modern hyraxes—animals that still survive throughout much of Africa today, even if they are ecologically inconspicuous. Modern hyraxes are small mammals—about the size of a large rabbit—that tend to cluster around rocky outcrops in groups of up to fifty individuals. In contrast to living hyraxes, those inhabiting the Fayum occupy a broad range of body sizes and ecological niches.[30] The largest of these, which is appropriately named *Titanohyrax ultimus,* approaches the size of a small rhino. A second species, *Antilohyrax pectidens,* looks as much like a gazelle as a modern hyrax. Both *Titanohyrax* and *Antilohyrax* browse on the abundant Fayum foliage. A third Fayum hyrax, *Bunohyrax major,* seems to have assumed the ecological role of pigs. Still others look and act like . . . well, hyraxes.

Despite their diversity and abundance, hyraxes are not the only herbivorous mammals living on the Fayum coastal plain. The largest of these other mammalian herbivores also qualifies as the oddest member of the Fayum fauna. Resembling nothing alive today, *Arsinoitherium* stands out primarily because of its colossal pair of nasal horns. Unlike those of modern African rhinos, the nasal horns of *Arsinoitherium* are arranged side by side, rather than front to back. Arsinoitheres are specialized browsers, like some of the Fayum hyraxes. Their high-crowned cheek teeth allow them to masticate especially tough, fibrous vegetation, which they consume in massive quantities. The only other Fayum mammals that approach *Arsinoitherium* in size are the archaic elephants, *Palaeomastodon* and *Phiomia.* Their small tusks and primitive cheek teeth notwithstanding, these Fayum elephants are far more recognizable than their distant evolutionary cousins the arsinoitheres.

Regardless of the absence of true mammalian carnivores—animals that are closely related to living cats, dogs, and bears—the Fayum's herbivores can't afford to let down their guard. A variety of predatory mammals lurks among the forests and glades of the Fayum coastal plain. These species vary in size and habits, but all belong to an extinct group of car-

Figure 18. Artist's rendering of *Arsinoitherium,* one of the extinct mammals that formerly inhabited the Fayum coastal forest during the early Oligocene. Adapted from Morgan and Lucas 2002.

nivorous mammals called hyaenodontids. They differ fundamentally from true carnivores in having a whole series of cheek teeth that are specialized for cutting and slicing through flesh. Living mammalian carnivores restrict this function to a single set of occluding teeth on either side of their jaws.

In contrast to the exotic mammals we observe on the ground, when we look up into the trees, we see creatures that seem far more familiar. All of the Fayum primates can be readily identified as monkeys, even if they look more like their living South American cousins than modern African monkeys. We rapidly discern four or five species, based on obvious differences in size, coat color, and vocalizations. They can all be segregated into two main groups. The first of these, which we recognize as parapithecids, are small, agile primates who prefer to bridge small gaps in the forest canopy by leaping from branch to branch. From our vantage point on the ground, we can spot several species of parapithecids, exemplified by the very abundant *Apidium phiomense.* Less numerous but hardly rare are the propliopithecids, of which *Aegyptopithecus zeuxis* is the most conspicuous example. At roughly fifteen pounds (6.7 kilograms), *Aegyptopithecus* outweighs all other Fayum primates. *Aegyptopithecus* is even big by the standards of modern South American monkeys, but the largest living African monkeys are substantially larger yet. Many of these modern African monkeys—like baboons, mandrills, and patas monkeys—spend much of their time on the ground. The Fayum monkeys, like their New World counterparts, rarely leave the safety of their arboreal domain. A closer look at their anatomy shows why this is so.

The forelimbs of living terrestrial monkeys are specialized in a number of ways that allow them to move swiftly and efficiently on the ground.[31] Some of these anatomical modifications stabilize the shoulder and elbow during quadrupedal locomotion. In contrast, arboreal monkeys emphasize climbing over fast quadrupedal running. Their shoulders are built to maximize mobility rather than stability. As a result, arboreal monkeys lack the stabilizing features at the shoulder and elbow found in ground-dwelling monkeys. Anatomically, this is reflected by the more globular shape of the head of the humerus in tree-dwelling monkeys (this is the "ball" part of the ball-and-socket joint at the shoulder). In baboons and other terrestrial monkeys, the front part of the humeral head is eclipsed by a bony protuberance called the greater tuberosity, which projects above the level of the "ball." Arboreal and terrestrial monkeys also differ from one another in the stance they maintain at the elbow joint. The elbows of arboreal monkeys are habitually flexed, possibly in order to lower their center of gravity and decrease the risk of falling. In contrast, the elbow can be almost fully extended in terrestrial monkeys, so that the entire arm lies nearly in a straight line. Predictably, this difference in elbow posture is reflected in the bony structure of the elbow region. For example, the pit on the back of the humerus known as the olecranon fossa is much more deeply excavated in terrestrial monkeys than it is in arboreal species. This gives the ulna (one of the two bones of the forearm) more room for extension, which partly explains the difference in elbow mobility between terrestrial and arboreal monkeys.

So long as we're investigating the ancient Fayum ecosystem, we decide to see how the behavior of *Apidium* and *Aegyptopithecus* matches their anatomy. It's relatively easy to watch *Apidium* and *Aegyptopithecus*, since both species are active in broad daylight. But to examine their anatomy in any detail we need access to their skeletons. It takes time, but we eventually locate recently deceased representatives of both species. The specimens have been dead long enough that insects and other scavengers have mainly consumed the soft tissues, but ligaments continue to hold the major skeletal elements together. For a paleontologist, the greatest benefit of a time machine is the luxury of finding biological specimens that would rarely, if ever, get preserved in the fossil record.

As soon as we begin to examine our prize specimens of *Apidium* and *Aegyptopithecus,* we see that their shoulders and elbows resemble those of modern arboreal monkeys, yet differ substantially from those of terrestrial species like baboons. Neither *Apidium* nor *Aegyptopithecus* has the enlarged greater tuberosity or the deep olecranon fossa of a ground-dwelling monkey.[32] This explains why both of these Fayum primates are so reluctant to leave the safety of the trees. Yet we've also noticed that *Apidium* and *Aegyptopithecus* move quite differently through the lush foliage of the Fayum forest. *Apidium* tends to pace frenetically along the tops of branches, hardly even pausing at the end of a branch before launching itself to its next arboreal avenue. *Aegyptopithecus* is slower and more deliberate, spending more of its time moving vertically between successive stories in the canopy. Its powerful arms and legs even allow *Aegyptopithecus* to hang briefly beneath branches as it reaches for hard-to-get pieces of fruit.

What anatomical features underlie these different patterns of locomotion? Turning back to our skeletal specimens, we note that the humerus of *Aegyptopithecus* bears extremely well-defined areas for muscle attachment, while that of *Apidium* is slightly more gracile. The distinctly curved finger bones of *Aegyptopithecus* sport clear flexor sheath ridges, reflecting the importance of prolonged and powerful grasping in its mode of locomotion. In contrast, the finger bones of *Apidium* are straighter, because its preferred mode of running along the tops of branches demands less in the way of powerful grasping.[33] *Apidium* and *Aegyptopithecus* further differ in the anatomy of their hindlimbs. The most obvious distinctions lie in the ankle and lower leg. Just above the ankle joint in *Apidium,* we see that the two bones of the lower leg—the tibia and fibula—are tightly joined and nearly fused. This part of the leg is missing in our skeletal specimen of *Aegyptopithecus,* but a very different condition occurs in the closely related genus *Propliopithecus.*[34] Here, the tibia and fibula make only casual contact near the ankle joint, allowing a wide range of motion at this joint. Most primates have a tibiofibular articulation similar to that of *Propliopithecus.* The unusually tight apposition between tibia and fibula that we see in *Apidium* limits rotational movement at the lower leg and ankle joints, while it stabilizes the hingelike flexion and extension that is so important during leaping. The same patterns emerge from our comparison of the ankle bones of *Apidium* and *Aegyptopithecus.* Those of *Apidium* are built to enhance stability during flexion and extension, while those of *Aegyptopithecus* facilitate a much wider range of motion. Clearly, the hindlimbs of *Apidium* bear the stamp of a primate that is committed to leaping, while those of *Aegyptopithecus* support its more varied repertoire of movement.

Although our brief observations of Fayum monkeys reveal significant differences in their locomotion, we can easily document similarities in other aspects of their behavior. For example, virtually all of the Fayum monkeys—parapithecids and propliopithecids alike—eat fruit.[35] Only the smallest and least common parapithecids, like *Qatrania wingi,* round out this mainly frugivorous diet with saps and gums. Oddly, none of the Fayum monkeys shows a strong predilection for eating leaves. Only *Parapithecus grangeri,* a uniquely specialized species that completely lacks lower incisors, consumes a few leaves along with its daily intake of fruit.[36] By the same token, the social organization of Fayum monkeys seems fairly monotonous. Both *Apidium* and *Aegyptopithecus* live in large, socially complex troops composed of a few adult males, multiple adult females, and their offspring. A rigid social hierarchy among the males is maintained by an endless series of threats, punctuated by brief bouts of actual combat. Both the male threat displays and the short-lived fights rely upon the sexually dimorphic canines that are ubiquitous among the Fayum monkeys.[37]

As we wrap up our stint as ecologists in this ancient forest in the Sahara, the impression that lingers most in our minds is how surprisingly modern the Fayum monkeys appear to be. If we were to capture some parapithecids and propliopithecids and transport them back to the future with us in our time machine, they would hardly seem out of place in the monkey house of any large zoo. For that matter, if we set them free in the Amazonian rain forest, only experienced primatologists would likely become excited. Even among such avid professionals as these, the resulting buzz would focus more on the discovery of "new" living species of South American monkeys than any major rift in the fabric of time.

Looking around the exotic Fayum scenery for the last time, we realize that the answers we seek cannot be found here. The Fayum monkeys look and act so much like their modern relatives that they provide only modest insight into the origin and early evolution of the anthropoid lineage. Instead of bridging the biological gap between modern anthropoids and their prosimian cousins, the Fayum anthropoids establish an unambiguous baseline for anthropoids in the early Oligocene. Both their advanced anatomy and their teeming diversity emphasize one point un-

Plate 1. A tarsier, the only living primate that consumes nothing but live animal prey.
Photograph courtesy of and copyright by David Haring/Duke University Primate Center.

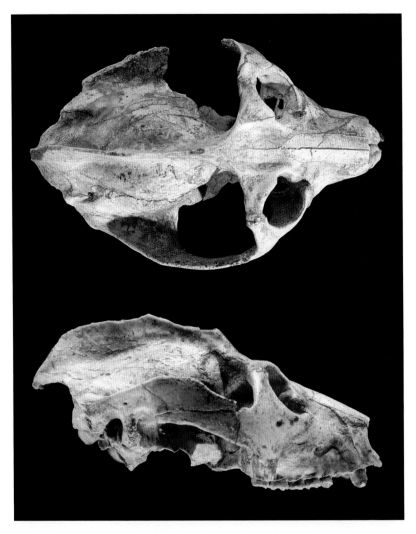

Plate 2. A well-preserved skull of *Adapis parisiensis,* a leaf-eating adapiform primate once common in the Quercy region of southern France. Photograph by D. Serrette, courtesy of and copyright by Muséum National d'Histoire Naturelle, Paris.

Plate 3. Artist's rendering of *Adapis parisiensis* in the karstic limestone environment of southern France about thirty-eight million years ago. Original art by Mark Klingler, copyright Carnegie Museum of Natural History.

Plate 4. Artist's rendering of *Notharctus venticolus* in the vicinity of the Buck Spring Quarries, Wind River Basin, Wyoming, about fifty million years ago. Original art by Mark Klingler, copyright Carnegie Museum of Natural History.

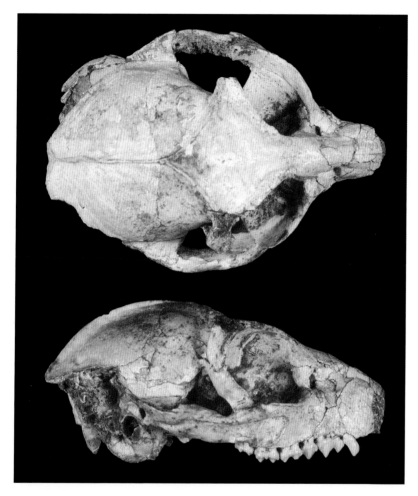

Plate 5. A well-preserved skull of *Necrolemur antiquus*, a microchoerid primate once common in the Quercy region of southern France.

Plate 6. Artist's rendering of *Necrolemur antiquus* in the forests of southern France roughly thirty-eight million years ago. Original art by Mark Klingler, copyright Carnegie Museum of Natural History.

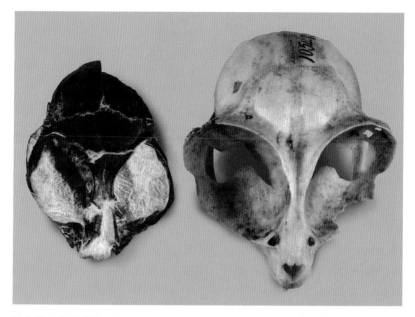

Plate 7. A skull of *Shoshonius* compared with that of a living tarsier (right).

Plate 8. Artist's rendering of *Shoshonius cooperi*, a nocturnal omomyid primate that lived in central Wyoming roughly fifty million years ago. Original art by Mark Klingler, copyright Carnegie Museum of Natural History.

Plate 9. The L-41 site in the Fayum region of Egypt, where *Catopithecus*, *Proteopithecus*, *Arsinoea*, and other early anthropoids were discovered. Photograph courtesy of and copyright by Marc Godinot.

Plate 10. The Shanghuang fissure-fillings, where *Eosimias* was first discovered.

Plate 11. Prospecting for fossils at Locality 7 on the southern bank of the Yellow River in Henan Province, Yuanqu Basin, central China.

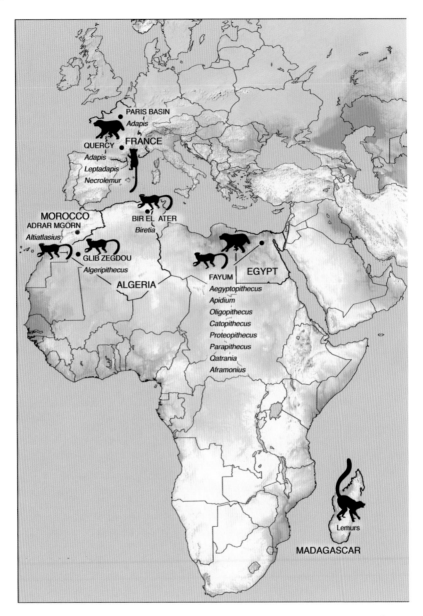

Plate 12. Geographic distribution of various living and fossil primates in Eurasia and Africa. Original art by Mark Klingler, copyright Carnegie Museum of Natural History.

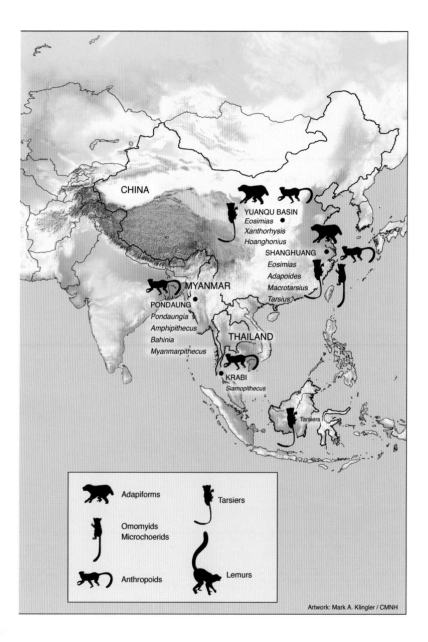

CHINA

YUANQU BASIN
Eosimias
Xanthorhysis
Hoanghonius

SHANGHUANG
Eosimias
Adapoides
Macrotarsius
Tarsius

MYANMAR

PONDAUNG
Pondaungia
Amphipithecus
Bahinia
Myanmarpithecus

THAILAND

KRABI
Siamopithecus

Tarsiers

Adapiforms

Omomyids
Microchoerids

Anthropoids

Tarsiers

Lemurs

Artwork: Mark A. Klingler / CMNH

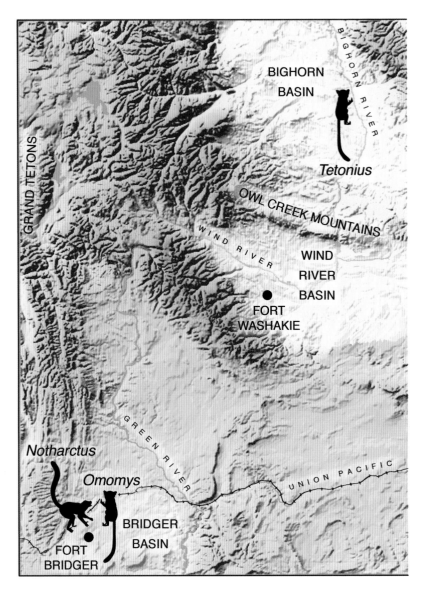

Plate 13. Map of Wyoming, showing the location of some of the key fossils, fossil sites, and other geographic features mentioned in the text. Original art by Mark Klingler, copyright Carnegie Museum of Natural History.

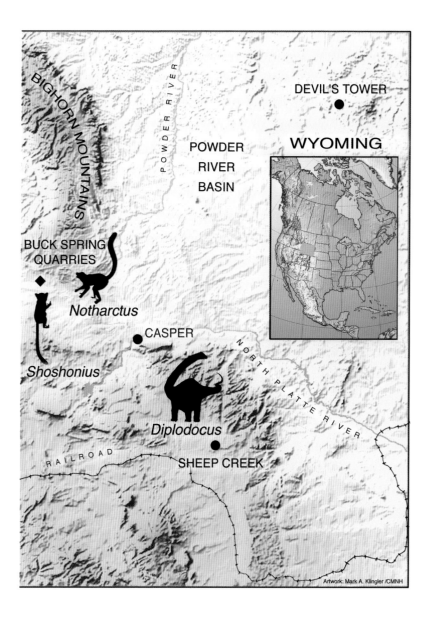

BIGHORN MOUNTAINS

POWDER RIVER

POWDER
RIVER
BASIN

DEVIL'S TOWER
●

WYOMING

BUCK SPRING
QUARRIES

Notharctus

Shoshonius

● CASPER

NORTH PLATTE RIVER

Diplodocus

RAILROAD

SHEEP CREEK
●

Artwork: Mark A. Klingler /CMNH

Plate 14. A troop of *Eosimias sinensis* encounters a group of smaller eosimiid primates near the central coastline of China roughly forty-five million years ago. Original art by Mark Klingler, copyright Carnegie Museum of Natural History.

equivocally—by the early Oligocene in northern Africa, the major features of anthropoid evolution have already taken place. If we want to learn about the earliest and most critical phases of anthropoid evolution, we have no choice but to plunge back into the mysterious void known as the Eocene. As we've already seen, creatures like *Adapis, Notharctus,* and *Tetonius*—primitive primates that differ conspicuously from anthropoids—reigned over this prolonged interval of time.

Like most good science, our Fayum odyssey has generated more questions than answers. How far back in the Eocene must we go to chart the origin of the anthropoid lineage, and will we be able to recognize it when we see it? Should we concentrate our fossil-collecting efforts on other sites in Africa, or should we extend our search elsewhere? How do the well-known Eocene primate groups—the adapiforms, omomyids, and microchoerids—fit into the chronicle of anthropoid origins? Is it possible to use the rapidly expanding database regarding the evolutionary relationships among living primates—built mainly on the back of DNA sequences—as a framework for interpreting fossils? Finally, from a more humanistic perspective, what can be said about the legacy of Elwyn Simons, whose work in the Fayum has been so integral to our story?

Simons has dedicated most of his professional life to recovering fossils that bear on the most important milestones in primate and human evolution—critical transitions like those from prosimian to anthropoid and from ape to human. In many ways, Simons has enjoyed phenomenal success. No one has unearthed more fossil primates from a wider variety of times and places than he. Never lacking in ambition, Simons continually sought to secure his place as a leading authority on the origins of anthropoids and humans alike by having the oldest relevant fossils. Inevitably, this simple strategy was doomed to fail, but for very different reasons in each case.

The controversy over *Ramapithecus* placed Simons in the middle of the debate over human origins for two decades, from the early 1960s, when Simons revived *Ramapithecus* as a likely human ancestor, until he was forced to abandon this view in the early 1980s. Throughout this period, Simons steadfastly defended the status of *Ramapithecus* as the earliest known hominid. He beat back challenges from competitors like Louis Leakey, whose similar claim for *Kenyapithecus* Simons neutralized by subsuming *Kenyapithecus* within *Ramapithecus*. He repeatedly denounced the emerging field of molecular systematics and its "molecular clock," which routinely predicted that hominids diverged from apes too recently for *Ramapithecus* to be an early member of the hominid lineage. Only the discovery of relatively complete fossils showing detailed and compelling resemblances to orangutans rather than humans caused Simons to recant his opinion that *Ramapithecus* was a hominid. In the end, Simons had staked a claim for the earliest hominid that was not only too old, but also based on inadequate anatomical evidence.[38]

Similarly, Simons's long series of expeditions to the Fayum made him the undisputed custodian of the oldest fossils relevant to anthropoid origins for most of his career. Exclusive access to these Fayum fossils allowed Simons to develop and promote numerous theories on anthropoid origins, even if these often conflicted with his own previous ideas of how early anthropoids evolved. Despite his constant pruning and grafting of the anthropoid family tree, Simons consistently argued that anthropoids originated sometime near the Eocene-Oligocene boundary. Of course, this timing happened to reinforce the pivotal status of the early Oligocene anthropoids from the Fayum. As we shall see, when substantially older anthropoid fossils began to be described in the early 1990s, Simons defended the significance of his Fayum fossils by arguing that the competing fossils weren't as old as they were purported to be, or else that they weren't anthropoids at all.

The long-standing hegemony that Simons maintained over the issue of anthropoid origins eventually collapsed under the weight of its own success. As more and more anthropoids showing virtually modern anatomy emerged from the Jebel Qatrani Formation, it became increasingly obvious that speculating about anthropoid origins on the basis of Fayum fossils was akin to using Neanderthals to determine how humans evolved from apes. Instead, we needed to find the anthropoid equivalent of the australopithecines—fossils that were clearly transitional in both time and anatomy between the undoubted anthropoids of the Fayum and the horde of primitive prosimians known from the Eocene.

Resisting for the moment the temptation to heed the Eocene's beckoning call, we step back into our time machine and head for home.

5

Received Wisdom

By the late 1980s, paleontologists had been successfully recovering fossil primates for more than 150 years. From the pioneering efforts of Georges Cuvier and Jacob Wortman to the more recent expeditions of Elwyn Simons, a burgeoning inventory of Greek and Latin names charted the latest revelations from the fossil record. Despite the confusing proliferation of new species and genera, the signal concerning our own deep evolutionary history seemed easy enough to decode. The earliest fossil primates from the Eocene all belonged to one of two major groups. The first of these consisted of vaguely lemurlike adapiforms such as *Adapis* and *Notharctus*. The second group—omomyids and microchoerids—included smaller, more tarsierlike primates such as *Tetonius* and *Necrolemur*. Yet nowhere among this vast assemblage of Eocene primates was there clear and persuasive evidence for the lineage leading to modern monkeys, apes, and humans. Instead, the oldest fossils documenting the emergence of anthropoids hailed from early Oligocene sites in the Fayum region of Egypt. The simple fact that anthropoids followed prosimians in the grand succession of the fossil record reinforced the notion that life evolves in ladderlike fashion from primitive to advanced—or from tree shrew to lemur to tarsier to monkey to ape to human in the widely adopted scheme of Sir Wilfrid E. Le Gros Clark.

Around this same time, paleoanthropologists settled on a broad consensus about the major features of anthropoid evolution, even if occasional fights still broke out regarding specific details. In a nutshell, the party line went as follows. The anthropoid lineage originated somewhere in Africa near the Eocene-Oligocene boundary, about thirty-four million years ago. These early African anthropoids evolved from anatomically advanced prosimians that differed only marginally from their anthropoid offspring. Therefore, as the fossil record improved, it should become increasingly difficult to separate the earliest anthropoids from their immediate prosimian forebears. Nevertheless, at some point along this prosimian-anthropoid continuum, our distant ancestors crossed a critical threshold—a basic shift in anatomy and ecology that would force everyone to agree that the evolutionary product was an anthropoid rather than a prosimian. One of the last remaining hurdles was to identify the prosimian ancestors of these earliest anthropoids. Here, all bets were off, though, because different scientists championed one or the other of the major Eocene prosimian groups as anthropoid ancestors. Some argued that anthropoids descended from an advanced adapiform, while others thought that anthropoids evolved from some unknown omomyid.

Even these persistent—and often bitter—debates over which prosimian group evolved into early anthropoids represented genuine progress, however. By the late 1980s, no serious scholar questioned the common ancestry, or monophyly, of all living anthropoids. According to this view, South American primates ranging in size and behavior from pygmy marmosets to spider monkeys sprang from the same ancestral stock that—on the other side of the planet—evolved into macaques, baboons, orangutans, and humans. This grand unification of New World monkeys with their brethren from across the Atlantic was an old idea whose time had finally arrived. It supplanted an earlier consensus in paleontology that viewed the two major groups of living anthropoids—South American monkeys, on the one hand, and Old World monkeys, apes, and humans, on the other—as distant relatives, having evolved along parallel lines from separate prosimian ancestors. Let's briefly consider some of the factors behind this dramatic shift of scientific opinion.

To some extent, the newfound support for anthropoid monophyly was contingent upon more earth-shaking developments in a separate field of science. A German scientist named Alfred Wegener began promoting the concept of continental drift as early as 1912. At the core of Wegener's model was a truly radical idea—that the continents themselves were capable of moving about the surface of the globe like a fleet of rubber ducks in a bathtub. Although Wegener highlighted a variety of evidence—including the geographic distribution of certain fossils—as support for continental drift, his views were largely dismissed by the scientific community of the time. Among other problems, Wegener could never explain the natural forces that might cause continents to drift. Decades later, geologists showed that seafloor spreading was the driving force in plate tectonics, confirming Wegener's theory of continental drift and his place in the history of earth sciences (although he himself did not live to see it). As far as geology was concerned, the plate tectonic revolution transformed almost everything. Geological phenomena that had previously been thought to be unrelated—from volcanoes to earthquakes to mountain building—suddenly shared a common theoretical foundation. At the same time, brand-new possibilities emerged for interpreting how plants and animals happened to live where we find them today.

Prior to the plate tectonic revolution, the rock-solid stability of the continents posed an obvious problem for the idea that all living anthropoids share a common origin. If South American monkeys evolved from the same source that also gave rise to Old World monkeys, apes, and humans, how did these ancestral monkeys manage to cross the Atlantic Ocean? Faced with this dilemma, early paleontologists downplayed the anatomical evidence supporting a common origin for anthropoids in favor of the geographic evidence suggesting that a single origin was unlikely. Instead of trying to explain how the ancestors of New World monkeys reached South America from the Old World, most paleontologists felt more comfortable with the idea that New World monkeys evolved from Eocene prosimians that already had the advantage of living on this side of the Atlantic. After all, paleontologists dating back to Joseph Leidy had noted that the teeth and jaws of *Notharctus*—whose fossils were so abundant in Wyoming—required only minor evolutionary tinkering to be transformed into those of a South American monkey.[1]

Shored up by geography and such plausible New World ancestors as *Notharctus,* the notion that South American monkeys evolved separately from their Old World counterparts hardly suffered from lack of support. Yet this is precisely what the idea received, and it could hardly have come from a more credible source. As part of the initial backlash against the radical views of Wegener, the renowned paleontologist William Diller Matthew decided to erect his most synthetic contribution to evolutionary biology on the terra firma of the anti–continental drift movement. Matthew's deeply influential 1915 article "Climate and Evolution" suggested that ordinary processes like climate change adequately explained how organisms came to live where we find them today. For Matthew, to speculate about continents drifting was not merely untenable, it was also unnecessary, and his views swayed several generations of paleontologists against the notion of a common origin for living anthropoids.

Matthew's basic premise held that cyclical changes in global climate, along with a prevailing tendency for mammals to disperse from north to south, account for most of the odd geographic patterns shown by living mammals. For example, Matthew noted that many groups of mammals inhabit far-flung patches in the tropics or on the southern continents. Tapirs (distant relatives of rhinos with long, flexible proboscises) provide a case in point, because three species live in Central and South America, while a fourth occurs in southeastern Asia. Camels and their close relatives, such as llamas and alpacas, occupy a similarly discontinuous range. Llamas, alpacas, and related species live in South America, while "true" camels are confined to the Old World. Living primates occupy the same basic parts of the globe, although they obviously prefer more humid and forested regions than camels do. Matthew showed that in many instances, the fossil relatives of living mammals had been more widespread (especially across North America and Asia), closing the gaps in their modern distribution. Hence, Matthew saw little need to invoke submerged land bridges or continental drift to explain the patchy geographic ranges of living tapirs, camels, primates, and other mammals. Instead, he thought, their modern distributions represent small fractions of what were once continuous realms.

But why were so many of these problematic occurrences restricted to tropical or southerly parts of the globe? Matthew explained the prevalence of spotty geographic ranges in the south as the result of alternating global climatic regimes. When the Earth's climate was significantly warmer than it is today, fairly uniform environments extended from the poles to the equator. Among other things, this accounts for the presence of primates like *Notharctus* in such unlikely places as Wyoming during the Eocene. These warm, homogeneous periods were punctuated by times when the higher latitudes differed from the tropics by a strong environmental gradient, as they do today. Each shift in climatic regime triggered important evolutionary responses among plants and animals. Natural selection was relaxed during the warm intervals because of the abundant supply of food and the idyllic climate. The onset of cooler episodes resulted in harsher and more variable environmental conditions, which were initially restricted to higher latitudes. These more extreme climatic conditions intensified natural selection, causing adaptive radiations of

Figure 19. William Diller Matthew, preeminent mammalian paleontologist of the early twentieth century, whose theoretical work on the relationship between climate and evolution influenced the study of anthropoid origins for decades. American Museum of Natural History archival negative 116557, copyright American Museum of Natural History Library. Reproduced by permission.

new and improved kinds of mammals. Matthew believed that humans and many other groups of modern mammals first evolved in the northern reaches of the globe—especially central Asia—under these shifting climatic circumstances.[2] Animals that failed to adapt to the severe environments in the north had but two options—to follow the shrinking tropical forests toward the equator or to succumb to extinction. The southward retreat of these tropically adapted mammals led to patchy modern distributions, because their descendants became isolated in South America, Africa, or southeastern Asia, depending on which part of the Northern Hemisphere their immediate ancestors had happened to call home.

Most scientific efforts to bring a wide variety of data under the umbrella of a single theory offer both pluses and minuses. Matthew's views on climate and evolution were no different. The model worked surprisingly well for many mammals, tapirs and camels being prime examples. Although neither camels nor tapirs occur naturally in North America today, both groups boast long and extensively documented fossil records there. Just as Matthew's model predicted, these North American fossils indicate that camels and tapirs once lived far beyond their current, highly disjointed ranges in South America and the Old World. Yet as much as camels and tapirs supported Matthew's model, monkeys conflicted with it. For Matthew, the major problem was that no one had ever found a fossil monkey in North America. At that time, only a handful of early anthropoids were known to science. These included the Fayum specimens described by Osborn and Schlosser and a poorly known creature from Patagonian Argentina that had been named *Homunculus*.[3] None of these fossils helped bridge the geographic gap separating modern anthropoids from the New and Old Worlds.

Faced with such an obvious contradiction to his model, Matthew knew of only three options for explaining away the discrepancy. The first was to blame it all on the inadequacy of the fossil record. In this scenario, monkeys had once inhabited North America, even though their fossils had never turned up there. Perhaps these elusive North American monkeys had lived in specialized environments that were rarely sampled in the fossil record, or maybe they had just passed through briefly on their way to more attractive regions in the south. After all, Darwin himself had frequently cited an incomplete fossil record as the source of apparent conflicts between his theory and the various types of evidence available to him. But this approach held little appeal for Matthew, if only because he and his colleagues had done so much to improve the North American fossil record through the years. Furthermore, Matthew found it hard to explain how fossil monkeys could have eluded generations of paleontologists scouring the American West, while considerably less effort led to their rapid discovery in places like Argentina and Egypt.

If Matthew's monkey problem wasn't caused by deficiencies in the North American fossil record, it could only be because monkeys had never lived there. But how had early anthropoids made their way from the Old World to the dense jungles of the Amazon Basin without traversing North America? Under Matthew's theory of a stable Earth, early anthropoids couldn't have walked to South America across an ancient land bridge, and they certainly hadn't arrived there on a drifting continent. At the same time, Matthew knew that various types of terrestrial animals had colonized islands, some of which were quite remote. Another possibility was thus that early monkeys had somehow crossed the Atlantic Ocean on natural rafts of floating vegetation. As unlikely as this might seem at first glance, Matthew thought that given sufficiently long intervals of time, such extraordinary events might occur.[4] He even went so far as to promote rafting as the most probable explanation for the origin of the peculiar living mammal fauna of Madagascar.

In contrast to the teeming biodiversity of the nearby African continent, Madagascar harbors an impoverished mammal fauna consisting of only a few major groups. Prominent among these are more than twenty species of lemurs. Matthew regarded Malagasy lemurs as the descendants of one or more immigrant species from the African mainland. Like castaways from a shipwreck, these ancestral lemurs would have been fortunate to survive the dangerous voyage across the Mozambique Channel, the body of water that separates the island from southeastern Africa. But merely surviving a treacherous passage across open water isn't enough to colonize an island like Madagascar. Even after their raft washed ashore on some ancient Malagasy beach, these ancestral lemurs had to establish a viable population in unfamiliar territory. To do so, at least one pregnant female had to be on board the lucky African raft. Alternatively, the raft had to support two or more individuals of the same species, including at least one representative of each sex. Given the relatively short distance across open water separating Madagascar from Africa, Matthew regarded rafting as a viable means of ancestral Malagasy mammals getting onto the island. But the idea that a pregnant monkey—or multiple monkeys of the same species—could survive a much longer trip by raft across the South Atlantic Ocean strained the limits of credibility. Matthew therefore rejected this possibility for transporting ancestral monkeys to South America. Only one alternative remained that didn't violate Matthew's overarching resistance to continental drift.

Instead of positing that the North American fossil record was oddly skewed, or that pregnant monkeys had succeeded in navigating the Atlantic Ocean millions of years before Columbus, Matthew settled on the least objectionable option available to him. South American monkeys must have evolved separately from and independently of Old World monkeys, apes, and humans. The idea was appealing because it performed the intellectual equivalent of killing two birds with one stone. Not only did it bring monkeys back into the fold of Matthew's grand theory of climate and evolution, but it also upheld the integrity of the North American fossil record. By arguing that Old and New World anthropoids evolved from different groups of Eocene prosimians, the distribution of anthropoid lineages through space and time began to resemble the patterns Matthew had already established for camels and tapirs. Furthermore, it was no longer true that the North American fossil record had inexplicably failed to yield the ancestors of South American monkeys. The fossils had been right there all along, perhaps in the form of the well-known adapiform *Notharctus.*

Of course, the biggest problem with Matthew's notion that South American monkeys evolved separately from their counterparts in the Old World was the simple fact that the two groups of anthropoids look so much alike. Regardless of whether a monkey hails from Borneo or Brazil, they all share many more features in common than they do with Eocene prosimians like *Notharctus.* Compared to *Notharctus,* all living monkeys have bigger brains, complete bony eye sockets, a reduced snout, and forelimbs and hindlimbs of similar length. If Matthew was correct, the fundamental resemblance of all living anthropoids seemed to defy a basic principle of evolution—that similar features indicate descent from a common ancestor. How could such an apparent contradiction be explained?

Fortunately for Matthew, a pair of British anatomists soon offered a possible answer. Sir Grafton Elliot Smith and Frederic Wood Jones each sought to explain primate evolution by focusing on the critical adaptations that propelled the group's obvious biological success. In contrast to paleontologists like Matthew, Elliot Smith and Wood Jones were not overly concerned with the fairly tractable issues of when, where, and how humans and other primates evolved. Instead, they were motivated by the quasi-philosophical question of *why* humans and other primates diverged from their more primitive mammalian ancestors. As a likely solution, they settled on a basic aspect of biology that virtually all primates share—life in the trees. Appropriately, the evolutionary scenario that Elliot Smith and Wood Jones constructed came to be known as the "arboreal theory" of primate and human origins.

The arboreal theory insisted that life in the trees led inexorably to a steady evolutionary progression—the now familiar series from tree shrew to lemur to tarsier to monkey to ape to human. Natural selection in an arboreal setting produced higher intelligence, increased visual acuity, enhanced hand-eye coordination, and a degeneration of the sense of smell. From the neck down, life in the trees promoted functional differentiation between the forelimbs and hindlimbs, allowing the hands to become progressively freed from the mundane requirements of quadrupedal locomotion. Of course, most of these supposed trends simply anticipated the evolution of bipedal, big-brained humans. By this standard, all other living primates were biological failures, having fallen off the wagon of steady evolutionary progress toward humanity.[5]

Despite its absurd proposition that the vast majority of living primates somehow failed to benefit fully from life in the trees, the arboreal theory of primate evolution provided a possible explanation for the apparent similarity of New World and Old World anthropoids. Instead of reflecting their close common ancestry, many of the features shared by the two groups simply resulted from their long exposure to natural selection in an arboreal environment. Thus, South American monkeys and their Old World counterparts look alike because both groups had arrived at a similar evolutionary stage—one that stood about midway between those occupied by tree shrews and humans. Weighed against the geographic difficulties posed by anthropoid monophyly, the idea that anthropoids arose twice, because both groups had survived for so long in an arboreal habitat, seemed compelling to paleontologists working under the assumption of a stable Earth.[6]

However, as soon as it became acceptable to move continents about Earth's surface, all of these calculations changed. Some of Matthew's more troublesome patches of tropical organisms demanded a fresh look. New World monkeys ranked high on this list of suspects, and biologists began studying them with renewed vigor. The resulting analyses showed repeatedly that New World monkeys resemble their Old World cousins in ways that the arboreal theory simply couldn't explain. For example, a survey of how developing primate embryos become implanted in the uterus and sustained by the placenta demonstrated that New World and Old World anthropoids differ from all other primates in these critical aspects of reproduction.[7] It was hard to imagine how life in the trees might cause anthropoids to evolve a certain type of placenta, while similarly arboreal lemurs required a different method of nourishing their young in utero. Sometimes, when it walks like a duck and quacks like a duck, it really is a duck—or in this case, an anthropoid. The idea that monkeys evolved independently in the New and Old Worlds collapsed like a house of cards.

Although the plate tectonic revolution reunited South American monkeys with their Old World cousins, it did not fully erase the geographic issues that Matthew and other stable Earth paleontologists found so problematic. Ideally, continental drift would have completely resolved the question of how South American monkeys happen to live where they do.

This would have been the case if the common ancestors of New World and Old World anthropoids inhabited an ancient landmass that once encompassed both Africa and South America. As the two continents drifted apart, those anthropoids that became stranded to the west would have evolved into South American monkeys, while those to the east would have given rise to humans and our Old World relatives. This biologically passive mode of placing similar animals on distant terrains goes by the technical name of vicariance. Under vicariance models, the animals themselves do not move. Instead, they are transported on drifting blocks of continental crust, or else their populations become fragmented by novel geographic features, like a rising mountain range or a major river that shifts its course. Vicariance contrasts with the biologically active mode of moving closely related animals to distant lands, which is known as dispersal.

Plate tectonics teaches us that an ancient landmass subsuming both Africa and South America—a supercontinent known as Gondwana—existed in the remote past. Unfortunately, it is abundantly clear that ancestral anthropoids didn't live there. The problem is one of timing. Africa and South America were last connected by dry land roughly a hundred million years ago, during the middle part of the Cretaceous Period. At this early date, modern groups of placental mammals like primates had yet to evolve. As a result, even if we factor continental drift into the equation, we must still explain how early anthropoids crossed the Atlantic Ocean. Our only consolation is that—depending on how far back in time we believe early anthropoids made their fateful voyage—the distance would have been appreciably less than at present. Because Africa and South America have been steadily drifting apart since they parted ways in the Cretaceous, a voyage across the South Atlantic thirty million years ago would have totaled about 1,250 miles (2,000 kilometers); forty million years ago, the same excursion would have covered only 940 miles (1,500 kilometers). Today, the shortest possible route across the South Atlantic spans approximately 1,825 miles (2,920 kilometers).[8]

While such long distances across open water may seem daunting, there are reasons to believe that early anthropoid stowaways on natural rafts of vegetation could have tolerated the trip. First, mariners have reported a number of chance encounters with large floating islands on the high seas. One particularly compelling anecdote describes such a structure off the coast of North America in the summer of 1892.[9] The floating mass was said to comprise an area of roughly 9,000 square feet (1,000 square meters), and it supported trees that stood thirty feet (nine meters) high, making it visible from a distance of seven miles (eleven kilometers). Apparently, the structure possessed enough physical integrity to drift far from its point of origin. The natural raft was first documented off the coast of New England, about 250 miles (400 kilometers) east of Cape Cod. Two months later, the same mass of vegetation was cited far to the northeast, about 375 miles (600 kilometers) east of Newfoundland. By this time, the floating island must have been transported more than 1,000 miles (1,600 kilometers). Its ultimate fate is unknown, but smaller pieces of flotsam and jetsam are known to have conveyed living animals far from their natural range.[10] Floating islands that are large enough to support stands of trees can mimic sailboats, enabling prevailing winds to accelerate their progress across long distances. During the middle Cenozoic, geologists believe that both the prevailing winds and ocean currents in the South Atlantic trended from east to west. Taking into account the contributions of ocean currents and winds, estimates of the time required for a floating island to reach South America from Africa thirty million years ago vary from ten to fifteen days.[11] Because mammals about the same size as early anthropoids have been documented to survive more than thirteen days without water, it seems feasible that early anthropoids could have survived a transatlantic voyage. Given millions of years and a sufficient number of trials, feasibility translates into a significant probability.

Not only does continental drift narrow the oceanic barrier separating South America from Africa, but it also decreases the possibility that South American monkeys arrived there from the north, as Matthew envisioned. Today, the narrow isthmus of land known as Panama forms a bridge linking the Americas. But this connection is a geologically recent development, dating to no more than about two and a half million years ago. For most of the Cenozoic, South America was an island continent similar to Australia. It remained isolated for millions of years after the first monkeys show up in the South American fossil record about twenty-five million years ago. As a result, no matter what route the first South American monkeys followed, they had to cross a significant expanse of open water. If they came from the north as Matthew believed, they would have needed to overcome the added disadvantage of rowing against the tide, because geologists reconstruct ancient ocean currents as flowing from South America toward North America at this time.

Regardless of how the first South American monkeys arrived, the plate tectonic revolution and its biological fallout binds them inextricably with similar anthropoids living in the Old World. Only after this fundamental issue of anthropoid monophyly was settled could science move forward to investigate when, where, and how this common stock of early anthropoids evolved. Predictably, as soon as the question of whether anthropoids evolved once or twice was put to rest, new disputes emerged over the broad outlines of primate evolution. The most contentious point of all was determining which prosimian group lay in or near the ancestry of the lineage leading to monkeys, apes, and humans. Solving this puzzle had major evolutionary implications, because it dictated the types of modifications that were required to make a monkey out of a prosimian, not to mention when and where this pivotal transition occurred. At least three opposing factions soon set up camp across the scientific landscape, each touting its own theory of where anthropoids fit on the primate family tree.

Ever since Leidy's early work on *Notharctus*, various paleontologists have supported a version of anthropoid origins that views them as descendants of Eocene adapiforms. The modern revival of this theory is the handiwork of Philip Gingerich, whose research on *Adapis* is discussed in chapter 2. A paleontological wunderkind, Gingerich rapidly completed a doctoral degree at Yale under the supervision of Elwyn Simons. Like his mentor, Gingerich became fascinated by the major anatomical transitions that occurred during primate evolution. For his doctoral research, Gingerich decided to address the long-standing problem of primate origins. He focused on an extinct group of Paleocene mammals known as plesiadapids—vaguely squirrellike animals whose most familiar representative, *Plesiadapis,* was about the size of a modern groundhog *(Marmota).* At the time, plesiadapids were widely regarded as archaic primates—creatures that helped to bridge the gap between primitive mammals and more advanced prosimians like omomyids and adapiforms.[12]

Paleontologists who study fossils that pertain to the base of vast evolutionary radiations run the risk of developing novel ideas about how the entire group evolved. In focusing on *Plesiadapis,* Gingerich was no exception. Toward the end of his dissertation, Gingerich proposed a new family tree for primates. According to Gingerich's version of primate phylogeny, anthropoids evolved from Eocene adapiforms, while omomyids and tarsiers descended from archaic primates like those he studied in his dissertation.[13] On the face of it, Gingerich marshaled an impressive body of evidence supporting his contention that adapiforms gave rise to anthropoids.

First, Gingerich listed numerous ways in which adapiforms—but not omomyids—resembled early anthropoids from the Fayum. Most of these

Figure 20. Philip D. Gingerich, the modern architect of the adapiform theory of anthropoid origins.

similarities are restricted to the jaws and teeth, a pattern that is not too surprising, considering that these notoriously durable elements form the lion's share of the primate fossil record. For example, Gingerich noted that the lower incisors of adapiforms and anthropoids are small, vertically oriented, and spatulate (more or less shovel-shaped), while those of omomyids are larger, more protruding, and pointed. Like Fayum anthropoids, several different lineages of adapiforms, including *Notharctus* and *Adapis*, fused the two halves of their lower jaws into a single bony element known as the mandible. Similar bony fusion at the chin has never been documented among omomyids. Farther back in the mouth, Gingerich pointed out, adapiforms possess fairly large canines like those of Fayum anthropoids, while the canines of omomyids are uniformly reduced in size. At least some adapiforms—particularly *Notharctus*—

display sexual dimorphism in the size and shape of their canines, another similarity to early anthropoids. In a few adapiform species, the upper canine develops a "honing" facet with the lower premolar with which it occludes, again recalling what is seen in anthropoids. Finally, Gingerich noted that the lower molars of many adapiforms resemble those of anthropoids in having lost a primitive cusp known as the paraconid, a feature that is virtually always retained in omomyids. Beyond the similarities in their teeth and jaws, both adapiforms and Fayum anthropoids tend to be much larger animals than omomyids. As we have already seen, this difference in body size means that both adapiforms and anthropoids could exploit the potentially lucrative ecological niche presented by leaf eating. Omomyids, on the other hand, were forced by their diminutive size to consume food that was richer in calories, mainly insects and fruit. Postcranially, the ankle bones—especially the calcaneus or "heel" bone—of adapiforms and anthropoids lack the elongation typically found in omomyids.

While Gingerich emphasized that adapiforms resemble anthropoids in many respects, he also determined that other Eocene primates are too specialized to serve as plausible anthropoid ancestors. Instead, both omomyids and their close European cousins the microchoerids share features with living tarsiers. In the case of the skull and jaws, Gingerich cited such important similarities as enlarged eye sockets, a bony tube lining the external opening of the ear, pointed incisors, and an upper jaw shaped like a bisected hourglass (using this analogy, the muzzle near the front of the upper jaw corresponds to the constriction at the midpoint of the hourglass). Although postcranial elements were known for only a few species of omomyids and microchoerids, Gingerich was equally impressed by the fact that omomyid limb and ankle bones consistently indicate a propensity for leaping. In contrast to adapiforms, omomyids and microchoerids have elongated ankle bones (especially the calcaneus), and the tibia and fibula of microchoerids such as *Necrolemur* become fused near their distal ends, just above the ankle joint. A more extreme version of these features occurs in tarsiers. Gingerich interpreted the similarities between omomyids and tarsiers as evidence that tarsiers, but not anthropoids, evolved from this group of Eocene primates.[14]

Additional support for Gingerich's thesis that anthropoids evolved from adapiforms came from the fact that certain fossils could not be easily assigned to one group as opposed to the other. For example, Gingerich pointed out that the molar anatomy of *Oligopithecus savagei*—a poorly known fossil collected by Elwyn Simons's first Egyptian expedition—is remarkably similar to that of *Hoanghonius stehlini,* a species widely regarded as an Eocene adapiform from China. For Gingerich, the correspondence between *Oligopithecus* and *Hoanghonius* suggested that either or both of these animals were transitional fossils connecting the more advanced Fayum anthropoids with their primitive adapiform ancestors. Likewise, Gingerich argued that the perennially controversial Eocene primates from Myanmar—*Amphipithecus* and *Pondaungia*—might also lie near the transition between adapiforms and anthropoids. Because all of these anatomically intermediate fossils date to the latter part of the Eocene or the early Oligocene, they also happened to be the right age to bridge the gap between undoubted anthropoids from the Fayum and their Eocene prosimian ancestors.[15]

Although Gingerich constructed his adapiform theory of anthropoid origins to reflect his reading of the fossil record, it runs counter to what the comparative biology of living primates says about their evolutionary relationships. As we saw in chapter 1, numerous aspects of biology indicate that, among living prosimians, tarsiers are the nearest evolutionary cousins of anthropoids. In 1918, a British anatomist named Reginald Innes Pocock published what would eventually become a classic study, showing that tarsiers resemble anthropoids in the structure of their noses and upper lips. In contrast, lemurs retain the primitive mammalian pattern—also found in domestic cats and dogs—in which a moist, naked patch of skin surrounds the nostrils, and the upper lip bears a vertical slit that is tightly attached to the underlying gums. On the basis of these anatomical distinctions, Pocock divided all living primates into two main groups, the Strepsirhini (which means "twisted noses") and the Haplorhini ("simple noses").[16] The Strepsirhini encompasses lemurs, lorises, and bushbabies, while the Haplorhini includes tarsiers and anthropoids. Long after Pocock's division of primates into strepsirhines and haplorhines, new types of data emerged—especially from the nascent field of molecular evolutionary biology—to reinforce his opinion. Gingerich's problem was to reconcile his ideas about primate evolution based on the fossil record with the very different signal that the comparative biology of living primates seemed to offer. How could adapiforms have given rise to both lemurs and anthropoids if tarsiers, which apparently evolved from omomyids, were the nearest living relatives of primates? Something in this equation had to be wrong.

For most other experts on primate evolution, the solution to Gingerich's paradox was that adapiforms had no bearing on anthropoid origins. Instead, omomyids seemed far more suitable candidates for this role.

Figure 21. Frederick S. Szalay, the leading proponent of the omomyid theory of anthropoid origins.

In contrast to Gingerich's adapiform theory, the idea that anthropoids descended from omomyids coincides with the biological evidence that tarsiers are the nearest living relatives of anthropoids. After all, even Gingerich admitted that tarsiers must have evolved from omomyids. Although a number of prominent paleoanthropologists have supported omomyids as potential anthropoid ancestors, no one has made this case more vigorously than Frederick S. Szalay, a morphologist and paleontologist based at the City University of New York.

For Szalay, the evidence supporting an omomyid origin for anthropoids took several forms. In certain critical respects, Szalay found that omomyids resemble anthropoids, while adapiforms look more like lemurs. Unlike the simple dental traits cited by Gingerich, major anatomical features of the skull appeared to link omomyids with anthropoids. For example, oxygenated blood is supplied to the brain in the same basic way in both omomyids and anthropoids, but a different anatomical route is followed in adapiforms and lemurs.[17] Most primates rely on the internal carotid artery—a major blood vessel that runs through the neck—to transmit oxygen-rich blood to the brain. However, the internal carotid artery can follow different paths as it ascends through the neck to enter the braincase in the vicinity of the ear. In omomyids and anthropoids the internal carotid enters the auditory bulla (the bony housing of the middle ear) from behind and toward the midline. In lemurs and adapiforms the internal carotid runs more laterally, so that it enters the auditory bulla from behind and toward the outside. In all primates, the internal carotid eventually divides into two smaller arteries, known as the promontory and stapedial arteries, once it is inside the auditory bulla. The promontory artery is large in omomyids and anthropoids, while the stapedial artery is small or even vestigial. The opposite is true in adapiforms and lemurs. Elsewhere in the skull, both omomyids and anthropoids show evidence for downplaying the sense of smell, while this primitive mammalian attribute continues to be emphasized in lemurs and adapiforms. Not only do omomyids and anthropoids possess narrower and shorter snouts than those of adapiforms and lemurs, but they also bear reduced olfactory lobes of the brain. Szalay regarded all of these important cranial features as positive evidence that anthropoids evolved from omomyids.

Identifying shared features of the skull is all well and good, but if omomyids were as distinctly tarsierlike as Gingerich claimed, they would still be too specialized to stand as suitable ancestors for anthropoids. To counter this argument, Szalay and his colleagues emphasized that not all omomyids possess the complete laundry list of tarsierlike features that Gingerich attributed to them. The supposed distinctions between adapiforms and omomyids in the anatomy of their front teeth serve as a case in point. Although the lower incisors of many omomyids are enlarged and pointed as Gingerich maintained, Szalay argued that this was not universally true. Unfortunately, the fossils Szalay could muster to support his contention were imperfect. Few omomyid specimens preserve the lower incisor crowns in place, and of those that do, most possess enlarged, forwardly protruding incisors, thus conforming to Gingerich's pattern. Despite this problem, Szalay was able to find some omomyid specimens in which the root sockets (or alveoli) for the lower incisors were as small as those of typical adapiforms, even though the delicate incisor crowns were missing.[18] The take-home message was that in at least some omomyids, the incisors remained primitive enough to have evolved into those of anthropoids. A similarly variable pattern emerged in the postcranial skeleton of omomyids and microchoerids. Although a few species resemble tarsiers in having a fused tibia and fibula, in other omomyids these bones remain separate, as they do in anthropoids and most other primates.[19] The variation that Szalay and his colleagues documented among omomyids and their relatives hinted that the group might be diverse enough to include the ancestors of both tarsiers and anthropoids.

If omomyids played such a central role in primate evolution, intermediate fossils should turn up that link anthropoids with their omomyid precursors. Much to the dismay of those who supported the omomyid theory of anthropoid origins, paleontologists seemed remarkably inept at finding such specimens. Even more embarrassing was the fact that Gingerich could recite a long list of fossils that potentially connected Fayum anthropoids with adapiforms. Inevitably, a scientific standoff developed. The omomyid camp emphasized how their views agreed with the comparative biology of living primates, while those who supported the adapiform theory of anthropoid origins laid claim to the mantle of the fossil record. The flip side of this equation also applied. Thus, the omomyid camp increasingly regarded the fossil record with skepticism, especially when it came to drawing simple, ladderlike evolutionary sequences, as Gingerich and his colleagues were prone to do. For their part, the adapiform contingent repeatedly dismissed the mounting evidence from the comparative biology of living primates. After all, there was no straightforward way to reconstruct soft anatomical structures like noses and placentas in extinct primates like omomyids and adapiforms. Nor could their DNA be sampled and sequenced.

In science, as in other human pursuits, gridlock breeds both innovation and exasperation. Gingerich and Szalay had each made compelling cases, but neither side had been able to vanquish its opponents from the field. Rather than resigning themselves to intellectual stalemate, a few researchers began to experiment with new ideas about anthropoid origins. Two prominent colleagues of Elwyn Simons's at Duke University, Matt Cartmill and Richard Kay, formed the most enduring and influential of these maverick factions. Their novel idea was that anthropoids were not closely related to any of the well-known Eocene primates. Instead, living tarsiers and anthropoids—the modern primates that Pocock had segregated as haplorhines—shared a more recent common ancestor with each other than they did with omomyids or adapiforms.[20] For simplicity, let's refer to Cartmill and Kay's version of anthropoid origins as the tarsier theory. Like those who favored a derivation of anthropoids from omomyids, Cartmill and Kay accepted the biological evidence that supported Haplorhini. But they went further in emphasizing a few features that were found only in tarsiers and anthropoids, and not in omomyids. Let's briefly examine the evidence for and against the tarsier theory.

As noted in chapter 1, tarsiers probably qualify as the weirdest primates alive today. They are the only primates that eat nothing but live animal prey. Tarsiers are active at night, yet their eyes lack the "glow in the dark" tapetum lucidum layer that enhances night vision in so many other nocturnal primates and mammals. Their postcranial skeleton—characterized by hindlimbs that are much longer than the forelimbs, fusion between the tibia and fibula, and uniquely elongated ankle bones—rivals the upright, bipedal skeleton of humans in terms of its distinctiveness among primates. Yet despite their pervasive oddity, tarsiers resemble anthropoids in a few bony features. For paleontologists, these hard structures are potentially more informative than the similarities in soft anatomical parts—noses, placentas, and eyeballs—that Pocock and other anatomists had delineated. As the adapiform camp so often stressed, soft anatomical structures are notoriously difficult to reconstruct and interpret in fossils. In contrast, the bony features that tarsiers share with anthropoids might serve as critical waypoints toward deciphering how anthropoids fit on the primate family tree.

Two idiosyncratic aspects of skull anatomy support the tarsier theory of anthropoid origins. The first of these concerns the bony eye socket—one of the most immediately diagnostic features of monkey, ape, and human skulls. The front margin of all primate orbits is completely surrounded by bone. In lemurs, adapiforms, and omomyids, this is the only part of the orbit that is so defined. Among these anatomically primitive primates, a narrow bony strut called the postorbital bar is all that separates the eyeball and the muscles that control its movement from the side of the head known as the temporal fossa. For whatever reason, anthropoids have segregated the eyes and related parts of the visual system from the powerful chewing muscles that fill most of the temporal fossa. Anthropoids achieved this partition by developing a more or less complete sheet of bone between the orbit and the temporal fossa. This highly distinctive bony plate, called the postorbital septum, produces the bony eye sockets that make monkey skulls look so remarkably human. Tarsiers resemble anthropoids and differ from omomyids and other primitive primates in having a partial postorbital septum.[21]

A second point of similarity between tarsier and anthropoid skulls lies in the ear region. In both groups of living haplorhines, the auditory bulla encompasses two spaces separated by a bony septum. The first of these spaces is the middle ear cavity itself, defined as the region that accommodates the three ear ossicles (malleus, incus, and stapes), whose function is to transmit sound waves from the eardrum to the inner ear. In mammals that possess a bony auditory bulla, this structure always encloses the middle ear, so the fact that tarsiers and anthropoids share this attribute is unremarkable. However, the second chamber that lies within the auditory bulla of anthropoids and tarsiers—a space often called the anterior accessory cavity—does not exist in other living and fossil primates. Furthermore, as the internal carotid artery enters the auditory bulla of tarsiers and anthropoids, it runs within the bony septum that separates the anterior accessory cavity from the middle ear.[22]

While these highly unusual similarities in skull anatomy bolster the tarsier theory, problems with Cartmill and Kay's view of anthropoid origins soon became evident. For one thing, the tarsier theory represents a nearly complete rejection of the fossil record. Although the adapiform and omomyid camps disagreed about how to draw the family tree, both groups believed that the ancestry of tarsiers and anthropoids could be found among the broad range of Eocene primates that paleontologists had already unearthed. Yet if the tarsier theory were correct, all of these Eocene primates would be relegated to the very base of the primate family tree, before the evolutionary split between the tarsier and anthropoid lineages. Paradoxically, while the base of the primate tree would then be adorned by a whole alphabet of characters ranging from *Adapis* to *Uintanius,* not a single specimen would document the major branch leading to tarsiers and anthropoids. How could the fossil record be so perversely biased?

Even if the fossil record is as skewed as the tarsier theory posits, the long list of similarities between omomyids and tarsiers remains to be discounted. Some of this evidence dates back to the days of Edward Drinker Cope, whose earliest description of Wortman's *Tetonius* skull noted that its large orbits recall those of tarsiers. In fact, the idea that tarsiers evolved from omomyids has enjoyed such broad and long-lasting support among paleontologists that this was the only major feature of the family tree held in common by the competing factions led by Gingerich and Szalay. Prompted by the radical denial of the tarsier camp, new studies of skull anatomy in omomyids and microchoerids revealed a number of additional ways in which these animals resemble tarsiers.[23] Among other features, omomyids and tarsiers both possess remarkably narrow openings between the nasal passages and the pharyngeal region—the anatomical continuity that allows us to breathe through our noses while our mouths are closed. Anthropoids and other primates have a much wider connection between the nose and the back of the mouth.

At the same time that the evolutionary link between omomyids and tarsiers was being reasserted, Elwyn Simons and his former student Tab Rasmussen began to question the reliability of the bony features that sustained the tarsier theory of anthropoid origins.[24] This dispute goes right to the heart of a long-standing debate in evolutionary biology. Various species often share traits that are similar to a greater or lesser degree. However, when substantial anatomical discrepancies exist, the possibility that these features evolved convergently must be considered. By definition, homologous traits are anatomically similar structures that were inherited from a common ancestor. Precisely because of this evolutionary continuity, whenever we map the distribution of homologous features across various species, we typically arrive at a pretty accurate family tree. Difficulties arise when two species share anatomically similar structures that were absent in their common ancestor. From an evolutionary perspective, such features are analogous but not homologous. They arose by a process known as convergent evolution. For instance, convergent evolution obviously produced the streamlined, fish-like bodies of sharks, dolphins, and ichthyosaurs, because we know that dolphins (which are mammals) and ichthyosaurs (which are extinct marine reptiles) evolved from separate quadrupedal ancestors that walked on land. Unfortunately, not all examples of convergent evolution are this straightforward, so that a fundamental problem in evolutionary biology is to distinguish homologous and convergent traits. In many ways, the features that Cartmill and his colleagues cited in support of the tarsier theory of anthropoid origins exemplify this debate. A closer comparison of the postorbital septum in tarsiers and anthropoids illustrates this point.

No living or fossil primate aside from tarsiers and anthropoids possesses a sheet of bone between the eye and the temporal fossa. Accordingly, the sheer rarity of this feature implies that it may be an ideal evolutionary milestone—one that evolved only once in the history of primates. On the other hand, anatomical distinctions between the partial postorbital septum of tarsiers and the complete postorbital septum of anthropoids open the possibility that the two versions arose independently, from ancestors whose eye sockets resembled those of a lemur or an omomyid. Not only is the partial septum in tarsiers less complete than that of anthropoids, but it is also made from different bones. In anthropoids, the zygomatic bone—roughly equivalent to what we colloquially refer to as "cheek bones"—makes up most of the postorbital septum. In tarsiers, the zygomatic makes only a minor contribution to the bony partition between the eye and the temporal fossa. Instead, most of the partial postorbital septum in tarsiers consists of the frontal and alisphenoid bones, which lie internal to the zygomatic. Tarsiers diverge further from anthropoids in possessing a unique bony flange that extends upward from the rear of the maxilla or upper jaw. While this structure causes the eye sockets of tarsiers to seem more complete than they actually are, the resulting similarity to anthropoids is more apparent than real. Depending on whether we emphasize the similarities or the differences between the postorbital septa of tarsiers and anthropoids, these structures can be regarded as homologous or convergent. Combined with its lack of congruence with the fossil record, such ambiguity poses a grave challenge for the tarsier theory of anthropoid origins.

This survey of the three major models of anthropoid origins in vogue roughly a decade ago has shown the strengths and weaknesses inherent to them all. None of the dominant personalities behind the three camps—Gingerich, Szalay, and Cartmill—succeeded in achieving a consensus around his preferred scheme of anthropoid origins. At the same time, none of the models has been summarily dismissed. To this day, all three factions have their own committed partisans.

Beyond the personal squabbles and scientific egos, larger issues were at stake behind the three models of anthropoid origins. In paleoanthropology, as in other areas of science, strong differences of opinion occasionally transcend the routine data that support one idea as opposed to another. Often, an underlying philosophy motivates scientists to champion a certain viewpoint. Because of its high profile, the controversy surrounding anthropoid origins eventually became a surrogate for broader debates in evolutionary biology. These ran the gamut from how to go about the general task of reconstructing evolutionary relationships—using the "ladder" or "tree" paradigm—to the importance of fossils versus biological evidence from living animals in doing so.

The strongest link between the debate over anthropoid origins and these more philosophical issues in evolutionary biology emanated from the adapiform camp. The strength of this connection derived from the fact that Gingerich's scientific interests went beyond reconstructing the primate family tree. He also wanted to defend his own method of outlining evolutionary history, a technique that he called stratophenetics.[25] Gingerich maintained that the best way to sort out relationships among extinct primates and other organisms was to trace evolving lineages over time. His method relied heavily on both stratigraphy (a proxy for time) and overall biological similarity (or phenetics), hence the term *stratophenetics*. What I have called the ladder paradigm includes stratophenetics as well as related methods of reconstructing evolutionary history. As we

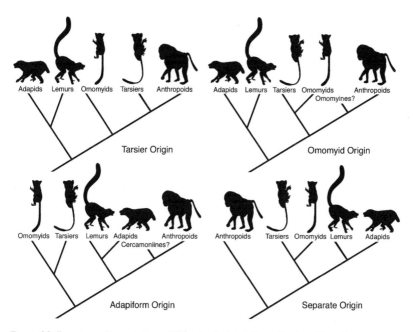

Figure 22. Four competing versions of the primate family tree, showing alternative ways in which anthropoids have been linked with other primates. The tarsier theory of anthropoid origins (upper left), originally proposed by Matt Cartmill and his colleagues, is consistent with a variety of biological evidence, although it conflicts with the fossil record. The omomyid theory of anthropoid origins (upper right), advocated by Frederick Szalay, strives to reconcile the biological evidence linking anthropoids and tarsiers with the fossil record. The adapiform theory of anthropoid origins (bottom left), resurrected by Philip Gingerich, depends on seemingly intermediate fossils, dismissing biological evidence linking anthropoids and tarsiers. Given the problems raised by each of these traditional theories of anthropoid origins, Susan Ford eventually proposed a radical alternative—that anthropoids were the first major group of primates to diverge from the ancestral primate stock in the early Cenozoic (bottom right). Original art by Mark Klingler, copyright Carnegie Museum of Natural History.

have seen, the ladder paradigm can work quite well within the narrow yet ideal confines of the Bighorn Basin, where the fossil record is exceptionally rich. But could it also function on a global scale, and across a broader expanse of time?

The origin of anthropoids provided a prominent test for the general usefulness of the stratophenetic approach. Although the vagaries of fossil preservation might mask the ladderlike signal of the record, Gingerich predicted that a dense succession of intermediate forms once connected adapiforms with their anthropoid descendants. The only thing standing in the way of scientific progress was a dearth of fossils. As things stood,

Gingerich and his scientific acolytes believed that they could identify several important rungs on the ladder leading from adapiforms to anthropoids. Key fossils like *Hoanghonius, Oligopithecus, Pondaungia,* and *Amphipithecus* each documented transitional steps, even though the gaps between the rungs might span thousands of miles and millions of years. If biochemistry and the structure of noses and placentas conflicted with the fossil record, Gingerich and the adapiform faithful knew which line of evidence they would trust.

On the other side of the intellectual divide lay Cartmill and his tarsier theory of anthropoid origins. By default, the tarsier theory found its philosophical home under the tree paradigm, also known as cladistics. Not a single fossil documented the lineage that hypothetically led to tarsiers and anthropoids. Rather, in agreement with cladistic principles, Cartmill and his colleagues had identified anatomical features in tarsiers and anthropoids that had obviously evolved after the origin of lemurs and other primitive primates. Such shared and derived traits as the postorbital septum and the oddly specialized ear region of tarsiers and anthropoids formed the most reliable guideposts to reconstructing evolutionary relationships. While the proponents of the tarsier theory did not deny the importance of fossils, they placed their faith in the utter complexity that could be gleaned from the biology of living animals. Long sequences of amino acids and DNA, as well as the minute details afforded by soft anatomy, upheld the close evolutionary affinity between tarsiers and anthropoids. With an overwhelming abundance of biological data like this, who needed a few fragmentary fossils? Although Cartmill himself did not advocate such extreme views, a few radicals argued that the disparity between the fossil record and the biology of living animals was so great that fossils should actually be ignored when it comes to reconstructing evolutionary relationships.[26]

Szalay's omomyid theory occupied the broad middle ground between these opposite poles. Like Cartmill, Szalay accepted the biological evidence for the haplorhine grouping among living primates. Even though similar biological evidence for omomyids was lacking, Szalay included them among the haplorhines as well, mainly on the basis of their reduced emphasis on the sense of smell. At the same time, the very lemurlike skeletons of adapiforms implied that these animals belonged with the living strepsirhine group of primates. Accordingly, the omomyid theory of anthropoid origins coincided reasonably well with both biological and paleontological data. The major shortcoming of the omomyid theory was its failure to yield transitional fossils that bridged the anatomical gap between omomyids and anthropoids. For this reason, the omomyid theory was more compatible with the tree paradigm than with the ladder.

Despite their underlying philosophical differences and the divergent family trees they implied, all three models of anthropoid origins agreed on several key points. Conspicuous among these was the idea that anthropoids originated fairly recently. Several lines of evidence supported this view. First, the fossil record suggested that anthropoids followed prosimians in time. As far as primates are concerned, the Eocene was dominated by prosimians, but most of these Eocene prosimian lineages became extinct by the end of that epoch. At the same time, the series of expeditions to Egypt's Fayum region led by Elwyn Simons showed that early anthropoids flourished there by the early Oligocene. Hence, the evolutionary transition between prosimians and anthropoids must have occurred sometime near the boundary between these two Cenozoic epochs, roughly thirty-four million years ago. In agreement with this interpretation, the most compelling candidates for anthropoid ancestry among the vast array of Eocene fossils—primates like *Hoanghonius, Amphipithecus*, and *Pondaungia*—all lived near the end of the Eocene. Even the proponents of the tarsier theory of anthropoid origins, who doubted that any of the known groups of Eocene prosimians were ancestral to anthropoids, accepted the conventional wisdom that anthropoids arrived late on the evolutionary stage. After all, if tarsiers and anthropoids had gone their separate ways far back in the Eocene, shouldn't their fossils be apparent among the vast number of Eocene primates that were already known?

The general consensus in paleoanthropology that anthropoids evolved sometime near the Eocene-Oligocene boundary proved to be auspicious, because this was no ordinary interval in Earth history. Following in the wake of the greenhouse world of the Eocene, the Eocene-Oligocene boundary was a time of dramatic global cooling and drying. Typical environments in the interior parts of continents like North America changed from warm, humid forests to cooler, drier, more open conditions. This pervasive environmental shift must have affected primates in profound ways. As we've already seen, the prosimians that previously inhabited Europe and North America were decimated by the events that transpired across the Eocene-Oligocene boundary. Presumably, the radically altered Oligocene ecosystems of North America and Europe could no longer sustain primates. In terms of scale, the environmental changes that took place across the Eocene-Oligocene boundary exceeded anything that had happened since the asteroid impact at the Cretaceous-Tertiary boundary. The asteroid impact is widely credited with ending the reign of the dinosaurs, thus paving the way for the "Age of Mammals." Did the extensive cooling and drying across the Eocene-Oligocene boundary exert a similar effect on primates—causing the extinction of most prosimians and opening the door for anthropoids?

For some, the answer to this question was a resounding yes. Susan Cachel, a paleoanthropologist at Rutgers University, developed the most prominent and wide-ranging hypothesis linking anthropoid origins to climate change.[27] In Cachel's view, most of the features that distinguish anthropoids from prosimians evolved as a response to deteriorating climatic conditions across the Eocene-Oligocene boundary. Her argument went as follows. As climate cooled significantly, ancestral anthropoids responded by becoming larger. Increasing body size enhanced the ability of early anthropoids to maintain a constant core body temperature—the hallmark of being "warm-blooded" or endothermic—by decreasing their body's ratio of surface area to volume. This change in body size cascaded across the biology of ancestral anthropoids, modifying their anatomy, behavior, and diet. For example, the larger size of early anthropoids allowed them to exploit diets that were unavailable to their smaller prosimian forebears. As noted earlier, small primates are forced to consume high-calorie diets that emphasize insects. As anthropoids evolved their larger body mass, they shifted to eating mainly fruits, abandoning their primitive insectivorous niche. Fortuitously, fruits became a more abundant and predictable resource in the more seasonal conditions of the Oligocene, because tropical trees tend to synchronize their production of fruit under seasonal climatic regimes. An increasing reliance on eating fruit provided the impetus for daytime activity patterns among early anthropoids, while insectivorous prosimians continued to be most active at night. Color vision evolved in these diurnal, frugivorous anthropoids to allow them to distinguish ripe and unripe fruits. As diurnal, color-sensitive anthropoids increasingly relied on vision rather than smell, their brains expanded and became reorganized, especially in areas related to neural processing of visual information. The dietary shift from insects to fruit caused the postorbital septum to evolve, providing an expanded area for the attachment of the front part of the temporalis muscles—the powerful muscles that assist in moving the lower jaw during chewing.

Scientists often refer to ideas that unify a wide variety of observations as elegant, in much the same way that a literary critic might praise a particularly evocative poem. If nothing else, Cachel's adaptive hypothesis about anthropoid origins ranks as elegant. Here, in one tidy package, Cachel had explained a major portion of what it means to be an anthropoid. In contrast to the unrealistically simple arboreal theory of primate evolution proposed by Elliot Smith and Wood Jones, Cachel developed her views in terms that were both plausible and internally consistent. But like Wood Jones's and Elliot Smith's, Cachel's goal went beyond the issues of how and when anthropoids evolved. She also wanted to address that most philosophical of all questions—why? In this case, the answers to why and when were much the same—climatic deterioration across the Eocene-Oligocene boundary spurred a sequence of evolutionary changes that culminated in the origin of our distant ancestors, the anthropoid primates.

Like so much sand before a gale-force wind, most of these ideas about our deep evolutionary history would soon be blown away.

6

The Birth of a Ghost Lineage

fell in love with Wyoming the first time I saw the place. Back then, I could never have predicted that Wyoming's vast, open basin and range country would play such a pivotal role in my career. But this is hardly surprising, since I also had no idea what type of career I might pursue. Like a lot of boys entering their awkward teenage years, I was far more interested in sports than science. Growing up in the 1970s in the Blue Ridge Mountains of western North Carolina, backyard basketball filled most of my free time. Dean Smith and his constant retinue of star athletes at the University of North Carolina earned my boyhood idolatry. Fossils seemed as remote as outer space. In fact, in the highly metamorphosed belt of rocks where I grew up, my chances of finding a fossil were roughly equivalent to finding a bit of meteorite or some other wayward souvenir from the cosmos.

Still, I'll never forget the first time I saw Wyoming. My family was on vacation in the summer of 1976, driving west toward the Rocky Mountains in a station wagon equipped with a dubious air conditioner. We had already made the obligatory stops along Interstate 90 in western South Dakota. Wall Drug, with its carpet-bombing approach to billboard advertising, was a bit of a disappointment for me. On the other hand, the endless sea of White River badlands on display in Badlands National

Park justified the long, hot drive across the Great Plains. A few days in the alpine splendor of the Black Hills allowed us to catch our breath before embarking for Yellowstone. Not long after we crossed the Wyoming state line, I saw the summit of an enormous column of rock off on the northern horizon. The road map informed us that the odd geological feature had to be Devil's Tower. Instead of detouring for a closer look, we continued west across the immense expanse of sagebrush and rolling prairie that makes up the Powder River Basin. Its starkly surreal landscapes captivated me. Abundant herds of pronghorn, with their eponymous headgear, looking like nothing else but a living fossil, caused my mind to wander. I imagined that our station wagon had been caught up in some type of time warp. As we drove on toward the distant crest of the Bighorn Range, we might just as well have been traversing the millennia back to the Miocene.

Years later, once it became clear that my aptitude for science far exceeded my prospects in basketball, I rediscovered the interest in fossils that I had first developed as a youngster. Like many kids, by the time I enrolled in first grade, I could name more species of dinosaurs than my teachers thought had ever existed. My father, a biology teacher at one of the local high schools, had instigated my early fascination with fossils by reading me bedtime stories featuring all kinds of prehistoric beasts. After repeated bouts of career-oriented soul-searching in college, I eventually admitted that what I really wanted to do was paleontology. Only later did I learn that I could combine my childhood fixation on fossils with my inordinate fondness for Wyoming. That, in a nutshell, is how I found myself following in the footsteps of Jacob Wortman.

Like Wortman, I began searching for fossils in central Wyoming's Wind River Basin at an early age. I spent my first field season in the Wind River Basin in the summer of 1990, exactly 110 years after Wortman first explored the region as an employee of Edward Drinker Cope. I was still in my twenties that summer—only a few years older than Wortman had been—and I was the proud holder of a freshly minted Ph.D. I even worked for the same museum that employed Wortman for a short, turbulent, but highly successful stint searching for Jurassic dinosaurs not too far to the southeast, at a place called Sheep Creek.

Beyond these basic parallels, my experiences in the Wind River Basin bear little resemblance to what Wortman must have endured there. Over the intervening century, science, society, and the American West have all changed dramatically. Technological advances allow us to enjoy a standard of living in the field that Wortman would find shocking, if not downright decadent. Rather than working solo as Wortman did, my field parties typically consist of six to eight persons, depending on the comings and goings of team members who can't sign on for the entire tour of duty. We divide the daily chores of camp life, and we socialize in the evenings—playing cards, telling jokes, or just passing time by the campfire. Being a one-man operation, Wortman had to perform all of the mundane tasks of fieldwork himself, not to mention enduring the psychological toll of working alone in the vast emptiness of the Wind River badlands. Thanks to plastic coolers and a propane-powered refrigerator, the members of my team drink a variety of cool beverages to stay hydrated. A hundred years ago, while scouting terrain where ephemeral drainages bear such enticing names as Poison Creek and Alkali Creek, Wortman was lucky to find potable water of any temperature. Virtually every evening, fresh meat, fruit, and vegetables grace the dinner table at our base camp. Without the benefit of rapid transportation and modern refrigeration, Wortman relied on dried and salted meat and other nonperishable foods. We speed across stretches of sagebrush in our air-conditioned, four-wheel-drive Chevy Suburban. Traversing similar distances cost Wortman precious hours, moving from outcrop to outcrop on horseback or on foot.

Despite the hardships he had to overcome, Wortman set high standards for measuring paleontological success. He unearthed the remains of so many new types of fossil mammals in the Wind River Basin that subsequent expeditions to the region have mainly served to fill in the gaps. But as we saw in chapter 3, Wortman's greatest scientific achievement did not take place in the Wind River Basin. Nor did it occur along the banks of Sheep Creek in southeastern Wyoming, where Wortman exhumed the nearly complete skeleton of the sauropod dinosaur known as *Diplodocus carnegii*. Rather, Wortman's most notable contribution to paleontology was his discovery of a skull of the omomyid primate *Tetonius* in the Bighorn Basin in 1881, during his first full season of fieldwork there. For more than a century, Wortman's prize specimen was the sole example of an omomyid primate skull known to science.[1] To this day, no one has ever found a second one in the Bighorn Basin, despite the fact that tens of thousands of fossils have been collected there since Wortman's time. Ironically, it took an intensive new round of fieldwork in the Wind River Basin—the region that Wortman abandoned in favor of more lucrative opportunities to the north—to end the century-long drought in finding omomyid skulls.

Any paleontologist who spends time in the field learns an important lesson early on. The fossil record is inherently capricious. Its treasures typically emerge sparingly, in fits and starts. You spend years looking for that special fossil with little or no success, then move to a different site and find precisely what you've been searching for, sometimes in remarkable abundance. Richard Stucky and Leonard Krishtalka, former colleagues of mine at the Carnegie Museum of Natural History, made exactly this type of breakthrough when they discovered the Buck Spring Quarries, an extraordinarily rich set of fossil localities in the Wind River Basin, in 1984. Alongside the jaws and partial skeletons of a wide range of mammals and other vertebrates, Krishtalka and Stucky uncovered the first omomyid skulls to be found in North America since Wortman's day. A diagnostic feature on their upper molars—an extra cusp known as the mesostyle—revealed that all of the omomyid skulls from the Buck Spring Quarries belonged to the same species, *Shoshonius cooperi*. Like most other omomyids, *Shoshonius* was previously documented only by its teeth and jaws. The new material from the Buck Spring Quarries suddenly made *Shoshonius* the gold standard for assessing how North American omomyids fit into the broader scheme of primate evolution.

I first learned of this amazing scientific advance while I was still enrolled in graduate school. I happened to be attending the annual meeting of the American Association of Physical Anthropologists when Richard Stucky gave the Wind River Basin *Shoshonius* specimens their public debut. As Stucky flashed images of what appeared to be the virtually complete skull of an omomyid primate on the screen, a palpable buzz swept through the standing-room-only audience at the conference hotel. At the time I had no way of knowing that I would eventually be entrusted with studying these precious specimens and be given the chance to add to the record that Stucky and Krishtalka had already amassed.

A series of lucky coincidences caused my path to cross that of *Shoshonius*. As a graduate student at Johns Hopkins University in Baltimore, I made frequent pilgrimages to Pittsburgh to visit the Carnegie Museum of Natural History. Then, as now, my interests overlapped broadly with those of Stucky, Krishtalka, and Mary Dawson, all of whom were curators at the museum and experts on various groups of Eocene mammals. At the time, I was part of a team of paleontologists working in the Bighorn Basin, which lies across the divide created by the Owl Creek Mountains from Stucky and Krishtalka's Wind River Basin sites to the south. Because both teams focused on mammals dating to the early Eocene, a good-natured sense of rivalry developed over which basin yielded the better fossils, as well as how those fossils should be interpreted. Both groups believed that the fossil record from their respective basins revealed that certain lineages of mammals evolve gradually (or "anagenetically") through time, but they differed on how this evolutionary pattern should be expressed. My mentors in the Bighorn Basin, Ken Rose and Tom Bown, highlighted the gradual accumulation of evolutionary change by dividing lineages of Bighorn Basin mammals into successions of different species. Stucky and Krishtalka disagreed with this approach, preferring to collapse all members of a single lineage into one species, no matter how many evolutionary changes might accrue. Tongue in cheek, Krishtalka derided the "typological" stance taken by his rivals in the Bighorn Basin, using a simple geographic analogy to considerable effect.

> Wyoming geography suffers from the same typological thinking. In central Wyoming, the Wind River becomes ("speciates anagenetically" into) the Bighorn River at the entrance to the Wind River Canyon. The canyon links the Wind River Basin (where I work) and the Bighorn Basin (where Bown, Rose, and Gingerich work). It is one river (one species) described as two only because different parties explored and named different sections of the river at different times. "Wind River" has priority.[2]

Despite the fact that I worked for the other side, every time I visited the Carnegie Museum of Natural History as a graduate student, I received a warm reception. Richard Stucky and his wife, Barb, usually invited me to stay with them in their home—a welcome relief given my meager budget as a graduate student. These acts of personal generosity were matched by opportunities for professional advancement. I was allowed to publish on fossils in the museum's collection. Once, the museum went so far as to provide partial funding for my personal fieldwork, on the stipulation that any fossils I recovered would be added to the museum's collections. Fortunately for me, by the time I finished my doctoral dissertation at Johns Hopkins, Richard Stucky had decided to leave Pittsburgh for a post at the Denver Museum of Nature and Science, where he still works. The resulting vacancy seemed tailor-made for me, and I jumped at the chance to join the curatorial staff of the Carnegie Museum of Natural History. My new position gave me a leading role in the museum's ongoing Wind River Basin project. It also placed those beautiful *Shoshonius* skulls in my hands.

I spent much of my first year at the museum becoming acquainted with the Eocene fauna of the Wind River Basin by reading the technical papers that my colleagues had published on the subject and by studying the large collection of fossils the museum had acquired from the region.

When early summer rolled around, I could hardly wait to assemble a team and head out west for the field. One of my primary goals was to enhance the already impressive record Stucky and Krishtalka had compiled for *Shoshonius* from the Buck Spring Quarries. By then, Stucky and Krishtalka had found four skulls of *Shoshonius*, but none of them was perfectly preserved. The specimens—about the size of a half-dollar coin in circumference—were so small and delicate that each of them had sustained postmortem distortion or breakage of one sort or another. Some were crushed from top to bottom, so that the eye sockets looked like narrow slits instead of rounded orbits. Another was relatively undistorted, but it lacked the base of the skull where the ear region—so critical for interpreting evolutionary relationships—should have been preserved. As a result, to get an accurate picture of what a pristine skull of *Shoshonius* should look like, we needed to find some additional specimens. To flesh out the emerging picture of *Shoshonius,* I also hoped to locate some of its limb and ankle elements. With a clear picture of the season's objectives in mind, I loaded my team into the museum's field vehicle and began the long cross-country excursion from Pittsburgh to Wyoming.

For me, the trip induced a strong feeling of déjà vu as we rolled past Wall Drug and White River badlands, herds of pronghorn and miles of sagebrush. Only this time, Wyoming's austere landscapes offered more than a sense of romantic nostalgia. Instead, the exposed rock strata promised a journey back to the Eocene, to a period more than fifteen million years before *Aegyptopithecus* and *Apidium* would haunt the lush, tropical coastal plain of northern Egypt. These and other early Oligocene primates from the Fayum region of Egypt had already demonstrated that, in order to trace the origins of the anthropoid lineage, we had to focus on fossils from the Eocene. Knowing this, I couldn't help but wonder whether the Wind River Basin might reveal something fundamental about our own distant ancestry.

As we continued west past the city of Casper, whose 40,000-odd inhabitants make it one of Wyoming's largest metropolitan areas, I began to see that fieldwork in the Wind River Basin would depart from my former experiences in the Bighorn Basin. The Wind River Basin is larger in terms of area, but its outcrops of Eocene badlands are less extensive and less continuous than those I had surveyed as a graduate student. Most of the Wind River Formation consists of relatively drab sequences of mudstones and sandstones. By and large, these strata lack the brilliant hues of purple, yellow, orange, and red that prevail in the Willwood Formation to the north. Mercifully, ambient summer temperatures in the Wind

River Basin tend to be several degrees cooler than in the Bighorn Basin, because the higher elevation more than compensates for its more southerly position. Turning north from the highway onto the gravel road leading to our field area, I also realized that our standard of living here might easily eclipse what I had grown accustomed to in the Bighorn Basin.

The reason for this inequality owed more to geology than economics. The Willwood Formation crops out mainly in the center of the Bighorn Basin, far from the mountains that virtually encircle the basin. Basing our operation miles from the nearest mountain front and any reliable source of fresh water forced us to run a dry camp. Among other things, this meant that each member of the Bighorn Basin field party bathed only once a week, whether we needed it or not. In contrast, in the Wind River Basin, some of the best Eocene outcrops occur near the southern flank of the Owl Creek Mountains, along the northeastern margin of the basin. Leveraging this natural advantage, we chose to camp in the foothills of the Owl Creeks, several hundred feet above the basin floor. Between our base camp and the best exposures of the Wind River Formation lay the drainage of Badwater Creek. Despite its name, the stream provided a steady supply of icy water, fed from melting snowpack in the nearby mountains. Working under the assumption that a happy field crew finds more fossils, I insisted that each member of my team bathe every day in its refreshingly brisk currents. Aided by a menu featuring such gourmet cuisine as roasted Cornish game hens and a generous selection of microbrews, camp morale remained remarkably high in the Wind River Basin. As word of our lifestyle filtered out across the Owl Creek Mountains, we sometimes offered temporary shelter to refugees from the more spartan Bighorn Basin camp to the north.

Our fieldwork in the Wind River Basin fell into two general categories. Throughout the long summer field season, we targeted the Buck Spring Quarries, where we had the best chance of finding additional skulls and postcranial elements of *Shoshonius*. But while the site served as our bread and butter, working there was tedious. First, the fossil-bearing stratum had to be made accessible by removing the overlying layers of rock, which bore few, if any, fossils. This entailed old-fashioned manual labor—the kind that convicted criminals once performed back in the days of highway chain gangs. We used picks, shovels, sledgehammers, and large metal wedges to remove the unwanted "overburden." Once the pay layer was exposed, we cast aside these heavy tools in favor of trowels, chisels, brushes, and dental picks. Many of the fossils we worked so hard to recover at Buck Spring were tiny and delicate. These specimens disintegrate

Figure 23. Sandra Beard and Alan Tabrum collecting fossils for the Carnegie Museum of Natural History at the Buck Spring Quarries in the Wind River Basin, central Wyoming.

rapidly once they are exposed to the elements through the natural erosion of the Wind River badlands. As a result, most of the smallest animals from the site had never been seen before and were unknown to science. These included relatives of living shrews, hedgehogs, rodents, and even bats—the last of which are among the oldest ever found in North America.[3] Given the nature of the fossils we sought, we had no choice but to proceed cautiously in order to limit any damage during their recovery to the barest minimum. Yet the slow pace of the work made it difficult to stay focused on the task at hand. To maintain the team's interest and enthusiasm, I launched into lengthy discussions about the significance of the latest fossil to be discovered at the site. Whenever that tactic failed, I placed a bounty on future specimens, typically in the form of ice cream treats at the local general store. Soon my crew began referring to fossil mammal jaws with two or more complete teeth—the minimum qualifying standard—in terms of this highly desirable commodity, asking each other how many "ice creams" they had collected at the end of each day.

To relieve the monotony at Buck Spring, I split my team into two groups. Each day, one group returned to Buck Spring, while the other surveyed for fossils at new or established sites elsewhere in the basin.

The two groups traded places nearly every day, so that everyone had a chance to find fossils using both techniques. Although the best fossils almost invariably turned up at Buck Spring, I confess that I've always enjoyed prospecting for fossils more than anything else in paleontology. For me, the thrill of exploring a new area of badlands, where no one may ever previously have looked, is what makes fieldwork so addictive. In a setting as lovely as the Wind River Basin, even when you strike out as a paleontologist, you can't help but be enthralled by the natural world around you. I've stumbled across dens of bobcat kittens; I've seen rattlesnakes engage in territorial combat (it resembles arm-wrestling); and I've come face-to-face with a startled golden eagle while climbing a ridge of badlands.

Despite the many diversions, however, nothing beats finding a new locality that is rich in fossils. On average, we located about one excellent new site each year that we worked in the Wind River Basin. In 1992, for example, we decided to scout out some new areas well to the west of Buck Spring. As we pushed deeper and deeper into new terrain, it became clear why our predecessors had ignored the badlands in this region. Most of the sites we visited were barren of fossils, and they were so far away from our base camp that the effort hardly seemed worthwhile. Just as I was about to give up and return to Buck Spring, I noticed an appealing patch of badlands about half a mile from where we had parked our field vehicle. There, the exposed strata stood out because they were striped with tinges of olive, distinguishing them from the drab beds we had been surveying with little success. As we walked across the sagebrush flat to reach the olive-colored strata, I noticed an extraordinary number of grasshoppers littering the ground. You could barely take a step without crunching them beneath your feet. This was partly because they were so numerous, but also because they were more focused on each other than the human interlopers on their domain. All around us, an orgy of paired-off grasshoppers seemed intent on ensuring the propagation of their species.

The utterly Darwinian example set by the grasshoppers proved a good omen for the olive beds in the distance. As soon as we reached our destination, we began to find fossils. These were not only abundant, but they were also well preserved. We found a nearly complete lower jaw of a small tapir called *Selenaletes*. Nearby was the lower jaw of a primitive brontothere known as *Eotitanops*. Millions of years later, during their evolutionary heyday, more advanced brontotheres would grow larger than the biggest living rhinos, whom they generally resembled but for their single, forked nasal horn, shaped vaguely like a wishbone. *Eotitanops* would have paled in comparison to its brutish younger relatives, but its lower jaws were nonetheless substantial—about the size of those of a large cow. Finding such large, conspicuous fossils right on the surface of the outcrop could mean only one thing—we had stumbled upon a tract of virgin badlands. The fossils we were finding had been weathering out of the surrounding rock strata for millennia, and no human—or at least no human with an interest in fossils—had been here previously. As we continued our happy fossil collecting, we unearthed treasure after treasure. My most memorable finds consisted of two partial skeletons of armadillolike mammals known as palaeanodonts. Not until the evening light began to falter did I call a reluctant halt to our frenzy of discovery. We all agreed that such a wonderful place demanded a distinctive name. To this day, in the computerized database of fossil localities at the Carnegie Museum, the site is officially known as "Grasshoppers in Love."

Our survey work in the Wind River Basin was fun, but it was also important from a scientific perspective. Not all fossil localities are rich enough—or concentrated enough—to be quarried like Buck Spring. Yet, by collecting fossils from a broad spectrum of sites, we accomplished several objectives. First, by taking a thorough census from multiple stratigraphic horizons, you develop a good picture of how the local fauna changes over time.[4] Certain species appear while others go extinct. Others simply evolve. In order to understand the significance of fossils from any single locality—no matter how good it might be—it's important to know something about the dynamic picture of change through time. A second reason to conduct broad paleontological surveys is that every fossil site is in many ways unique. They each sample a particular environment over a finite duration of time. For example, we know from geological studies of the Buck Spring Quarries that the fossils we unearth there accumulated in a swampy environment.[5] Swamps and ponds, because they occupy topographic low points, function as reservoirs for gathering animal bones and carcasses that may ultimately become fossilized. But the ancient ecosystem of the Wind River Basin was hardly confined to these restricted habitats. Understanding the true complexity of ancient landscapes hinges on surveying sites that sample different environments, even when these sites are precisely the same age.

Taphonomy—the wide range of factors that determines how living organisms become preserved as fossils—also requires us to cast a broad net during paleontological fieldwork. Comparing one of the richest sites in the Wind River Basin, a locality known as Sullivan Ranch, with the Buck Spring Quarries shows how deep the taphonomic imprint on the fossil record can be. Carnegie Museum field parties have collected thousands of fossil mammal specimens at Sullivan Ranch through the years. Despite their abundance, fossils from Sullivan Ranch lack the spectacular preservation found at Buck Spring. This difference is owing to the divergent taphonomic histories of the two sites. The vast majority of fossils from Sullivan Ranch derive from paleosols, or ancient soil horizons, which are well exposed over an area of about two and a half acres (one hectare). As these specimens slowly became incorporated into developing soil profiles, most were exposed to scavengers, the elements, and other destructive agents prior to burial. In contrast, at Buck Spring, many specimens sank directly to the bottom of a prehistoric pond or swamp immediately after death. In these cases, any damage the fossils sustained happened long after they were entombed in the underlying mud. Instead of being weathered and scavenged, Buck Spring fossils suffer mainly from deformation caused by being buried beneath layer upon layer of younger rock. The take-home message for a field paleontologist is that if you want to find lots of fossils in a short time, go to Sullivan Ranch. If you want to find exquisitely preserved specimens, go to Buck Spring.

These taphonomic factors directly affected the primate fossil record in the Wind River Basin. Fossils of *Shoshonius* occur at both Sullivan Ranch and Buck Spring. In terms of the absolute number of specimens, *Shoshonius* is more than three times as abundant at Sullivan Ranch. But no one has ever found a skull or a major limb element of *Shoshonius* there. On the other hand, the extraordinarily thorough sampling at Sullivan Ranch paints a vivid portrait of the full range of animals that inhabited the Wind River Basin at that time. Among other things, the Sullivan Ranch record shows that *Shoshonius* lived alongside several other species of omomyids, all of which were comparatively rare. In fact, three of the omomyid species that have been found with *Shoshonius* at Sullivan Ranch have rarely, if ever, been unearthed anywhere else. Only through exhaustive sampling of the Sullivan Ranch site were we able to find these cryptic species, which we named *Anemorhysis natronensis*, *Trogolemur amplior*, and *Trogolemur fragilis*.[6] All three of these rare omomyids were small, and they were apparently more closely related to one another than to *Shoshonius*. Judging from their remarkably hypertrophied and protruding lower incisors, the two species of *Trogolemur* must have been ecologically specialized gum-feeders. Perhaps this narrow dietary niche explains their poor representation in the fossil record.

The important point to be gleaned from comparing Sullivan Ranch with Buck Spring is that bias of one sort or another pervades the fossil record. Recognizing this, any full understanding of the history of life must be constructed from all available evidence. By diligently collecting thousands of specimens from Sullivan Ranch, my colleagues and I succeeded in fleshing out a few details about the evolutionary history of omomyids in North America. But to understand how omomyids as a group fit on the primate family tree requires fossils that are substantially more complete. Fortunately, our concurrent work at the Buck Spring Quarries was progressing nicely on this front.

During the three field seasons during which my team worked there, our efforts to find additional specimens of *Shoshonius* in the Buck Spring Quarries succeeded beyond my wildest expectations. We added two more skulls to the four that had already been found by Stucky and Krishtalka. Neither of the new *Shoshonius* skulls was immaculate, but one of them preserved the critical ear region in better shape than any of the previously collected specimens. We also uncovered many of the major limb and ankle elements of *Shoshonius,* thus revealing aspects of its postcranial skeleton for the first time. More than a century of fieldwork in the American West has established that postcranial remains of omomyids rival their skulls in terms of rarity. Complete forelimb and hindlimb bones of North American omomyids, such as the humerus and femur that we unearthed at Buck Spring, had simply never been found previously. In the span of less than a decade, Carnegie Museum excavations at Buck Spring had transformed *Shoshonius cooperi* from an obscure species whose only known remains consisted of some jaws and teeth to one of the most thoroughly documented Eocene primates in the world. Yet the more we learned about *Shoshonius,* the more it became apparent that prevailing theories of anthropoid origins were woefully incorrect.

In paleontology, as in many other human pursuits, timing can be everything. As I settled down to study the anatomy of *Shoshonius* in detail, I began to entertain a notion that most of my colleagues would have regarded as heretical. As far as our efforts to reconstruct anthropoid origins were concerned, it seemed to me that our timing was off. The cumulative wisdom gained from the primate fossil record, from the days of Cuvier right up to the present, held that anthropoids had evolved fairly recently. The oldest well-known and undisputed anthropoids came from Egypt's Fayum region, where fossils like *Aegyptopithecus* and *Apidium* dated to the early Oligocene. A few scrappy specimens—such as the perennially controversial Burmese primates, *Amphipithecus* and *Pondaungia*—

extended the record of possible anthropoids back to the waning days of the Eocene. But even if *Amphipithecus* and *Pondaungia* turned out to be anthropoids, they were only marginally older than their Egyptian cousins. *Aegyptopithecus* and *Apidium* dated to roughly thirty-three million years ago, and *Pondaungia* and *Amphipithecus* might push the record back as far as thirty-seven million years or so, a difference of little more than 10 percent. No one had ever claimed to have found a fossil anthropoid—no matter how primitive or controversial it might be—going back as far as fifty million years or more, which was the era of *Shoshonius*. By that measure, I had no reason to suspect that our work in the Wind River Basin would fundamentally alter our ideas about anthropoid origins. In fact, if the bounty of material from Buck Spring clarified nothing else, it proved that *Shoshonius* was not an ancestral anthropoid. Yet the difference would be moot, as we shall see in due course.

By carefully examining its skull and postcranial skeleton, I soon became convinced that *Shoshonius* is closely related to living tarsiers. Even the most cursory comparison between a tarsier skull and that of *Shoshonius* reveals an obvious similarity between these two primates—they both have enormous orbits. Recall that tarsiers stand out among living primates in having greatly hypertrophied eyeballs, each of which encompasses about the same volume as its brain. Although the orbits of *Shoshonius* don't quite measure up to tarsier standards, they go well beyond anything found in other primates, either living or fossil. At the same time, *Shoshonius* lacks the long, doglike muzzle seen in most prosimians. The overall result is a very tarsierlike face, one that is dominated by its huge, forward-facing orbits.

Detailed similarities to tarsiers also extend to the underside of *Shoshonius*'s skull. Like many other omomyids, *Shoshonius* possesses an upper dental arcade shaped like a bisected hourglass. While the left and right molars are widely separated, the short, narrow muzzle constricts the upper dentition in the region of the canines and incisors. The same unusual configuration of the upper dentition exists in tarsiers, which is one of the reasons that paleontologists have linked omomyids with tarsiers for so long. Oddly enough, while the molars on either side of the upper jaw diverge widely from each other, the connection between the nasal passages and the back of the mouth (technically known as the choanal region) is very narrow in *Shoshonius*. As a result, the upper molars appear to be isolated on bony platforms on either side of the aperture between the nose and mouth. The same condition occurs in tarsiers and those few omomyids and microchoerids that are documented by complete skulls, but not in other living or fossil primates.

Moving on to the ear region, traditionally regarded as the location of numerous features important for reconstructing evolutionary relationships, we find additional links between *Shoshonius* and tarsiers. For example, in both of these primates, distinctive bony flanges overlap the auditory bulla in two different spots. One of these bony flanges occurs on the outer side of the bulla, in front of the external opening that leads to the eardrum. There, a backward extension of the lateral pterygoid lamina—a thin sheet of bone that anchors some of the muscles involved in chewing—envelops the bulla's external surface. The second bony flange, derived from the basioccipital bone, lies on the internal side of the bulla, toward its rear. With the exception of the European microchoerid *Necrolemur* and possibly *Tetonius* (Wortman's *Tetonius* skull is badly damaged in this region), no other primates possess similar bony overgrowths on their auditory bullae. Tarsiers and *Shoshonius* also depart from the typical primate pattern in having a very narrow skull base. In other primates, the basioccipital and basisphenoid bones are much broader, yielding greater separation between their ear regions. Without belaboring the numerous similarities in the ear regions of *Shoshonius* and tarsiers, it seems worthwhile to point out that these extend to seemingly trivial features. The stapedius muscle, one of the smallest muscles in the primate body, provides a case in point. Its name comes from the fact that it attaches to the stapes, one of the three tiny bones that link the eardrum with the inner ear. In almost all primates, the narrow groove that houses the stapedius muscle lies within the auditory bulla. Tarsiers and *Shoshonius* depart from this nearly ubiquitous pattern in having the stapedius groove exposed on the sidewall of the skull, immediately adjacent to the bulla.[7]

In contrast to the numerous and profound ways in which their skulls resemble each other, the postcranial skeleton of *Shoshonius* differs appreciably from that of tarsiers. Still, important points of resemblance are evident. For example, the femoral head (the "ball" part of the ball-and-socket joint at the hip) of *Shoshonius* is not spherical as it is in anthropoids and most other primates. Rather, it approximates the more cylindrical shape characteristic of tarsiers and bushbabies. In both of these living prosimian primates, the asymmetrical shape of the femoral head reflects the importance of flexed and widely splayed hip postures, the stance that is required when these small animals cling to relatively vertical supports. The distal femur (the upper part of the knee joint) of *Shoshonius* also resembles that of tarsiers and bushbabies in being high from front to back and narrow from side to side. Just above the ankle joint, the tibia and fibula of *Shoshonius* are not fused as they are in tarsiers, but an extensive bony scar shows that the two bones lay in tight apposition for at least a quarter of their length.[8]

From this long series of anatomical comparisons among *Shoshonius,* tarsiers, and other primates, a coherent picture begins to emerge. *Shoshonius* shares many features with tarsiers that are lacking in most or all other primates. At the same time, tarsiers and *Shoshonius* differ in a number of important ways, which is hardly surprising given their separation in time and space. Interpreting the ways in which tarsiers diverge from *Shoshonius* is straightforward in an evolutionary sense. In practically every case, tarsiers have evolved novel or advanced features (such features are referred to as "apomorphic" in the jargon of cladistics) that *Shoshonius* lacks. The list of these features is fairly long, but a few of them are spelled out here, because they are important for interpreting the ancient lifestyle of *Shoshonius* as well as its evolutionary position. (1) Tarsiers have larger, more vertically oriented (or "frontated") orbits than *Shoshonius.* (2) Tarsiers possess a partial postorbital septum, while *Shoshonius* retains a simple postorbital bar. (3) The base of the skull is flexed in tarsiers, so that the foramen magnum enters the braincase from below. In *Shoshonius* there is little basicranial flexion, and the foramen magnum lies at the back of the skull. (4) The auditory bulla of tarsiers encapsulates both the middle ear space and a second chamber (the "anterior accessory cavity"), the latter of which is absent in *Shoshonius.* (5) The hindlimbs are much longer than the forelimbs in tarsiers, while this disparity is less pronounced in *Shoshonius.* (6) The tibia and fibula are fused in tarsiers, while a long scar demarcates the region where these bones lay in tight apposition in *Shoshonius.* (7) The calcaneus (or "heel" bone) of tarsiers is extremely long, while that of *Shoshonius* is only moderately long.

The fact that tarsiers are generally more specialized than *Shoshonius* makes a lot of sense intuitively. Because *Shoshonius* dates from more than fifty million years ago, we might expect it to diverge less from ancestral primates than living tarsiers do. After all, fossil hominids such as *Australopithecus* more closely approximate our apelike ancestors than we do. There is a general correlation between a fossil's age and how primitive its anatomy tends to be, but the relationship is hardly perfect. For example, in at least one respect—the presence of an extra cusp on the upper molars known as the mesostyle—*Shoshonius* is more specialized than living tarsiers. This seemingly trivial feature highlights a broader point. It hardly ever makes sense to refer to a given species—whether living or fossil—as being "more primitive" than another, for reasons that go beyond any value-laden connotations the comparison carries along with it. Tarsiers are more primitive than humans in having three premolars on either side of their lower jaws and in lacking a complete mandible formed by bony fusion at the chin. Humans are more primitive than tarsiers in retaining a separate tibia and fibula and in having much smaller eyes. The important distinction here is that, while entire species can rarely be arranged from primitive to advanced, individual features usually can be. In fact, paleontologists rely upon exactly these types of trait-by-trait comparisons to decipher the biology of extinct organisms, as well as to reconstruct how they fit on the evolutionary tree.

The quality of the Buck Spring Quarry sample of *Shoshonius* fossils allows us to paint a reasonably detailed portrait of its way of life. *Shoshonius* was a small primate, probably weighing no more than two or three ounces (sixty to ninety grams). As discussed previously, primates in this size range are forced to consume foods that are rich in calories—mainly insects, small vertebrates, fruits, and gums. In contrast to most other North American omomyids, the cheek teeth of *Shoshonius* bear sharp crests, suggesting that its diet emphasized small animal prey over fruits and gums.[9] The extremely large orbits of *Shoshonius* indicate that it was a nocturnal species, because primates that are active at night tend to have larger orbits than diurnal species. In fact, *Shoshonius* resembles tarsiers in having orbits that are substantially larger, relative to the length of its skull, than those of other primates, including such small nocturnal species as bushbabies and lorises. This extreme orbital hypertrophy suggests that *Shoshonius* may also have resembled tarsiers in lacking the reflective tapetum lucidum layer, so that its greatly enlarged eyeballs evolved as a means of compensating for this loss. While vision remained of utmost importance to *Shoshonius,* its reduced snout and narrow interorbital region suggest that the sense of smell—so critical to lemurs and many other mammals—was being diminished.

In most respects, from the neck up, *Shoshonius* is very tarsierlike. Its diet, activity cycle, and major sensory adaptations probably closely tracked these aspects of the biology of tarsiers. From the neck down, important differences emerge. These anatomical discrepancies indicate that *Shoshonius* traveled through the dense Eocene forests of central Wyoming in a different manner than tarsiers move through the jungles of Borneo and the southern Philippines today. Several factors indicate that *Shosho-nius* utilized more quadrupedal patterns of locomotion than tarsiers do. Tarsier anatomy recalls that of humans in the fundamental sense that the neck and torso are habitually situated below the head, rather than behind it. Primarily, this is because the foramen magnum, through which the spinal cord achieves its neural connections to the brain, is located underneath the skull in tarsiers and humans, while in most other primates (including *Shoshonius*) the foramen magnum lies toward the back of the skull. Just as adult humans feel more comfortable standing upright than crouching on all fours, tarsiers strongly prefer a vertical orientation for their torsos. Although *Shoshonius* could adopt vertical postures whenever the need arose, the primitive location of its foramen magnum indicates that, like most other primates, it frequently held its trunk more horizontally. Its locomotion must have been more varied than that of tarsiers as a result. Rather than being anatomically committed to vertical clinging and leaping, the postcranial skeleton of *Shoshonius* implies that it engaged in quadrupedal walking and running and powerful climbing as well as leaping. For example, *Shoshonius* lacks the hindlimb dominance and extreme elongation of the calcaneus characteristic of tarsiers.

Although *Shoshonius* differs from tarsiers in its habitual posture and mode of locomotion, these two primates seem to occupy adjacent branches on the family tree. To put this apparent contradiction in perspective, consider the disparity between knuckle-walking in chimps and bipedalism in humans. Despite the obvious differences in how chimps and humans move, it is abundantly clear that the two species are closely related. The differences in chimp and human locomotion evolved sometime after their evolutionary divergence, which probably dates to roughly seven million years ago. Tarsiers have been evolving independently from *Shoshonius* for at least fifty million years, and probably longer. Yet the differences in locomotion between tarsiers and *Shoshonius* were probably less substantial than those between chimps and humans. The evidence that chimps and humans are closely related includes the chimplike skulls of early fossil hominids as well as the close similarities in their DNA. How compelling is the evidence that links *Shoshonius* to tarsiers in an evolutionary sense?

Unless molecular biologists succeed in recovering DNA from Eocene fossils à la *Jurassic Park,* our understanding of how *Shoshonius* fits on the primate family tree will always rely on anatomical evidence. Thanks to the bounty of the Buck Spring quarries, we are on unusually solid ground when it comes to interpreting *Shoshonius* this way. As we have seen, *Shoshonius* uniquely resembles tarsiers in a number of important details. For the most part, these similarities lie in the skull. But similarities alone are not enough to conclude that *Shoshonius* and tarsiers are closely related. To take that step, we have to demonstrate that the features shared by *Shoshonius* and tarsiers are evolutionary novelties (also known as shared derived characters or "synapomorphies") that arose in their close common ancestor, after most other primate lineages had gone their separate ways. For example, tarsiers resemble *Shoshonius* and many other omomyids in having a cusp on their lower molars known as the paraconid. This cusp is absent in many adapiforms, living lemurs, and anthropoids. However, no serious paleontologist would cite the presence of paraconids as evidence that tarsiers and *Shoshonius* are closely related. The reason is that virtually all early primates—including basal adapiforms and anthropoids—retain this feature. Its occurrence in tarsiers and *Shoshonius* signifies that, in this particular respect, neither of them has diverged from the ancestral primate condition. The same could be said about the fact that most living primates have five fingers and toes. Similarities like these may be anatomically compelling, but because they were inherited from such a distant common ancestor, they fail to illuminate evolutionary relationships among its descendants. Evolutionary biologists refer to such features as shared primitive characters or "symplesiomorphies." They cannot be used to trace the complex pattern of evolutionary bifurcations that describes the primate family tree.

In practice, showing that shared anatomical features are derived rather than primitive is not always straightforward. Nevertheless, a number of criteria can help us make this determination. For example, features that are confined to just a few species (like the bony flanges that overlap the auditory bulla in tarsiers and *Shoshonius*) are more likely to be derived than features that are widespread (like the presence of five fingers and toes). Features that are anatomically complex (like the postorbital septum) are more likely to be derived than anatomically simpler structures (like the postorbital bar). Some features (like the evolutionary loss of a premolar or a toe) can be regarded as derived because it is difficult or impossible to imagine evolution proceeding in the opposite direction. Often, it is possible to ascertain whether a feature is primitive or derived by casting a wider biological net, and looking at animals outside the group of immediate interest. Today, powerful computer programs routinely sort through large volumes of anatomical and genetic data to determine the most likely arrangement of primitive and derived characters. None of these procedures is foolproof, but they all help to decode the evolutionary signal provided by fossils, genes, or other biological traits.

Returning to the issue at hand, it is clear that most of the skull features that link *Shoshonius* to tarsiers are derived rather than primitive. Prominent examples include hypertrophied orbits, a reduced snout, the narrow connection between the nasal passages and the back of the mouth, the bony flanges that overlap the auditory bulla, and the narrow cranial base. This imposing set of shared derived traits makes a solid case for a tight evolutionary relationship between *Shoshonius* and tarsiers. Many evolutionary relationships are proposed and gain wide acceptance on thinner evidence. Yet in this case, the features linking tarsiers with *Shoshonius* conflict with the evidence put forward by those who support the tarsier theory of anthropoid origins (see chapter 5). Recall that tarsiers and anthropoids each possess a postorbital septum. In contrast, *Shoshonius* resembles *Tetonius, Necrolemur,* and other primitive primates in having only a simple postorbital bar. Furthermore, while *Shoshonius* and tarsiers resemble each other in many aspects of ear anatomy, *Shoshonius* lacks the anterior accessory cavity that is present in both tarsiers and anthropoids. Like most of the features that link *Shoshonius* to tarsiers, the traits shared by tarsiers and anthropoids are obviously derived. Primitive primates lacked both the postorbital septum and the anterior accessory cavity. Hence, reconstructing where tarsiers lie on the primate family tree hinges on how we interpret this conflicting anatomical evidence. Apparently, evolution often works in such a way that the evidence it leaves behind can be misleading. The only explanation for the conflicting anatomical evidence regarding tarsiers is that one set of these features is the product of evolutionary convergence rather than inheritance from a common ancestor.[10] But which set of shared derived characters do we believe?

There are three basic methods of resolving evolutionary logjams caused by this type of character conflict. The simplest of these is to add up the derived features that support each of the evolutionary alternatives. The evolutionary tree backed by the most derived features wins. By this measure, a close evolutionary relationship between *Shoshonius* and tarsiers is preferred, by a margin of roughly six to two. (Here, I'm counting the two bony flanges over the auditory bulla as separate traits; otherwise, the vote is five to two). Unfortunately, the simplicity of this method belies some fundamental problems. The most obvious of these has been duly noted by Matt Cartmill, the driving force behind the tarsier theory of anthropoid origins:

The number of resemblances we can find between any two real objects—particularly objects as complex as organisms—is practically limitless. A sufficiently ingenious or perverse investigator can therefore go on for a surprisingly long time seeking out such resemblances and adding them to a favored list to make sure of securing a desired outcome. For instance, if we wished to argue for an elephant-human clade that excludes chimpanzees, we might advance such candidate elephant-human synapomorphies as a projecting nose, a bony chin, hairlessness, and an inability to gallop.[11]

A second, and in some ways preferable, way of resolving seemingly intractable evolutionary conflicts is to go back and reassess the source of the disagreement. Are any of the features suspect in some way? In chapter 5, the potential problems concerning the homology of the postorbital septum in tarsiers and anthropoids were discussed. Tarsiers have a partial postorbital septum, while anthropoids have a more complete sheet of bone separating their eye sockets from the large temporalis muscles on the sides of their heads. At the same time, different bones make up the bulk of the postorbital septum in each case. In anthropoids, the postorbital septum is formed mainly by the zygomatic bone, while the partial postorbital septum of tarsiers includes much larger contributions from the alisphenoid, frontal, and maxilla.[12] By the same token, the anterior accessory cavity that lies within the auditory bulla of tarsiers and anthropoids (but not *Shoshonius*) shows some interesting differences that warrant suspicion. In anthropoids, the anterior accessory cavity is filled with bony cellules. In tarsiers, the space is a simple hollow chamber. The differences between the anthropoid and tarsier versions of these derived characters immediately raise the possibility that they evolved convergently, rather than having been inherited from a common ancestor that possessed the same feature. Once again, a close evolutionary relationship between tarsiers and *Shoshonius* comes out on top.

In many respects, the best way to settle evolutionary disputes arising from clashing characters is to resort to additional lines of evidence. Since everything discussed so far relates to aspects of skull anatomy, what does the postcranial skeleton have to say about the two competing evolutionary reconstructions? Here again, the weight of the evidence comes down in favor of a connection between tarsiers and *Shoshonius*. While the locomotor skeleton of tarsiers is far more derived than that of *Shoshonius*, there are no obvious similarities between tarsiers and anthropoids from the neck down. For instance, tarsiers have cylindrical femoral heads, while those of anthropoids are spherical. *Shoshonius*, with its semi-cylin-

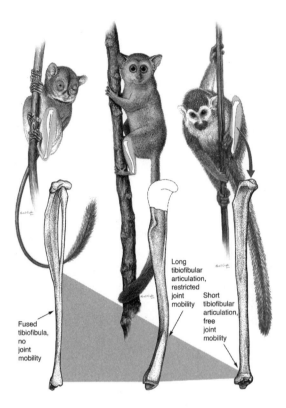

Figure 24. Variation in the anatomy of the tibiofibular joint reflects different adaptations for posture and locomotion in primates. In living tarsiers (left) the tibia and fibula are fused, stabilizing that joint as an adaptation for leaping. In living squirrel monkeys (right), the tibia and fibula remain unfused, allowing free joint mobility across the wide range of postures and modes of locomotion employed by these animals. An intermediate condition occurs in *Shoshonius*. Original art by Mark Klingler, copyright Carnegie Museum of Natural History.

drical femoral head, fits somewhere in between these two extremes. The tibia and fibula in tarsiers are fused, while in anthropoids these bones articulate with one another at a small joint just above the ankle. Again, *Shoshonius* lies somewhere in the middle, having a long and rugose area of tight tibiofibular apposition. While the calcaneus of tarsiers is markedly elongated, that of *Shoshonius* is moderately elongated. Anthropoids differ in having a calcaneus that is notably short.

After carefully considering all of the relevant evidence, my colleagues and I easily reached a consensus about how to draw this part of the primate family tree. The weight of the evidence all pointed in the same direction—*Shoshonius* is a close evolutionary cousin of living tarsiers that roamed the forests of central Wyoming between fifty and fifty-one million years ago.[13] By default, this also meant that some of the features that are shared by tarsiers and anthropoids—particularly the postorbital septum and the anterior accessory cavity—resulted from convergent evolution. If we were right, the tarsier theory of anthropoid origins had to be tossed onto the trash heap of scientific hypotheses that had been proposed, examined, and found to be wanting. But as we shall see, the other traditional theories of anthropoid origins—Gingerich's adapiform theory and Szalay's omomyid theory—would fare no better. For decades, paleoanthropologists had sought answers to the riddle of anthropoid origins in places that had yielded early anthropoid fossils—mainly in Africa, but also in Asia and even South America. Against all odds, our fieldwork in the Wind River Basin—on a continent renowned for never having yielded a fossil anthropoid and in beds that were millions of years too old—was fundamentally shifting the debate on this crucial segment of our remote evolutionary history. Jacob Wortman would have delighted in the multiple layers of irony.

Beyond demonstrating that the tarsier theory was fatally flawed, *Shoshonius* influenced the debate over anthropoid origins on the critical issue of timing. Previously, this was one of the few aspects of anthropoid origins that seemed more or less settled. All major theories of anthropoid origins posited that the prosimian-anthropoid transition took place near the Eocene-Oligocene boundary, roughly thirty-four million years ago. Gingerich had even nominated a few fossils—*Hoanghonius, Oligopithecus, Pondaungia,* and *Amphipithecus*—as hard evidence of this evolutionary milestone. All of them dated appropriately to within a few million years of the Eocene-Oligocene boundary. At the same time, the momentous environmental changes that transpired then offered a suitable context for the emergence of anthropoids. Hence, Susan Cachel explained most of the features that set anthropoids apart from prosimians as anatomical or behavioral responses to the cooler, more seasonal climates that prevailed across the Eocene-Oligocene boundary. As venerable and elegant as these ideas about anthropoid origins might be, *Shoshonius* showed that they simply had to be wrong.

To understand why this is so, bear the following three items in mind. (1) Among living prosimian primates, all available evidence indicates that tarsiers are the nearest evolutionary cousins of anthropoids (see

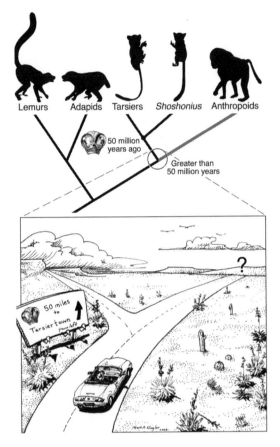

Figure 25. A cartoon depicting the logic behind the ghost lineage of early anthropoids suggested by our study of *Shoshonius*. Original art by Mark Klingler, copyright Carnegie Museum of Natural History.

chapter 5). (2) As we have just seen, numerous details of skull and post-cranial anatomy show that *Shoshonius* is an extinct relative of tarsiers. (3) *Shoshonius* dates to roughly fifty million years ago. If the origin of the anthropoid lineage is defined as the evolutionary split between ancestral anthropoids and ancestral tarsiers (in the same way that we equate the origin of the human lineage with the evolutionary dichotomy between chimps and humans), then the preceding factors require us to assert that anthropoids originated at least fifty million years ago, more than fifteen million years before the traditional theories would have us believe!

If you don't quite follow my argument, consider this analogy. You are driving down an unfamiliar road. The only thing you know for certain is that there is a single fork in the road, somewhere up ahead. One fork leads to Tarsiertown; the other goes to Monkeyville. Moreover, you know that the fork in the road lies midway between the two towns. No other roads lead to Monkeyville, which is your ultimate destination. Eventually, you arrive at the fork in the road. A sign points off to the left bearing the inscription "Fifty miles to Tarsiertown." No sign points to Monkeyville. Which fork do you take, and how much farther must you drive to reach your destination?

Shoshonius is equivalent to that sign pointing toward Tarsiertown. It allows us to predict that the anthropoid lineage extends back in time at least fifty million years, even if no anthropoid fossils come close to being that old. Paleontologists refer to such missing segments of the fossil record as "ghost lineages."[14] We are forced to admit their existence because of the shape of the evolutionary tree, along with the age and position of key fossils that adorn it. Thus, if *Shoshonius* lies on the same major branch of the family tree that includes tarsiers, and if anthropoids lie on a separate but adjacent branch, the continuity of evolutionary descent requires us to deduce that the anthropoid lineage is at least as old as *Shoshonius*. To argue otherwise is to deny the geometry of the evolutionary tree (for example, by claiming that *Shoshonius* is in the wrong place) or to advocate the spontaneous generation of anthropoids. Needless to say, my Wind River Basin colleagues and I were unwilling to concede either point.

Despite the impeccable logic behind our argument, the cool response I received after our first publications on *Shoshonius* told me that most of the reigning experts on anthropoid origins still weren't convinced. A graduate student of one prominent advocate of the view that anthropoids evolved fairly late in the game even accused me of being a traitor to the discipline of paleontology, since my interpretation of *Shoshonius* implied that the fossil record was so imperfect. I responded that if the fossil record were as complete as he and his academic advisor seemed to think, why were we spending so much time and money crawling around remote badlands looking for more specimens? Besides, I hardly felt like a traitor to the cause of paleontology. After all, I had used a fossil to make the novel case that the earliest phases of anthropoid evolution were still eluding us. In all scientific fields, new discoveries often highlight the gaps in our understanding that remain. For my money, paleontology is just like any other science.

Shoshonius spawned a ghost lineage of early anthropoids that threw virtually everything we thought we knew about their early evolutionary history out the window. Soon after our initial analysis of *Shoshonius* was published, I gave a lecture at Stony Brook University, on Long Island, one of the great centers of excellence in paleoanthropology in the United States. I challenged my audience to consider the ramifications of *Shoshonius* and the ghost lineage of anthropoids that it implied. What would a fifty-million-year-old anthropoid look like? Would we be able to recognize a fossil of such an early anthropoid if we saw it? And where on Earth should we go to find such elusive fossils? The only thing that was abundantly clear was that our old ideas about anthropoid origins had to be jettisoned. Everything else was up for grabs.

7

Initial Hints from Deep Time

Fertile fields and hillside vineyards whiz past the window in our compartment on the southbound TGV, the French version of a bullet train. My wife, Sandi, parcels out the food we've taken on board for lunch. Adhering to that well-worn maxim about "when in Rome," we share a freshly baked baguette, some fruit, and a rich assortment of cheeses. Sandi looks out at the passing scenery, seemingly oblivious to the blistering speed we've managed to achieve. She has seen this all before, having undertaken several archeological field seasons investigating how Paleolithic humans hunted Ice Age horses in France. It's a brand-new experience for me, though, and I marvel at our rapid progress through the Paris Basin toward our destination of Montpellier on the Mediterranean coast.

Our trip entails both business and pleasure. It is late in the spring of 1991, and we have just enough time to fit this excursion into our schedules before embarking on our summer field seasons. Afterward, I shall return to the Wind River Basin, but Sandi will stay in Europe to work on a new archeological project in Cyprus. Here in France, we plan to visit Marc Godinot, one of Europe's leading researchers on Eocene primates, who is perhaps best known for his work on *Adapis* (see chapter 2). Marc is an old friend in addition to being a valued colleague. We met for the first time at a scientific conference in Germany in 1985. A year or so later, Marc obtained a postdoctoral fellowship through NATO to study fossil primates at Johns Hopkins, in the same lab where I was conducting my doctoral dissertation research. Our similar interests in Eocene primates led to several collaborative publications, and we have kept in touch ever since. Knowing that my recent work on *Shoshonius* would benefit from detailed comparisons with one of its European cousins, Marc invited me to come over to have a look at several skulls of *Necrolemur* in his lab at the Institut des Sciences de l'Evolution at the Université de Montpellier. In addition to the opportunity to compare *Shoshonius* with *Necrolemur* side by side, Marc has generously offered us the use of his family's vacation house, located in the small resort town of Banyuls-sur-Mer, the last French outpost on the Mediterranean coast north of the Spanish border.

When our train pulls into the station in Montpellier, Marc is there to greet us. We exchange the usual pleasantries, but I can tell that Marc has something important to relate. Soon enough, our conversation veers from personal anecdotes about friends and family to the topic that has obviously been foremost in Marc's mind. He mentions that he has some small primate teeth from an Eocene site in Algeria that he wants to show me back in his lab. Since my purpose in coming to France is not to study isolated teeth but to examine complete skulls of *Necrolemur,* I know that these specimens must be special in some way. "What do you think they are?" I ask. Marc's excitement momentarily overwhelms his naturally conservative disposition as a scientist. "They seem to be tiny but remarkably advanced anthropoids!" he blurts out. "Really? How old are they?" "Middle Eocene at least, maybe older."

It takes a few moments before I fully grasp the significance of Marc's words. Less than a year previously, my Wind River Basin colleagues and I had achieved no small measure of scientific notoriety by positing that anthropoids originated fifteen million years or so before their first appearance in the fossil record. Our claim was based on *Shoshonius* being a close fossil relative of living tarsiers, an interpretation that most paleontologists regarded as eminently reasonable. Still, many experts disliked the concept of a ghost lineage of early anthropoids that persisted for millions of years without leaving behind hard evidence in the fossil record. If Marc's fossils held up under scrutiny, our prediction about the great antiquity of the anthropoid lineage would be vindicated. At the same time, our ghost lineage could move beyond its status as a purely theoretical construct. Even a few isolated teeth might transform our ghost lineage into something more tangible, something at least partly corpo-

Figure 26. Marc Godinot leads a team searching for fossil primates in France. Photograph courtesy of and copyright by Marc Godinot.

real. I insist that we go immediately to Marc's lab to have a look at these potentially groundbreaking specimens.

A glance under Marc's microscope reveals why he is so elated. The entire assemblage consists of only five isolated teeth, collected by washing and screening a fossiliferous layer of sandstone at a site called Glib Zegdou, near the border region between Algeria and Morocco. By any anatomical standard, the specimens are meager. Yet they come from a place and time that has previously been a void on the map of primate evolution. By this measure, the Algerian fossils are inherently interesting, no matter what type of primates they prove to be. Still, even my cursory perusal of the specimens forces me to agree with Marc's initial assessment. The complex pattern of cusps and crests on the tiny fossil teeth functions as an evolutionary fingerprint. For this reason, most species of living and fossil mammals can be distinguished on the basis of their teeth alone. At the same time, the intricate web of bumps and ridges of enamel provides a rich source of data for interpreting where Marc's Algerian primate lies on the family tree. Several diagnostic features on these miniscule teeth—with standard dimensions measuring no more than a tenth of an inch (2.5 mm)—ally them with anthropoids.

The two most instructive specimens in Marc's sample are an upper

Figure 27. The fossil site of Glib Zegdou, Algeria, where isolated teeth of *Algeripithecus minutus* have been found. Photograph provided by Rodolphe Tabuce, courtesy of and by copyright Mohamed Mahboubi.

second molar (abbreviated as M^2) and a lower third molar (or M_3). The crown of the upper molar is dominated by three main cusps, which are connected by crests. Together, these cusps and crests approximate an equilateral triangle, with the cusps corresponding to the apices of the triangle and the crests forming its sides. The internal (or lingual) side of the crown bears a broad and continuous shelf of enamel (technically known as a cingulum) that supports the crown's fourth major cusp, called a hypocone. The most important features that distinguish the M^2 from Algeria from those of Eocene prosimians are its complete lingual cingulum bearing a hypocone, the number and pattern of crests on its crown (most Eocene prosimians possess an extra crest known as the postprotocingulum), and the extremely low height of the molar crown as it is viewed from the side. Overall, the upper molar from Algeria closely resembles that of *Aegyptopithecus* from the Fayum region of Egypt, although the Algerian tooth is several times smaller and millions of years older.

The M_3 shows additional features that recall those of anthropoids, yet differ from Eocene prosimians. For example, the paraconid cusp, which is present in primitive adapiforms and virtually all omomyids, is absent in the Algerian tooth. At the same time, the back part of the tooth, known as the hypoconulid lobe, is strongly reduced—both from front to back and from side to side. Eocene prosimians typically bear an enlarged hypoconulid lobe on M_3, while later anthropoids diminish this structure.

As I raise my eyes from Marc's microscope, I sense that he is impa-

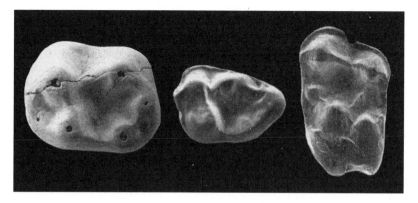

Figure 28. Isolated teeth of the early anthropoids *Biretia* (left) and *Algeripithecus* (right) from Algeria. Photograph of *Biretia* from Bonis et al. 1988, copyright Académie des Sciences; photographs of *Algeripithecus* molars from Godinot and Mahboubi 1994, copyright Académie des Sciences. Reproduced by permission.

tient to hear my verdict. I weigh my words carefully. "I can see why you think these teeth belong to tiny, primitive anthropoids, and I think you're right. But if you publish them without first finding better material, you know you'll be criticized by those who don't think anthropoids should exist this far back in time." I know from my own recent experience with *Shoshonius* that many leading paleoanthropologists dismiss the notion of middle Eocene anthropoids as something beyond the realm of scientific credibility. I also know that Marc doesn't care whether his views achieve consensus as soon as they are published. He and I share the belief that science only progresses by breaking out of the intellectual ruts that inevitably develop around long-standing ideas—even those that were never well substantiated in the first place. Still, it can be lonely being the sole champion of a new idea—one that even those colleagues whom you hold in high esteem regard as unlikely, if not downright wacky. It seemed only fair to remind Marc of the response he might face from the scientific community, while also letting him know that he had my support. Marc waves off these petty concerns, as I knew he would. "If you're going to cause a stir anyway, you may as well publish these specimens in *Nature*," I suggest.

When Marc and his Algerian colleague Mohamed Mahboubi published their fossils in *Nature* a few months later, the impact on theories of anthropoid origins was fairly predictable.[1] In a deliberate effort to evoke the better-known *Aegyptopithecus zeuxis*, Marc chose to name his new primate *Algeripithecus minutus*. I thought this was both savvy and appropriate, since the teeth of *Algeripithecus* and *Aegyptopithecus* resemble each other so closely. *Algeripithecus* went a long way toward confirming the ghost lineage model of anthropoid origins we had proposed on the basis of *Shoshonius*. At the same time, its small size (between five and ten ounces, or 150–300 grams) suggested that early anthropoids were no bigger than the smallest living monkeys—species like the pygmy marmoset of South America. If so, this conflicted with the widespread notion that the evolutionary transition from prosimians to anthropoids entailed a significant increase in body size. While some of these inferences were unconventional, in at least one respect *Algeripithecus* conformed to standard views of anthropoid origins. Like the abundant fossil anthropoids from the Fayum region of Egypt, *Algeripithecus* hailed from Africa. Marc and many others interpreted this limited geographic distribution of early fossil anthropoids as meaningful. Anthropoids, it seemed, must have originated in Africa tens of millions of years before the earliest hominids emerged on the same landmass. In fact, *Algeripithecus* was merely the latest in a series of discoveries indicating that the history of African anthropoids extended far back into the Eocene, long before the interval when *Aegyptopithecus* and *Apidium* scampered through the Fayum's coastal forests.

The first African anthropoid to break the Eocene-Oligocene barrier was found at the opposite end of Algeria, near its border with Tunisia. During geological and paleontological reconnaissance of the region in the 1980s, a team of French and Algerian scientists led by Jean-Jacques Jaeger discovered a new fossil site called Bir el Ater south of the Nementcha Mountains.[2] Jaeger, an enthusiastic field paleontologist who excels at finding new localities in unlikely places, had been exploring North Africa for many years. In this case, his persistence paid off. In contrast to most localities that yield the remains of terrestrial mammals, Bir el Ater samples an ancient deltaic environment, when shallow seas inundated much of the North African shoreline. As a result, at Bir el Ater fragmentary fossils of mammals occur alongside the remains of marine organisms such as sharks and bony fishes. Among the isolated teeth and bone fragments that were recovered by washing and screening fossiliferous sediment there, Jaeger's team found a single lower molar (M_1) of a primate.

Despite its unique and fragmentary nature, the Bir el Ater molar shows several features indicating its anthropoid affinities. For example, for a tooth of such small size (slightly larger than those of *Algeripithecus*), the molar crown is remarkably low, with inflated cusps and weakly developed crests. Paleontologists describe teeth having this combination of features as being bunodont. Early anthropoids tend to have cheek teeth that are more bunodont than those of Eocene prosimians, although there is some overlap between the two groups in this qualitative feature. Perhaps of more significance, the Bir el Ater molar lacks the paraconid cusp that is virtually ubiquitous among omomyids and primitive adapiforms. At the same time, the hypoconulid cusp at the back of the tooth is much larger than those of Eocene prosimians.[3] A small depression or fovea separates the hypoconulid from a neighboring cusp known as the entoconid. While a few omomyids or adapiforms are known to have one or two of these features, only anthropoids possess them all. Accordingly, Jaeger's team described the Bir el Ater molar as a new genus and species of fossil anthropoid, *Biretia piveteaui*.[4]

Although what little we know of its anatomy supports its anthropoid status, the scientific significance of *Biretia* owes more to its geological context than to any unique or unusual features it might possess. Today, the Bir el Ater locality occurs far to the south of the Mediterranean coastline, along the northern margin of the Sahara Desert. Yet we already noted the ancient deltaic environment that the Bir el Ater site samples. In this case, large-scale changes in sea level explain the paradoxical occurrence of fossil shark teeth in the Sahara Desert. During the late Eocene, global sea level was relatively high, and much of the North African coastal margin was inundated. This shallow sea receded across the Eocene-Oligocene boundary when sea level dropped precipitously, partly as a response to the formation of large ice sheets on Antarctica. The lower sea level of the early Oligocene exposed dry land where shallow seas and deltas prevailed a million years earlier. The abrupt change in environment corresponded to an equally dramatic shift in sedimentation. So long as the North African coastal plain lay beneath a calm, shallow sea, sediment accumulated there, as rivers deposited sand and mud from inland regions into deltas and nearby coastal estuaries. Yet once the sea receded, the freshly exposed coastal plain itself became subject to erosion as rivers dissected the newly emergent landscape on their rush to the sea. Geologically, this event manifests itself as an unconformity (i.e., a gap in the stratigraphic column caused by erosion) between late Eocene marine strata and younger rocks that are often continental in origin.

The Bir el Ater locality in Algeria occurs below the unconformity caused by the drop in sea level at the Eocene-Oligocene boundary, meaning that *Biretia* dates to the late Eocene, thirty-five to thirty-seven million years ago. At the time, this made *Biretia* the oldest fossil anthropoid on the African continent—proof that anthropoids originated sometime prior to their early Oligocene heyday in the Fayum. Indeed, as we saw in chapter 4, this much could have been deduced from the Fayum record itself. The remarkable diversity and virtually modern anatomy of Fayum anthropoids demanded some earlier interval of anthropoid evolution, if only to generate the number of anthropoid species that lived there during early Oligocene time. For a while, *Biretia* provided the only concrete support—limited though it may be—that this was indeed the case.

Far more compelling evidence concerning African Eocene anthropoids soon emerged from an unexpected yet familiar source—the Fayum region itself. As part of his wide-ranging investigation of Fayum geology, Tom Bown stumbled across a new fossil site there in 1983. Bown's broad expertise in geology and paleontology had already given him a leading role in documenting omomyid evolution in the Bighorn Basin and in reconstructing the Fayum's early Oligocene environment (see chapters 3 and 4). Despite these pivotal scientific advances, Bown's discovery of what at first appeared to be a modest new Fayum locality surely ranks as his most enduring contribution to the study of anthropoid origins. The new site, rather ascetically dubbed L-41, was situated well below the early Oligocene levels that yield *Aegyptopithecus* and *Apidium*. This stratigraphic fact alone made L-41 a promising addition to the Fayum's suite of fossil-bearing localities. Its potential was further enhanced when subsequent geological studies demonstrated that L-41 dates to the latest part of the Eocene, from thirty-five and a half to thirty-six million years ago.[5] However, because the only fossils Bown located during his original survey at L-41 were two small hyrax jaws and the kneecap of an arsinoithere, several years passed before further exploration revealed that the site was rich in fossil primates as well.[6]

By 1989, additional work at L-41 had uncovered two new species of anthropoids, which Elwyn Simons soon described as *Catopithecus browni* and *Proteopithecus sylviae*.[7] Despite their co-occurrence at such an early Fayum site, *Catopithecus* and *Proteopithecus* are remarkably divergent anatomically. Striking similarities in their dentition demonstrate that *Catopithecus* is closely related to the younger Fayum primate *Oligopithecus*.[8] Both *Catopithecus* and *Oligopithecus* possess two, rather than three, premolars in each jaw quadrant (a derived feature suggesting possible ties to *Aegyptopithecus* and the anthropoid group that includes Old World monkeys, apes, and humans), yet their lower molar morphology resembles that of Eocene prosimians in retaining the paraconid cusp (at least on M_1) and tall trigonids that project well above the level of the talonids.[9] These primitive features led Philip Gingerich to argue that *Oligopithecus* belonged to a group of transitional fossils linking adapiforms with anthropoids. However, the discovery of a crushed skull of *Catopithecus* at L-41 showed that Gingerich's proposal about the intermediate evolutionary position of *Oligopithecus* was unfounded. *Catopithecus* possesses numerous derived cranial features in common with anthropoids, including the presence of complete bony eye sockets and modifications to the ear region.[10] Rather than being a missing link between anthropoids and adapiforms, *Catopithecus* and *Oligopithecus* document a third major group of Fayum anthropoids—the oligopithecids—in addition to the parapithecids and propliopithecids from sites a few million years younger.

The second anthropoid to be described from L-41, *Proteopithecus*, hardly simplifies the picture of Fayum primate evolution. *Proteopithecus* differs from *Catopithecus* in retaining three premolars in each jaw quadrant. This primitive feature also occurs in parapithecids and New World monkeys (also known as platyrrhines), making it a hallmark of basal anthropoids. Yet the cheek teeth of *Proteopithecus* are far more primitive than those of parapithecids and platyrrhines. For example, *Proteopithecus* retains the paraconid cusp on M_1, its lower molar trigonids are substantially taller than their corresponding talonids, and its upper molars are transversely broad—much like those of certain omomyids. Nothing about the dentition of *Proteopithecus* links it strongly with any of the other three groups of Fayum anthropoids. Significantly, even as additional elements of *Proteopithecus* emerged—including crushed skulls and limb bones—they failed to clarify its relationships with other Fayum primates. The obvious solution placed *Proteopithecus* in yet a fourth group of early anthropoids, and Simons soon proposed the new family Proteopithecidae to signify this outcome.[11]

Spurred on by the discovery of *Catopithecus* and *Proteopithecus* at L-41, Simons redirected the focus of fieldwork in the Fayum to concentrate on this single locality. Eventually, three more species of anthropoids turned up there. These include a second proteopithecid, named *Serapia eocaena*, which is slightly larger than *Proteopithecus* but otherwise very similar to it; a small and very primitive anthropoid named *Arsinoea kallimos* that does not fit easily within any of the four main groups of Fayum anthropoids; and a small parapithecid named *Abuqatrania basiodontos*.[12] Aspects of the biology of the early anthropoids from L-41 can be discerned to the extent that relevant parts of their anatomy are preserved. *Catopithecus* and *Proteopithecus*, the two most common anthropoids at L-41, each appear to have been diurnal animals, based on their relatively small orbits. They also appear to have possessed sexually dimorphic canines, implying that at least some of the L-41 anthropoids lived in complex social groups like those of many monkeys. Judging by the anatomy of their teeth, all of the early anthropoids at L-41 ate fruits. Some or all of these species likely supplemented their diet with insects, seeds, and other types of food. Skeletal remains indicate that the locomotion of *Proteopithecus* was similar to that of many living platyrrhines, consisting mainly of quadrupedal walking and running, climbing, and occasional leaping. *Catopithecus* appears to have been more flexible in its locomotor abilities, emphasizing climbing and hanging at the expense of more basic quadrupedal activities.[13]

For those of us who thought that the sheer diversity of anthropoids in the classic Fayum localities implied an earlier interval of anthropoid evolution, the varied ensemble of late Eocene anthropoids at L-41 provided welcome vindication. Yet the same logic applies to L-41. Indeed, the diversity of anthropoids at L-41—including a parapithecid, an oligopithecid, and two proteopithecids in addition to the enigmatic *Arsinoea*—actually exceeds that of most early Oligocene sites in the Fayum! Obviously, it took some time to produce this wide variety of late Eocene anthropoids. What's more, at least some of the L-41 anthropoids already possess such typically anthropoid features as complete bony eye sockets, modified ear regions, and limb and ankle bones that are distinctly like those of other anthropoids, especially platyrrhines. If speculating on anthropoid origins using Oligocene primates from the Fayum is akin to using Neanderthals to reconstruct the origin of hominids, the late Eocene anthropoids from L-41 are only marginally more useful. Continuing the previous analogy, we've now reached the stage where we can envision human origins with the benefit of *Homo erectus* fossils instead.

While the mystery of anthropoid origins remained unresolved, no one could deny the dramatic progress that had been made in improving the record of early anthropoids in Africa. Among other things, the rash of new North African discoveries solidified the widespread opinion that Africa served as the birthplace of the anthropoid lineage. Aside from a few follow-up discoveries of *Pondaungia* and *Amphipithecus* in Myanmar, the Asian fossil record yielded nothing during the 1980s and early 1990s that might dissuade anyone from hopping on the "Out of Africa" anthropoid bandwagon. Significantly, not a shred of evidence had ever emerged from China—the Asian country with the most extensive Cenozoic fossil record of all—to counter the African juggernaut.

To make matters worse, the Chinese fossil record hinted that eastern

Asia was a distinctly marginalized theater of primate and human evolution. The historic discovery of "Peking Man" *(Homo erectus)* had played a major role in the development of paleoanthropology, but it was becoming increasingly clear that the earliest and most primitive hominids were confined to Africa. Similarly, although excavations in late Miocene strata (about seven to nine million years old) in southwestern China's Yunnan Province turned up an extraordinary collection of fossil apes, known as *Lufengpithecus,* the world's oldest and most primitive apes all hailed from Africa. In many cases, even the official names of these early apes and hominids—duly enshrined according to the international rules of zoological nomenclature—advertised their African heritage. Species like *Proconsul africanus, Kenyapithecus wickeri,* and *Australopithecus africanus* populated the paleoanthropological literature. It seemed that most, if not all, of the major evolutionary transitions in primate evolution—the origin of hominids, the origin of apes, and quite possibly the origin of anthropoids—took place in Africa. In contrast, Asian fossils appeared to represent fairly apical branches of the family tree, or else they belonged to animals so odd that they could only document extinct and evolutionarily insignificant side branches.

Aside from my work on *Shoshonius,* which impacted the debate over anthropoid origins obliquely, I watched most of this academic wrangling from the sidelines. I considered myself an expert on Eocene prosimians, and I was perfectly content to pursue my studies of omomyids and adapiforms without ever entering the fray over anthropoid origins. At the same time, other issues in paleontology—such as how and why a dramatic turnover in North American mammals took place across the Paleocene-Eocene boundary—captured my interest and began to occupy much of my fieldwork. With so many engaging projects to pursue in Wyoming and other parts of North America, I had little interest in initiating an international field project. Even if I had wanted to join the hunt for anthropoid origins, China would have been far down my list of places to go to address the issue.

My first introduction to Chinese fossil primates only served to reinforce my preconceptions. One day while I was working in my office at the Carnegie Museum of Natural History, my colleague Mary Dawson walked in to introduce me to Wang Banyue, a visiting Chinese paleontologist who was working with our collections. An energetic and cheerful woman by nature, Wang Banyue immediately began telling me about the nature of her visit to our museum. She had just collected a small assemblage of Eocene mammals from a new site in northeastern China's

Jilin Province—not far from the North Korean border—and wanted to compare them with North American fossils in the Carnegie Museum collection. I was initially skeptical that fossils from so far away could be usefully compared with North American specimens. Indeed, Mary informed me that the fossil rodents Wang Banyue had found were quite different from any North American species that she had ever seen. "But you should have a look at the nice primate jaw Banyue has from the same locality," Mary said. "Maybe it will help pin down the age of the site."

As soon as Wang Banyue unwrapped her specimen from the cotton she had packed around it for protection, I could see that its distinctive anatomy looked incredibly familiar. The lower jaw of the Chinese primate, preserving the third premolar and the last two molars (P_3 and M_{2-3}), shared numerous features in common with a poorly known omomyid from the middle Eocene of southern California known as *Stockia*. Both species possess an unusual lower molar architecture, characterized by trigonids that are strongly compressed from front to back, coupled with deeply basined talonids sporting a distinct lingual notch. Indeed, the degree to which Banyue's new Chinese primate—which she had named *Asiomomys*—matched the anatomy of *Stockia* was startling. Point by point, nothing else in North America comes this close, although a variety of related primates lived in western North America during the middle Eocene.[14] What this implied for primate evolution was inescapable— the same types of omomyid primates inhabited eastern China and the western United States during the middle Eocene. This unexpected distribution indicates that primates dispersed across the Bering land bridge roughly forty-five million years ago, during an interval when experts had previously thought that primates were evolving separately in North America and Asia.[15]

About the same time, I read a short paper in the Chinese paleontology journal *Vertebrata PalAsiatica* announcing the discovery of fossiliferous fissure-fillings dating to the middle Eocene in southern Jiangsu Province, not far west of the city of Shanghai.[16] The paper was remarkably unobtrusive, describing fossils of a primitive Eocene carnivore called *Miacis* and an early relative of rabbits, hares, and pikas known as *Lushilagus*. What caught my eye was a casual statement that the same site had yielded an abundance of other small mammals, including insectivores, bats, rodents, and even primates. The lead author of the paper was none other than Qi Tao, an expert on Eocene perissodactyls and other mammals who had spent time in residence at the Carnegie Museum of Natural History as a visiting foreign scholar in the late 1980s. Anxious to see whether the similarities between North American and Asian middle Eocene primates extended beyond Wang Banyue's remarkable find, I wrote to Qi Tao to see whether he was interested in having Mary Dawson and me come over to China to have a look at his fossils. Qi Tao responded immediately, inviting Mary (whom he affectionately referred to as his "American sister") and me to come to Beijing, see the fossils, visit the site, and explore our options for collaboration.

On New Year's Day, 1992, Mary and I boarded a flight to begin the long intercontinental trip to Beijing. Mary had visited China once before, being one of the first American paleontologists to be invited there as China began opening its doors to the West. I was a novice, but I had made the effort to learn some pidgin Mandarin from my friend Sun Xianghua, a native of Shanghai who works in the Carnegie Museum's library. At the time, direct flights on American carriers from the United States to China did not exist, and our connections were less than ideal. We had to transfer in Tokyo's massive Narita Airport, but we arrived so late in the evening that we had to spend the night at an airport hotel before getting up early the next morning to catch the final leg to Beijing. The night's rest did little to staunch the tide of jet lag that accumulates from traversing so many time zones so quickly.

We landed in Beijing on a cold, wintry morning. Both of us felt relief that the long journey was finally over. We were also excited by the scientific prospects that awaited us. After clearing customs and immigration at the airport, we were thrilled to find that Qi Tao had come out to meet us and give us a lift to our hotel. Qi Tao, an extroverted and gregarious man in his early fifties, greeted us warmly. We loaded our baggage into the vehicle Qi Tao had arranged for us and headed into the city. Along the way, Qi Tao related stories about the latest fossils found by him and his colleagues at the Institute of Vertebrate Paleontology and Paleoanthropology (IVPP), changes in the academic administration of the institute, and his read on late-breaking political developments in China and how they seemed to bode well for our planned collaboration. Qi Tao and Mary had developed a very cordial relationship during his tenure as a visiting foreign scholar at the Carnegie Museum, and it was clear that he was just as excited as we were about the possibility of working together on a field project in China.

As Mary and Qi Tao traded news, I gazed out at the exotic landscapes. Beijing sits on the margin of the North China Plain, a region that in many ways forms the breadbasket of the nation. The mountains north of the airport look oddly surreal from an American perspective. They are more rugged than the Appalachians but less imposing than the Rockies, distinctly unlike any mountains I had ever seen. On that day the mountains were also shrouded in blue haze, a sign of China's heavy reliance on coal for heating and cooking. The image reminded me of a common theme in traditional Chinese paintings, depicting tall mountains immersed in a heavy mist or fog. I had always chalked such representations up to artistic license—embellishments of reality in the same tradition as the dramatically overwrought landscapes rendered by earlier generations of artists back home. Now, I could see that the Chinese painters were more firmly grounded in realism than I had ever thought possible.

After checking in at our hotel, Qi Tao escorted us across the street to the IVPP, a scientific research institute with a global reputation built on its superb collections and trailblazing scientists. Proving once again that you can't judge a book by its cover, I was surprised to see that the internationally renowned institute was housed in a nondescript building that failed to hide the wear and tear of decades of use. Although it has since moved to a new building offering far more modern facilities, at the time, the IVPP occupied the same space that saw it through the social and political turmoil of Mao Zedong's Cultural Revolution. Qi Tao led us through a dimly lit corridor on the ground floor to a stairwell that gave access to the upper levels of the building. True to its scientific roots, the institute's architectural plan followed the stratigraphic logic of the fossil record. The lower levels of the institute were devoted to the study of early fishes and the first land vertebrates. Cenozoic mammals, being fairly high on the evolutionary tree of vertebrate life, were consigned to the upper reaches of the building. After hiking up several flights of stairs, we finally reached the level dedicated to China's extensive fossil record of mammals that lived after the demise of the dinosaurs.

Following the rules of traditional Chinese etiquette, we made the rounds from office to office, greeting old friends and new colleagues alike. Within an hour or so, Qi Tao had introduced us to the entire cadre of IVPP specialists on early Cenozoic mammals, a group that approximated a who's-who list for Asian paleontology. I already knew Wang Banyue, with whom I had just published a paper laying out the similarities between *Stockia* and *Asiomomys*, and Li Chuankuei, an expert on the early evolution of rodents and rabbits whom I had met at a scientific conference a couple of years earlier. For the first time, I was also introduced to Tong Yongsheng, Wang Jingwen, and Huang Xueshi—all of whom would subsequently collaborate with Mary and me on a separate field project in central China's Yellow River valley (see chapter 1). With such formalities out of the way, Qi Tao, Mary, and I finally settled down in his office to examine the cache of Eocene fossils Qi Tao had collected from fissure-fillings near a village in southern Jiangsu Province called Shanghuang.

In an effort to save time, Qi Tao had already booked our airline tickets for the field trip south to Jiangsu Province for the next morning. This meant that we had only part of a day to get a first impression of Qi Tao's treasure trove of fossils. Mary and I soon agreed that its significance far exceeded the introductory blurb Qi Tao had published in *Vertebrata PalAsiatica*. The vast majority of Qi Tao's specimens were collected by washing and screening the gooey red mudstone that fills the limestone fissures near Shanghuang. The process efficiently yields fossils in abundance, but at the cost of damaging many of the more delicate specimens. Once the fissure-filling matrix is harvested, it must be soaked in dilute solutions of bleach to break down the chemical bonds that hold the viscous mud together. The resulting slurry is sieved through a fine-mesh screen, which separates fossils and any other fairly coarse debris from the dissolved mud that previously entombed them. The overall technique has a lot in common with washing laundry by hand. And, just as physical agitation and exposure to certain detergents can damage fine clothes, pristine fossils often get broken during the process of washing and screening. Unfortunately, no other method is capable of quickly extracting specimens from the Silly Putty–like matrix at Shanghuang. Hence, most Shanghuang specimens are pretty fragmentary, and articulated skeletal material is unknown. Qi Tao's haul mainly consisted of isolated teeth, some jaws, bits and pieces of skulls, fairly complete hand and foot bones, and broken limb bones.

As advertised, fossils of small mammals dominated the assemblage, although a reasonable number of medium-sized ungulates like artiodactyls and perissodactyls were present as well. Mary's expertise on rodents and rabbits drew her to focus on those groups, while I naturally gravitated toward the primates. Fortunately, Qi Tao had two microscopes in his office, because most of his specimens could only be interpreted under low magnification. Qi Tao's work on the assemblage had progressed to the stage where the washing and screening was finished, but the fossils had not yet been segregated according to the major animal groups they represented. Mary and I each settled down to perform this task, first by sorting out the fossils of rodents, rabbits, and primates from the hodgepodge of bones, teeth, and jaws Qi Tao had stored in rectangular containers not much bigger than a cigar box. Soon enough, I spotted dozens of isolated primate teeth in addition to a few primate jaws and some ankle, toe, and finger bones. After a couple of hours of squinting through Qi Tao's microscope, I looked over at the substantial pile of Eocene primate fossils I had "high-graded" from his collection of Shanghuang bones and teeth. I had amassed an assemblage of primate fossils the likes of which had never been seen. In little more than two hours of picking and sorting through Qi Tao's screen-washed concentrate, I had found ten times more Asian Eocene primate specimens than the previous century of paleontological exploration on that continent had unveiled! With so little time left, I resisted the urge to continue sorting in order to try and figure out which major Eocene primate groups were represented by this bounty of fossil material.

I could readily identify a few of the Shanghuang primate specimens, but these were distinctly in the minority. One or more small adapiforms were certainly present in the sample. I was mildly surprised to see that the Shanghuang adapiforms resembled European species like *Adapis* and *Microadapis,* rather than Asian sivaladapids or North American notharctids. I also recognized an omomyid in the assemblage, but its fossils were extremely rare, consisting of only two isolated teeth. Nevertheless, these teeth were completely diagnostic. They belonged to a primitive new species of *Macrotarsius,* a primate fairly common in middle Eocene sites in western North America, but which had never been found before in Asia. Although its biogeographic affinities conflicted with those of the Shanghuang adapiforms, finding *Macrotarsius* at Shanghuang fit the pattern forged by Wang Banyue's discovery of *Asiomomys* in northeastern China. But the vast majority of the primate specimens I sorted from Qi Tao's magnificent box of treasures defied my attempts to classify them. I had simply never seen anything else like them, despite my working familiarity with all of the major groups of Eocene primates that had been described to date worldwide.

Not only were these mysterious primate specimens abundantly represented in the Shanghuang sample, but they were also obviously diverse. All were remarkably small—broadly overlapping the size range of omomyids, yet smaller than all but the smallest known adapiforms. Different size classes of anatomically similar specimens appeared to correspond to small, tiny, and truly minute species. Additional species stood out on the basis of their distinctive anatomy. For example, some species had cheek teeth that were very bunodont, while the molars of other species bore sharply defined crests emphasizing shearing over crushing and grinding. By segregating the Shanghuang primate fossils into groups based on size and dental anatomy, I estimated that the sample included between five and ten species of these puzzling new primates.

Just as I was beginning to focus on where the enigmatic Shanghuang primates might fit with respect to their Eocene cousins, Qi Tao announced that it was time for him to catch the IVPP shuttle bus to the part of town where he lived. Knowing that this meant that our workday was over, Mary and I packed up our things and headed back across the street to our hotel. Over dinner that night, we compared notes on what we had seen. Mary's impressions of the Shanghuang rodents ran parallel to mine. She noted that some Shanghuang rodents belonged to the squirrellike ischyromyids, a group well known in North America but rarely encountered in Asia. Primitive mouselike cricetids and ctenodactyloids, groups that were common in Asia during the Eocene, were also abundantly represented.[17] Still another group of Shanghuang rodents remained an enigma. After Mary finished her quick synopsis of what she had seen, she looked up from her bowl of Chinese noodles. "What about your primates?"

I quickly summarized the straightforward part about *Macrotarsius* and the adapiforms that so clearly resembled European primates like *Adapis* and *Microadapis*. Then I confessed that I was at a loss as to how to interpret most of the Shanghuang primates. "They have a lot of features in common with omomyids, and they're the right size," I said, "but there's something peculiar about them. I've never seen omomyids with molars like these, and their premolars have roots that are incredibly stout and long. If they're omomyids, they're the weirdest omomyids I've ever seen." Still suffering from the lingering effects of jet lag, we finished our meals and took the hotel elevator up to our rooms to rest for the trip south to Jiangsu Province the next morning.

The flight from Beijing to Nanjing— slightly further than from Pittsburgh to Atlanta—was uneventful, but the trip from the airport into Nanjing highlighted some of the differences between traveling in China and the United States. As part of the ongoing Chinese economic boom, local officials had decided to upgrade the road connecting the airport with downtown. But unlike on any American construction project of this size and scope, I saw little in the way of heavy equipment. Nothing larger than a small bulldozer was anywhere in sight. Instead, the task of building up the bed of the road so that it stood above the level of the surrounding fields was being performed by hundreds of manual laborers with picks and shovels—brute force of the sort that ancient pharaohs once mustered to build the pyramids of Giza. The scene struck me as decidedly anachronistic, especially since I had just arrived by Boeing. The ongoing battle between primitive and modern technology intensified on the road itself. There, Mercedes sedans competed with heavily laden donkey carts on the congested two-lane highway into town. For the most part, the primitive technology prevailed, because the slowest vehicles on the road generally dictated the flow of traffic. Even when sporadic gaps in oncoming traffic allowed the impatient drivers of foreign sedans and SUVs to pass the donkey cart that immediately impeded their progress, the thrill of speed was fleeting, being halted by the next medieval form of transportation obstructing the road ahead.

After surviving the harrowing trip from the airport, we checked in at a small hotel located near the center of town. Nanjing is a beautiful city built on the banks of the broad, lazy Yangtze River, or Chang Jiang. A former capital during various Chinese imperial dynasties, Nanjing still retains parts of the protective wall that once defended it. We had little time to savor Nanjing's charms, however, since our goal was to tour the site of the Shanghuang fissure-fillings to the south. Qi Tao informed us that the one-way trip typically takes several hours, depending on the number of donkey carts you have to pass along the way. We would depart for Shanghuang immediately after breakfast the next morning, giving us plenty of time for a site tour before returning to Nanjing late the same evening.

Our approach to the Shanghuang limestone quarry offered none of the signs I normally associate with an impending world-class fossil site. Instead of extensive badlands, vegetable fields dominate the surrounding terrain. On local geological maps, these fields are designated as Quaternary alluvium. The rocks that outcrop there, if they can be called that, consist of silt and mud that date back only slightly farther than the most recent flood. They are far too young to play any role in the quest for anthropoid origins. On the other hand, the only significant topographic feature in the immediate vicinity—a limestone hill that the same geological maps assign to the Triassic—is more than 150 million years too old to yield fossils of early anthropoids. However, as we got closer to the hill I saw that the monotony of the limestone was disrupted by reddish tubes of mudstone that snaked through the hill like the roots of some giant plant. Qi Tao pointed to these imperfections in the beige-colored limestone. "Look! Those are the Shanghuang fissure-fillings," he exclaimed.

Finally arriving at our destination, we emerged from our vehicle to behold a beehive of activity. The hill was actively being devoured by a commercial quarry operation, whose aim was to convert the limestone into cement. Workers swarmed across the nearly vertical quarry face, as well as the gaping hole in the ground in front of it. Some drilled holes to implant dynamite. During breaks in human activity, the detonations dislodged large chunks of limestone, which tumbled into the hole in the foreground. Other quarry workers specialized in breaking these limestone boulders down into more manageable pieces, using sledgehammers and chisels like those we employed to remove the overburden from the Buck Spring Quarries in the Wind River Basin. Only here, the scale of the operation was orders of magnitude larger, and it ran more or less permanently rather than during a few weeks each summer field season. Down in the hole, a few wheeled contraptions that looked like garden tillers hooked up to small homemade truck beds were being driven around, apparently to gather up the rendered limestone for transport to the nearby cement factory. The scene could easily have been edited from Mel Gibson's post-apocalyptic movie *The Road Warrior.*

Qi Tao gave hard hats to Mary and me, and we descended into the hole. Near the base of the quarry face, we stopped at one of the larger fissures within the limestone and began to explore its contents. The blood-red mud that filled the fissure was soft, wet, and amazingly sticky. I soon discovered that gobs of it would adhere to your fingers, hand tools, or virtually anything else that came into contact with it. Qi Tao explained that this, along with the fact that the stuff has no apparent value as a commodity, made the fissure-fillings a nuisance to the quarry workers. The quarry operators were therefore happy to have Qi Tao remove the fossiliferous mud from the fissure-fillings, but if this were not done quickly enough, the work of the quarry could not be held up, and whatever fossils they might contain would be forever lost to science.

The situation called for constant monitoring, but it required little in the way of technical expertise. None of us could afford to visit the site as frequently as the rapid progress of the quarry operation demanded. Even if that weren't the case, the brisk pace at which the limestone hill was being devoured meant that meticulous excavation of the fissures by hand—a technique that might yield fossils in better condition than Qi Tao's program of vigorous screen-washing—was out of the question. Instead, we had to devise a fast, simple, and low-cost strategy of salvaging the contents of the Shanghuang fissure-fillings. Recalling the primitive but effective method of road construction I had witnessed on the ride from the Nanjing airport, it seemed obvious to me that we needed to harness the raw manpower that flourished around us in such abundance.

We settled on a strategy of hiring quarry workers to extract the fossil-bearing mud from the fissure-fillings as soon as they were exposed. Like most of the work in the limestone quarry, this could be performed with simple hand tools. We made it clear that we did not want the quarry workers to find fossils for us. On the contrary, we discouraged them from doing so. Their job was simply to shovel the fossiliferous mud into heavy burlap sacks, each with a label noting the exact fissure from which the contents had come, which would be stored nearby until Qi Tao could return to supervise their shipment back to Beijing. We designated one of the quarry workers to oversee the entire process and promised to return to Shanghuang at regular intervals to check on the status of the work and pay the workers for their labor.

The best way to transport the bulk Shanghuang matrix back to Beijing was by train. There, it could be washed, screened, and sorted under Qi Tao's supervision. Because this was the most labor-intensive and technically demanding part of the whole operation, we agreed that roughly half of the Shanghuang matrix would be sent directly from Beijing to Pittsburgh, where Carnegie Museum technicians could perform the same tasks under our supervision. At the same time, the Shanghuang fissures made us optimistic that additional fissure-fillings might be found in other limestone quarries in the surrounding region. With luck, these might differ in age from those at Shanghuang, allowing us to sample fossils from a more extensive interval of time. Of course, the cost of all this labor, transportation, and additional fieldwork would require significant funding. We needed to write a grant proposal to obtain the money to carry the project forward.

Having satisfied our curiosity about the site itself and its potential for further scientific work, we decided to return to Beijing as soon as possible. This would give us time to make further observations on the fossils Qi Tao had already collected and to hammer out an agreement between the IVPP and the Carnegie Museum of Natural History regarding our future collaboration. All the way back, I couldn't stop thinking about the mystery primates from Shanghuang. How should I go about determining where these "weird omomyids" fit on the primate family tree? I decided to approach the problem systematically. I would make a list of the anatomical features that appeared to be characteristic of the group and hope that some consistent pattern emerged. I could hardly wait to get back to Qi Tao's office in the IVPP.

The morning after we flew back to Beijing, I woke up early, exhilarated by the scientific challenge posed by Shanghuang's weird omomyids. After Mary and I finished our meager breakfast of imported instant oatmeal and Starbucks coffee (neither of us has ever grown accustomed to traditional Chinese breakfast fare, which often consists of a watery rice gruel, vegetables pickled in salt brine, and steamed buns), we hurried across the street to meet Qi Tao in his office. I found the pile of Shanghuang primate fossils exactly as I had left them, inches away from the microscope I had used to sort them before the trip south to Jiangsu Province. Like a small child on Christmas morning, I eagerly began to examine each of my precious gifts.

Regardless of whether the mystery primates turned out to be weird omomyids, I decided to analyze them as if they were. Years before, during my stint as a graduate student in the Bighorn Basin with Ken Rose and Tom Bown, I had learned how to assess the evolutionary relationships of small primates like these, more by the process of osmosis than any formal coursework. In reconstructing how Bighorn Basin omomyids evolved, Ken and Tom had always stressed the anatomy of the front dentition. Characteristics like the size and number of the incisors and premolars, as well as their morphology, generally outweighed the structure of omomyid molars. For whatever reason, omomyid molars tended to be evolutionarily conservative, and they therefore offered only modest guidance in piecing together the family tree. Hence, I began my systematic study of Shanghuang's mystery primates at the very front of their lower jaws.

Two fragmentary lower jaws provided evidence about the front dentition, although neither of them preserved the front teeth in place. One of the two jaws supported the crowns of the last lower premolar (P4) and the first two molars (M$_{1-2}$), as well as showing the root sockets or alveoli for all of the front teeth except the incisors. The other jaw held only the P4 crown, but in this specimen the sockets for all the front teeth could be observed. To the extent that they were present in both specimens, the size and arrangement of the sockets for the front teeth were identical. In the single specimen that preserved them, the incisor sockets were small and vertically oriented. Significantly, the socket for the central incisor (I$_1$) was smaller than that for the lateral incisor (I$_2$). Already, I could show that Shanghuang's mystery primates differed from the vast majority of omomyids. As Philip Gingerich and others had emphasized, most omomyids possess central incisors that are substantially enlarged, and their sockets are strongly inclined toward the front of the jaw. I$_2$, in contrast, is frequently reduced in size or suppressed outright during the course of omomyid evolution. In omomyids like *Tetonius* and *Trogolemur,* these unusual incisor modifications probably represent adaptations for bark-

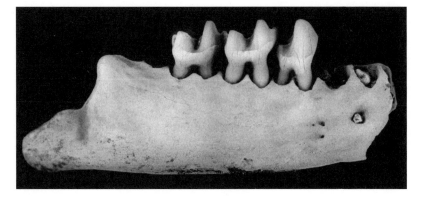

Figure 29. The holotype lower jaw of *Eosimias sinensis* from the Chinese Shanghuang fissure-fillings.

gouging and gum-feeding behavior (see chapter 3). Only a small minority of omomyids, including *Shoshonius* and its close relative *Washakius,* retain small, relatively vertically oriented lower incisors like those of the mystery primates from Shanghuang.

If the incisors of the mystery primates distinguished them from most omomyids, their canines set them apart from all of them. In both of my Shanghuang specimens, the girth of the root socket for the canine exceeded those of the incisors by a wide margin. At the same time, there was a significant discrepancy between the two Shanghuang specimens in this measurement. Clearly, the mystery primates had canines that were enlarged—certainly in diameter, and probably in height as well. In contrast, omomyids uniformly possess small canines, and some species lack them altogether. Given the sample size of two, I could not reasonably infer that the mystery primates were sexually dimorphic, but this possibility certainly remained open.

Goaded by the clear disparities between the mystery primates and omomyids, I moved on to see what could be deduced regarding their premolars. My first task was to determine how many premolars would have been present in a pristine lower jaw of one of these small Shanghuang primates. The most primitive known primates, including the majority of adapiforms and a few omomyids, possess four premolars (P_{1-4}) in each jaw quadrant. As various primate lineages reduced their dependence on the sense of smell, they evolved shorter snouts. This in turn affected jaw structure, requiring the teeth to become increasingly compacted from front to back as snout length decreased. Hence, a recurring theme in primate evolution is the loss of certain teeth—especially premolars—and the progressive crowding and compression of others. As a result, the front premolar (P_1) was independently lost several times during the course of primate evolution, leaving some adapiforms, most omomyids, tarsiers, and primitive anthropoids (including proteopithecids, parapithecids, and living platyrrhines) with three premolars in each jaw quadrant. A few omomyids, as well as more derived anthropoids (including oligopithecids, propliopithecids, Old World monkeys, apes, and humans), lost P_2 as well, leaving them with only two premolars in each jaw quadrant. Among the large number of primates that have three premolars, an important distinction exists regarding the number of roots that support P_2. In all but a few adapiforms (including all of those that retain P_1), P_2 is double-rooted. In contrast, omomyids, tarsiers, and anthropoids uniformly possess a single-rooted P_2, even in those rare cases among omomyids where P_1 is retained.

In my mystery primates from Shanghuang, three root sockets intervened between the crown of P_4 and the enlarged canine. These varied in size and position. A relatively small socket followed the canine. The two remaining root sockets were larger, and the more forward of them was staggered toward the labial side of the jaw. I noticed that the roots underneath P_4 occupied sockets that were similarly disposed. I interpreted the three empty root sockets in the Shanghuang specimens as having supported a single-rooted P_2 (the socket immediately behind the canine) and a double-rooted P_3 (the two sockets immediately in front of P_4). No other reasonable interpretation was possible, since virtually all Eocene primates have double-rooted P_3s, and since the oblique orientation of the two rear sockets faithfully mimicked the condition in P_4. Although I had never seriously entertained the possibility that the mystery primates from Shanghuang were weird adapiforms, had I ever done so, their lower premolar configuration would have caused me to reconsider my opinion. The single-rooted P_2 in the mystery primates matched the condition in all of the omomyids, tarsiers, and anthropoids that retain this tooth, while virtually all adapiforms have a double-rooted P_2.

Having pushed an analysis of empty root sockets about as far as I could go, I determined to have a closer look at the anatomy of the cheek teeth. The structure of the crown of P_4 in the mystery primates did not depart dramatically from that of primitive omomyids. For that matter, even some early notharctid adapiforms showed basic similarities in P_4 structure. The mystery primates differed mainly in having a shorter, wider P_4 crown, whose outer covering of enamel bulged out noticeably onto the labial margin of its front root. A more obvious distinction lay in the oblique orientation of the P_4 crown as a whole. This odd orientation was caused by the fact that the front root of P_4 was situated near the outer (labial) margin of the lower jaw, while its rear root lay more internally or lingually. Though its crown was missing, I could infer that P_3 would have shown the same oblique orientation in the mystery primates, based on the similar configuration of its root sockets noted earlier. The oblique orientation of P_{3-4} in the Shanghuang mystery primates struck me as their most peculiar attribute so far. I didn't know of any omomyids or adapiforms with similarly oblique lower premolars.

In keeping with the trend established by their front dentition, the lower molars of the mystery primates combined an odd mix of features, giving them a unique structure. Like most omomyids and adapiforms, the mystery primates possessed the primitive cusp known as the paraconid on M_{1-2}. However, the large size of the paraconid and its broad separation from the cusp behind it (called the metaconid) set the Shanghuang specimens apart from omomyids and adapiforms. A nearby cusp known as the protoconid was enlarged and set far apart from the metaconid. In omomyids and adapiforms, the protoconid and metaconid are similar in size, and their bases are crowded against each other. Finally, the mystery primates stood out because of the structure of the back part, or talonid, of their lower molars. Instead of having their molar talonids squared off in back like those of omomyids and adapiforms, the Shanghuang specimens showed a more arched structure. An internal cusp called the entoconid was surprisingly forward in position, while the hypoconulid cusp jutted out behind the rest of the talonid.

To complete my analysis of the mystery primates, I needed to consider certain characteristics of the lower jaw (or dentary bone) itself. The smooth texture of the bone forming the front midline of the dentary showed that its two halves were not fused at the symphysis to form a single lower jawbone or mandible. In this respect, the Shanghuang specimens were simply primitive, resembling omomyids, many adapiforms, and such early anthropoids as *Catopithecus* and *Arsinoea* from the Fayum's L-41 site. Primates showing solid fusion at the chin, such as the adapiforms *Notharctus* and *Adapis* and the vast majority of anthropoids, evolved several times from ancestors lacking this trait (see chapter 2). Despite the lack of fusion, not everything about the symphysis was primitive in the mystery primates. In particular, its general robustness and surprisingly vertical orientation contrasted with all omomyids and adapiforms—even those with fused symphyses. As a result, the root sockets for the lower incisors were long and vertically oriented in the mystery primates, while in adapiforms and omomyids the incisors were shallowly procumbent. Looking back toward the rear of the jaw, I could see that the robustness at the symphysis continued in that direction. The mystery primates therefore differed from omomyids and adapiforms of similar size in having deeper dentaries, in which the roots of the canines and premolars were strongly anchored.

Going back over the many differences I had enumerated, I was forced to conclude that the Shanghuang mystery primates were neither omomyids nor adapiforms—at least not like any that had ever been seen before. They certainly failed to fit my expectation—based on Wang Banyue's discovery of the *Stockia*-like omomyid in Jilin Province—that Chinese middle Eocene primates would strongly resemble those from North America. Their features ranged from being extremely primitive (like the unfused symphysis and the presence of paraconids on the lower molars) to obviously derived (like the robust vertical symphysis and obliquely oriented lower premolars). In order to assess where the mystery primates fit on the family tree, I had to focus on their derived characteristics alone. As I began to reel these off one by one in my mind, a possibility that I had not previously considered suddenly dawned on me. I looked over at Qi Tao, who was working at his desk. "Does anyone in the IVPP have a copy of Szalay and Delson's book *Evolutionary History of the Primates?*" I asked. This handy reference work lays out the fossil evidence for primate evolution in encyclopedic detail, with photographs and illustrations of most extinct primate species. Qi Tao replied that he thought someone did, and he ran down the corridor to see if he could get hold of it.

A few minutes later, Qi Tao returned with Szalay and Delson's compilation in his hands. I opened the book about two-thirds of the way through and gazed down upon a jarringly familiar combination of features. Jaw after fossilized jaw—from sites ranging from the Fayum to South America—showed small, vertically oriented incisors rooted in a deep, robust symphysis, followed by an enlarged canine, a single-rooted P_2, and obliquely oriented P_{3-4}. The molars in the photographs differed from those of my mystery primates in lacking any trace of the large paraconid cusp that the Shanghuang specimens displayed so prominently. However, I knew that even this distinction had been erased by paleontological discoveries made after the publication of Szalay and Delson's treatise.

The alternative that I had somehow avoided taking into consideration turned out to be correct. And I, more than anyone else, should have

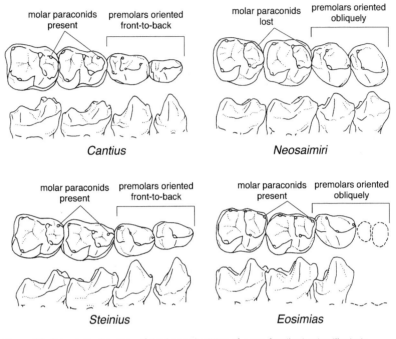

Figure 30. Schematic drawings of the lower dentition of some fossil primates illustrate some of the similarities and differences between *Eosimias* and other primates. Although *Eosimias* resembles primitive adapiforms and omomyids (such as *Cantius* and *Steinius,* respectively) in retaining large paraconids on its lower molars, its obliquely oriented premolars resemble those of anthropoids, including the early South American monkey *Neosaimiri.* Based on its unique combination of primitive and advanced features, I had no alternative but to interpret *Eosimias* as an extremely primitive anthropoid. Reprinted by permission from *Nature* (Beard et al. 1994), copyright 1994 Macmillan Publishers, Ltd.

been prepared to see it right from the start. The Shanghuang specimens weren't mystery primates at all. Instead, they conformed to the long ghost lineage of early anthropoids implied by my recent study of *Shoshonius.* The photographs of early anthropoids in Szalay and Delson's book duplicated most of the unusual features that I had been puzzling over for days. And while none of them had paraconids on their lower molars, I knew that many of the recently collected L-41 anthropoids—including *Catopithecus* and *Arsinoea*—retained this primitive structure. The main thing that set my mystery primates apart was their smaller size and the overwhelming preponderance of primitive features they retained. But given that the Shanghuang fissures are roughly ten million years older than the Fayum site of L-41, the startlingly primitive anatomy of the basal anthropoids from Shanghuang seemed imminently feasible. Here at last was a major source of information to address the long-standing riddle of anthropoid origins. If the Fayum's L-41 site yielded early anthropoids that were analogous, in the hominid scheme of things, to *Homo erectus,* the Shanghuang fissure-fillings were providing us with our first glimpses of australopithecines.

Unable to contain my enthusiasm any longer, I immediately told Qi Tao and Mary Dawson about my breakthrough. Qi Tao's face morphed from incredulity to exhilaration as I explained that most of the primate fossils he had found at Shanghuang belonged to tiny, very primitive anthropoids. Mary clapped her hands in excitement, fully recognizing the significance that our work now assumed. Not only did this remarkable discovery virtually guarantee that our collaborative research project would be funded; it also resonated strongly with Qi Tao's sense of national pride. China's historic role as the cradle of one of the world's great and enduring civilizations might now be extended tens of millions of years back in time, to an interval when the earliest members of the most diverse and successful branch of modern primates—the anthropoids—were just beginning to evolve the diagnostic features (like bigger brains, robustly constructed jaws, and associated changes in behavior and ecology) that would ensure their biological success. And because evolution—at least as it is viewed in hindsight—is a series of historical contingencies, had these earliest anthropoids failed to develop their unique set of characteristics, we humans would never have been granted our moment on the evolutionary stage.

More committed than ever to our collaborative research project, we began to strategize and make firm plans. All three of us agreed that the best way to press our agenda was for me to bring a selection of the Shanghuang specimens back to the Carnegie Museum of Natural History for further analysis that would eventually lead to publication. This would also make it easier for Mary and me to prepare our grant proposal to help support our joint activities.

A couple of days later, Mary and I boarded an eastbound Boeing 747 at the Beijing airport, carrying a deep sense of accomplishment in our hearts and several tiny fossilized jaws and teeth in our hand baggage.

8
Ghost Busters

ess than a month after my return to the United States, I received a telephone call from Rich Kay, a well-known professor of biological anthropology and one of the world's leading authorities on anthropoid origins. Rich told me that he, Elwyn Simons, and John Fleagle—another expert on primate anatomy and evolution, based at Stony Brook University, on Long Island—had decided to organize an international conference and workshop at Rich and Elwyn's home institution, Duke University. According to Rich, the rationale for the conference was obvious. Given the multiple theories of anthropoid origins that were being espoused, as well as the recent spike in the discovery of new fossils, it seemed like a good time to bring all of the scientific players together under one roof, along with many of the relevant fossils. If such a gathering couldn't achieve that most elusive of goals in paleoanthropology—a broad scientific consensus—at least it would serve as a venue to debate the competing viewpoints and give everyone a chance to examine some of the most important fossils.

Rich explained that all of the various factions in the debate over anthropoid origins would be represented and that ample time would be provided for discussion of each presentation. Philip Gingerich would be on hand to make the case for the adapiform theory of anthropoid origins,

Fred Szalay would anchor the omomyid camp, and Matt Cartmill would stand up for the tarsier theory. Specialists on the anatomy of early anthropoid skulls, teeth, and postcranial elements would discuss the evolutionary implications of their favored body parts. Last but certainly not least, those who had recently discovered key fossils would showcase their latest specimens and defend their interpretation of how the new fossils affected our constantly shifting understanding of anthropoid origins. Thus, Marc Godinot would travel from France along with his assortment of teeth of *Algeripithecus,* and Elwyn Simons would disclose some new specimens recently unearthed by his team at the Fayum's L-41 locality.

"Would you be willing to present your ideas about *Shoshonius* and how it bears on anthropoid origins?" Rich asked. I readily agreed, and I promised to bring along the best *Shoshonius* skulls we had collected in the Wind River Basin for purposes of show and tell. If my "ghost lineage" theory of anthropoid origins was to gain credibility in the scientific community, I needed to make a compelling case at the Duke conference. Besides, I would have found it hard to turn Rich down, since he had once served as my undergraduate honors thesis advisor. Although I was technically enrolled at the nearby University of North Carolina, as an undergraduate I spent much of my free time hanging around the Department of Anatomy at Duke, craving anything having to do with fossils or anatomy. Rich took me under his wing, involving me in a lab project aimed at estimating the age of monkeys by looking at microscopic growth rings in the cementum layer that encircles the roots of their teeth. The experience exposed me to the thrill of research for the first time, and our results formed the basis for one of my earliest scientific publications. I also benefited by developing a special rapport with Rich, although we had had less opportunity to interact after I left North Carolina for graduate school.

While I was glad of the chance to discuss my earlier work on *Shoshonius,* the excitement generated by my recent visit to China had eclipsed my North American field research, at least temporarily. Mary Dawson and I had agreed to maintain a low profile with respect to the Shanghuang primates, hoping to defer any publicity until we had had the chance to publish a formal paper announcing their discovery. But the opportunity posed by the anthropoid origins conference seemed too good to pass up. Impulsively, I asked Rich whether he might be interested in having me give a second presentation about some newly discovered fossils I had just brought back from China. I told Rich that I believed the specimens were extremely primitive anthropoids, although we were still making detailed analyses and comparisons. Timing would prevent us from publishing the specimens prior to the Duke conference, thereby raising the stakes considerably for my oral presentation. Since none of the assembled experts at the Duke conference could be prepared to digest their wide-ranging implications, the Shanghuang specimens had the potential to hit the congregation of paleoanthropologists like a bomb detonated without warning. Unveiling the Chinese fossils in such dramatic fashion guaranteed that they would be thrust immediately into the scientific spotlight. On the positive side, this instant celebrity might translate into funding for our collaborative fieldwork in China. However, in catching everyone off guard, I also had to prepare for the scientific backlash that might follow.

Nothing lends excitement to an academic conference like an unexpected controversy. Knowing this, Rich happily agreed to allocate two slots on his working schedule for the Duke conference to me—one to talk about *Shoshonius* and its implications for anthropoid origins, and another to introduce Qi Tao's Shanghuang fossils to the scientific community. Because the Chinese specimens promised to attract so much attention, I asked Rich whether it might be feasible for Qi Tao to attend the conference as well. Without hesitating, Rich offered to invite my Chinese colleague to the Duke conference too. Not only would this allow Qi Tao to meet many of the scientists involved in the debate over anthropoid origins, but it also gave us precious time to work together on either side of the conference in Pittsburgh. Qi Tao, Mary, and I could use the extra time to flesh out our plans for joint fieldwork in China. Having settled everything with Rich, I set about the task of preparing my two presentations and arranging for Qi Tao's imminent return to the United States.

Qi Tao was thrilled to be invited to the Duke conference and to spend a few weeks the Carnegie Museum. In the short time since Mary and I had visited him in Beijing, he had redoubled his efforts to wash, screen, and sort through the fossiliferous matrix from Shanghuang. When I greeted him at the Pittsburgh airport, he proudly displayed a medium-sized duffel bag containing the latest riches from the site. I could tell that he still couldn't quite believe that the tiny fossils he had salvaged from a limestone quarry in an unremarkable part of rural China were having an impact on one of the most contentious issues in paleoanthropology. At the same time, Qi Tao seemed more eager than ever to press forward with Mary and me on our scientific collaboration. He avidly read every technical paper I could give him on the anthropoid origins debate and told me that he could hardly wait to meet some of the leading researchers in the field at the Duke conference. As it happened, Qi Tao wouldn't even have to wait that long. Marc Godinot had arranged his transatlantic itinerary to fly straight from Paris to Pittsburgh in order to compare his *Algeripithecus* teeth directly with our new Chinese fossils.

Marc's arrival in Pittsburgh brought all of the earliest anthropoid fossils—those that were significantly older than the late Eocene L-41 site in the Fayum—together for the first time. For a couple of days before the Duke conference, the three of us held our own mini-summit on anthropoid origins at the Carnegie Museum, debating minor points of anatomy and discussing how the new specimens affected the evolving picture of anthropoid origins. Our deliberations soon revealed that Marc and I were drawing very different conclusions about the fossils before us. As is often the case in paleontology, a few trivial disagreements over basic anatomy cascaded into irreconcilable differences of interpretation. Much of the divergence between our points of view could be attributed to the paucity and fragmentary nature of the fossils themselves. The entire lot— including Marc's collection of teeth from Algeria and our assortment of jaws and teeth from China—fit comfortably in the palm of one hand. However, if the three of us couldn't agree on the meaning of our new material, there was no hope for any broad consensus among the much larger cast of scientists at Duke.

Our disagreement hinged on how to interpret the Chinese fossils, which Qi Tao and I now referred to informally as *Eosimias* ("dawn monkey"). According to our best estimates, the Chinese and Algerian fossils were roughly the same age, although they came from sites that were thousands of miles apart. Because neither site was associated with volcanic rocks suitable for standard radiometric dating, even the age of the specimens left room for reasonable people to disagree. In both cases, the sites were "dated" on the basis of biostratigraphy—a method that uses "guide fossils" to estimate the age of the entire assemblage. Guide fossils are typically abundant and widespread organisms whose age is established by their presence at multiple sites that have been dated by conventional means. Qi Tao and I felt confident about the age of *Eosimias,* because the Shanghuang fissure-fillings had yielded a wide variety of fossils that dated to the middle Eocene, roughly forty-five million years ago. For example, we had discovered several teeth of the omomyid primate *Macrotarsius* at Shanghuang. In North America, fossils of *Macrotarsius* occur at sites ranging from California to Utah, Wyoming, Montana, and Saskatchewan. All of these North American specimens of *Macrotarsius* come from a narrow interval of time in the middle Eocene, suggesting that the Chinese specimens are similar in age. Comparable evidence from a wide variety of other Shanghuang mammals—including rodents, carnivores, lagomorphs (rabbits and their kin), and perissodactyls (relatives of horses, rhinos, and tapirs)—reinforced this result.

If the age of *Eosimias* was reasonably well constrained, deriving a similar estimate for *Algeripithecus* was far more complicated. For one thing, only a small fraction of the fossil mammals found alongside *Algeripithecus* at the Glib Zegdou fossil site in Algeria had been thoroughly studied and described. Most of those that were already published belonged to species that were otherwise unknown to science, rendering them useless as guide fossils. Thanks to the vagaries of continental drift, Africa was geographically isolated from nearby landmasses in the Eocene. African mammals responded by evolving into a rich array of exotic forms that simply didn't exist elsewhere. To make matters worse, the Eocene fossil record of Africa is notoriously poor, so that few other African sites could be meaningfully compared with Glib Zegdou. Given these limitations, the fossil mammals from Glib Zegdou merely indicated that *Algeripithecus* was older than anything known from the Fayum. How much older was difficult to say, although fossil charophytes (algal cysts) from nearby pond deposits suggested that *Algeripithecus* might date back as far as the end of the early Eocene (about fifty million years ago).[1] If so, *Algeripithecus* could be as much as five million years older than *Eosimias*. This possibility troubled Marc profoundly.

Marc's reticence reflected his view of how evolution unfolds. Based on the few isolated teeth of *Algeripithecus* that were known, we all agreed that *Algeripithecus* appeared to be closely related to later African anthropoids, including *Apidium* and *Aegyptopithecus*. We also agreed that *Eosimias* had to occupy a more basal position on the family tree—regardless of whether or not it was an anthropoid. Yet *Algeripithecus* and *Eosimias* seemed to be roughly the same age, and it was even possible that *Algeripithecus* might be substantially older. Marc interpreted this to indicate that *Eosimias* was simply too young to be such a primitive anthropoid—in his words, *Eosimias* would already have been a "living fossil" in the middle Eocene. For Marc, the only viable explanation was that *Eosimias* was not an anthropoid at all. This conclusion coincided with Marc's opinion that anthropoids originated in Africa. If so, anthropoids as primitive as *Eosimias* should never be found in Asia, no matter how old they might be.

My own grasp of the mysterious workings of evolution embraces a broader range of possibilities. I felt confident that both *Eosimias* and *Al-*

geripithecus were early anthropoids, regardless of any difference in age between them. Anyone familiar with the range of anthropoids inhabiting the world today knows that they pertain to widely divergent branches of the family tree. Some living anthropoids are "more advanced" than others, at least from the narrow perspective of us humans. For example, the anatomy of New World monkeys departs from that of Old World monkeys, apes, and humans in a number of important ways, yet all are bona fide anthropoids that happen to live at the same time. Fossil anthropoids must have embodied a similar range of diversity during the remote past, especially when geographic isolation allowed groups to evolve independently in their own separate corners of the globe. I therefore had no difficulty imagining that anatomically primitive anthropoids like *Eosimias* might have lived in Asia at the same time that more advanced species like *Algeripithecus* were thriving in Africa. And if that made *Eosimias* a "living fossil" during the middle Eocene, so much the better!

While our different evolutionary philosophies prevented us from seeing eye to eye on *Eosimias,* Marc and I shared a common outlook on the broader issue of anthropoid origins. Both of us believed that anthropoids originated much farther back in time than traditional theories dictated. We also agreed that Gingerich's adapiform theory of anthropoid origins had little chance of being correct, despite the fact that it was being highly touted by Elwyn Simons and his former graduate student Tab Rasmussen. Simons and Rasmussen had revived the adapiform theory by declaring that many of the new anthropoid fossils from L-41 in the Fayum supported an evolutionary connection with those Eocene primates. Still, our older and more primitive Algerian and Chinese fossils shared little in common with adapiforms, and Marc and I were confident that science would ultimately reject this version of anthropoid origins. The only serious question was how long it would take before our views gained broad acceptance. If nothing else, the Duke conference gave us a chance to check the pulse of the paleoanthropological community on this matter.

Energized by the promise of intellectual give and take at the next day's gathering at Duke, Marc, Qi Tao, and I piled into my car for the eighthour drive from Pittsburgh to North Carolina. As we sped south through West Virginia on Interstate 79, Qi Tao remarked on how little congestion there was on American roads. The absence of slow-moving donkey carts enabled me to drive my cheap sedan much faster than a Mercedes sporting twice the horsepower on a typical Chinese highway. As he gazed out across dramatic vistas of the Mountain State, Qi Tao was moved to song. From the back seat of my car came John Denver's familiar refrain:

"Country road, take me home / To the place I belong." I knew from working with Qi Tao in China that he had a soft spot for ballads. Years before, his colleagues at the IVPP had recognized this personality trait by giving him the nickname *da jiao liu* ("big braying donkey"). Now his deep, baritone voice filled my little sedan with melodies ranging from the utterly exotic to the oddly familiar. Once Marc decided to join in, my car was transformed into a mobile stage for a United Nations sing-along. The miles swept past to the rhythm of Chinese, French, and Mongolian folk songs, punctuated by the occasional country and western classic. Before we knew it, we were pulling into our hotel in Durham, just across the street from the leafy campus of Duke University.

Early the next morning, we all made our way to the venue for the "Anthropoid Origins" proceedings, a medium-sized conference room on campus. Just as Rich had promised, the format of the sessions permitted speakers enough time to present and interpret their evidence to the audience of twenty-five other specialists, each of whom had been selected to participate because of his or her individual expertise on anthropoid origins. The talks were organized into thematic groups, and the floor was opened for discussion after every three presentations or so. This provided for freewheeling debate, and it soon became clear that a significant fraction of the group was divided along familiar lines (see chapter 5). As expected, a number of experts continued to maintain that anthropoids evolved from either adapiforms or omomyids. Yet as more and more presentations were given, I realized that a razor-thin majority of the scientists in attendance favored the idea that anthropoids must have evolved from some poorly known "third group" of early primates that differed from both adapiforms and omomyids. In practice, this nebulous third group was equivalent to the ghost lineage of early anthropoids I had proposed on the basis of *Shoshonius*. I was stunned to see that so many colleagues were willing to jettison the traditional adapiform versus omomyid duality that had dominated research on anthropoid origins for decades. I was also impressed by the diversity of evidence that supported the third group alternative.

This evidence took several forms. First, a number of presentations stressed that Eocene adapiforms and omomyids were already too specialized to have been anthropoid ancestors. My own presentation on *Shoshonius* fell into this category, since I made the case that *Shoshonius* shared so many derived features with tarsiers that it must belong somewhere on the tarsier branch of the primate family tree.[2] Along similar lines, several scientists argued that adapiforms shared derived features with living lemurs that excluded them from anthropoid ancestry.[3] If both adapiforms and omomyids were too specialized to have given rise to anthropoids, the only choice left was to posit that anthropoids evolved from some poorly known third group of early primates.

Analyses of the postcranial anatomy of living and fossil anthropoids reinforced and even extended the third group option. Susan Ford, an expert on New World monkeys at Southern Illinois University, argued that in many aspects of their postcranial anatomy, living and fossil anthropoids appear to be more primitive than prosimians. According to Ford, this meant that anthropoids must have been the first major group to diverge from the ancestral stock of primates in the early Cenozoic.[4] By positing such a basal divergence between anthropoids and all other primates, Ford effectively turned Le Gros Clark's venerable "ladder" of primate evolution—in which anthropoids followed prosimians in the great chain of being—on its head.

Although the most conservative scientists in the audience readily dismissed Ford's provocative new twist on primate evolution, it seemed worthy of serious consideration to me. After all, if anthropoids were the first major group to sprout from the primate family tree, the resulting ghost lineage might be only marginally longer than the one I supported on the basis of *Shoshonius*. But Ford's modest proposal raised its own set of problems, not least of which being that it denied that anthropoids could be specially related to tarsiers. Other experts on postcranial anatomy replicated Ford's results, but cautioned that other types of evidence—such as cranial anatomy, the loss of the reflective tapetum lucidum layer in the eyeball, and details of biochemistry including DNA—favor a more conventional account of primate evolution.[5] In either case, the distinctiveness of the anthropoid postcranial skeleton suggested that anthropoids did not evolve from adapiforms or omomyids. Instead, they must have originated from some mysterious third group of early primates. As more and more scientists lined up to endorse this idea, the nature and identity of the third group became a major focus of the whole conference.

The rapid migration away from traditional theories of anthropoid origins and toward the third group alternative was driven partly by news of the Chinese and Algerian fossils that Marc Godinot, Qi Tao, and I had brought to the conference. None of our specimens had been formally published, although Marc's paper describing *Algeripithecus* was in press at *Nature* and due to appear within the month. Still, the existence of our fossils was an open secret by the time the conference started. Everyone wanted to know what our new fossils looked like, how old they were, and the potential for finding more specimens where these came from. But until Marc and I made our formal presentations, we both agreed to hold our cards close to our vests.

At last, the time for full disclosure arrived. According to Rich's schedule, my slot came up first. I walked up to the speaker's podium and asked the projector to display my first slide, an image of one of the *Eosimias* jaws Qi Tao had recovered at Shanghuang. I structured my presentation in much the same way that I had originally gone about analyzing the Shanghuang primates in Qi Tao's office in Beijing. First, I pointed out what we could discern about the anatomy of the front dentition of *Eosimias,* based on empty root sockets alone. I stressed that, although we had no information about the shape of the incisor crowns, their root sockets showed that both incisors were small and fairly vertically oriented, and that the central incisor was smaller than the lateral incisor. All of these features conformed to the anthropoid pattern, although many adapiforms and a few omomyids showed similar incisor proportions. Next I showed that the empty canine socket indicated that this tooth was large and stout in *Eosimias,* an obvious difference from omomyids. Once again, *Eosimias* resembled anthropoids, although adapiforms also bore this feature. Then I spelled out the important features shown by the lower premolars of *Eosimias:* the presence of three (rather than four) premolars, the fact that P_2 was single-rooted rather than double-rooted, the oblique orientation of P_{3-4}, and the generally primitive structure of the crown of P_4, aside from the slight bulging of enamel over the external side of its front root. I emphasized that this suite of lower premolar characteristics was confined to anthropoids. Although some of these features occurred individually in omomyids and adapiforms, no living or extinct prosimian was known to have the whole package. Finally, I discussed the peculiar architecture of the molar crowns in *Eosimias.* These teeth, with their large paraconid cusps and high trigonids, differed fundamentally from the pattern that most experts had come to regard as typical of anthropoids. Still, I showed that in a few details, even the molars of *Eosimias* diverged from those of omomyids and adapiforms in ways that recalled anthropoids.

I concluded my presentation by charting a new and improved evolutionary tree of primates. Just above the critical juncture where the branch leading to living and fossil anthropoids diverges from that leading to omomyids and tarsiers, I placed *Eosimias.* A murmur swept across the room as various members of the audience turned to whisper to their neighbors. Above the muffled sound, I reiterated what was already clear to most of the authorities in the room. *Eosimias* deviated substantially from the myriad species of Eocene primates that had been unearthed and described previously. Mainly, *Eosimias* was remarkable for its unique combination of primitive and derived features. There was no denying that *Eosimias* resembled omomyids in certain respects. Like most omomyids, *Eosimias* was small, and its molars retained large paraconid cusps. But all of the similarities between *Eosimias* and omomyids were best explained as primitive holdovers from some remote common ancestor. None of them indicated that *Eosimias* and omomyids were close evolutionary cousins. On the other hand, *Eosimias* was hardly primitive in every respect. Its vertically oriented incisors, the depth and robustness of its lower jaw, and the significant compaction of its premolars from front to back all pointed in the same direction. All of these derived traits indicated a close evolutionary connection between *Eosimias* and anthropoids.

Finally, I reminded everyone that my interpretation of *Shoshonius*, along with separate evidence put forward by others at the conference, suggested that anthropoids evolved from some poorly known third group of early primates. *Eosimias* fulfilled many of the predictions of the ghost lineage theory that I had formulated on the basis of *Shoshonius*. "I'll put my head on the chopping block," I said. "I believe these animals are anthropoids, and that they are very different from omomyids and adapiforms."[6] With that, I rhetorically asked whether anyone had any questions or comments.

The backlash rolled up from the audience like a tidal wave. Elwyn Simons stated that he couldn't see why anyone would think that the Chinese fossils were anthropoids. Fred Szalay's assessment matched my own first impression of the Shanghuang primates—he suggested that they were simply weird omomyids. But the harshest critique of all came from Philip Gingerich, the modern architect of the adapiform theory of anthropoid origins. Gingerich, one of the few paleontologists in attendance whose professional expertise extended beyond primates to include other mammals, steadfastly refused to admit that the Shanghuang fossils were primates at all. Rising up from his chair for emphasis, Gingerich turned to face the audience. "I think these things are hedgehogs, broadly speaking," he pronounced.[7] If true, Gingerich's opinion would relegate *Eosimias* to the ranks of the lowly Insectivora, a motley assemblage of mammals whose other living members include such creatures as shrews and moles.

It was true that the basic molar pattern of *Eosimias* displayed some vague resemblances to that of primitive hedgehogs. In fact, the same could be said for virtually any primitive primate. Decades earlier, paleontologists had relied on precisely these similarities to argue that primates must have evolved from insectivores that resembled hedgehogs in many ways. But the hedgehog theory of primate origins had long since fallen out of fashion, partly because of the pervasive differences between hedgehogs and primates in their skulls and skeletons. Moreover, I knew that no insectivore had jaws and front teeth bearing the distinctly primatelike anatomy that was so evident in my Chinese specimens. Primitive hedgehogs possess long, slender lower jaws that uniformly retain more teeth than the short, stocky lower jaws of *Eosimias* I had just described. I challenged Gingerich to cite any hedgehog that failed to conform to this rule. He could not do so, but still he adamantly refused to cede his position.

Having reached a stalemate with Gingerich, I still hoped to persuade others in the audience about my point of view. For several minutes I argued with Szalay, who believed that our Chinese fossils belonged to an odd group of Asian omomyids. I agreed with Szalay that there were real similarities between *Eosimias* and omomyids. However, since all of these points of resemblance seemed to be primitive, none of them allowed us to conclude that *Eosimias* and omomyids share a close common ancestry. I asked Szalay to cite a single derived feature that *Eosimias* held in common with omomyids. He couldn't. Simons then jumped into the fray, arguing that *Eosimias* was simply too primitive to be an anthropoid. I responded that I'd be happy to refer to *Eosimias* as a "protoanthropoid" if that made him feel any better. For me, the critical issue was reconstructing where *Eosimias* fit on the primate family tree, not what we should call it. But Simons refused to acknowledge that *Eosimias* had anything to do with anthropoid origins, preferring to link anthropoids with adapiforms instead. After what seemed like an eternity of squabbling with Simons, Szalay, and other members of the audience, it was time for the next presentation to begin. I returned to my seat, frustrated and disheartened by my failure to convince the room full of experts that *Eosimias* forged a pivotal new link in the search for anthropoid origins.

A short while later, it was Godinot's turn to make the case for his Algerian fossils. Marc showed image after image of his small isolated teeth, emphasizing their close resemblance to teeth of *Aegyptopithecus* and other Fayum anthropoids. Just as Marc had rapidly convinced me that *Algeripithecus* was a tiny anthropoid in his office in Montpellier, many in the audience at Duke nodded in agreement as Marc enumerated his litany of anatomical features. Although Marc's fossils were substantially less complete than our Chinese specimens, two critical factors worked in his favor. First, *Algeripithecus* genuinely looked like a miniature version of fossils that everyone accepted as anthropoids. *Eosimias,* on the other hand, differed so radically from other primates that experts like Gingerich could claim that it belonged to a different group of mammals altogether. Second, *Algeripithecus* enjoyed the geographical advantage of having been found in North Africa, not so far from that familiar epicenter of early anthropoid evolution known as the Fayum. In contrast, the Asian fossil record of potential anthropoids was decidedly thin—consisting of *Amphipithecus, Pondaungia,* and little else. As a result, once Marc summarized his reasons for interpreting *Algeripithecus* as an early anthropoid, he found a far more receptive audience than I had faced. Even so, some questioned whether *Algeripithecus* was as old as Marc claimed it to be.

Throughout the rest of the conference, it became increasingly clear that any broad scientific consensus on anthropoid origins would have to wait. The debate had clearly shifted in favor of the ghost lineage version of anthropoid origins. Many who had previously viewed omomyids as the most plausible anthropoid ancestors crossed over to support the ghost lineage model instead. Yet several of the most prominent scientists in the field continued to defend more traditional theories. Elwyn Simons and his followers adamantly proclaimed that the latest discoveries at the L-41 site in the Fayum upheld Gingerich's assertion that anthropoids evolved from adapiforms. According to Simons, *Eosimias* could not be an anthropoid because it was too old and because it lacked the adapiformlike features that were so apparent in his L-41 primates.

On the other end of the theoretical spectrum, a few scholars maintained that tarsiers and anthropoids must have diverged fairly recently, making most Eocene fossils irrelevant to the issue of anthropoid origins. This faction, led by Matt Cartmill and Callum Ross, doubted the ghost lineage version of anthropoid origins because they dismissed any close evolutionary relationship between *Shoshonius* and tarsiers (see chapter 6). Without *Shoshonius* to anchor an early divergence date between the tarsier and anthropoid lineages, there was no reason to posit that anthropoids evolved from any poorly known third group of Eocene primates, regardless of whether *Eosimias* belonged to such a group or not. Hence, for very different reasons, supporters of the tarsier and adapiform theories of anthropoid origins continued to oppose the third group or ghost lineage model. I suspect that they fancied themselves as ghost busters.

Elwyn Simons had ample motive to lead the charge against the ghost lineage model and the ethereal fossil record that it implied. It was bad enough that the new model conflicted with the idea that anthropoids evolved from adapiform ancestors, a position that Simons had repeat-edly and emphatically endorsed. After suffering the indignation of hav-ing to abandon his *Ramapithecus* theory of human origins, Simons clearly wanted to be on the winning side of the anthropoid origins debate. Worse yet, the new fossils that bolstered the ghost lineage model threatened to trump Simons's long-standing monopoly on the world's earliest anthro-poids. Through the decades, Simons had leveraged his exclusive access to the Fayum fossils into numerous academic accolades. Early on, he was elected to the National Academy of Sciences, one of only a handful of paleoanthropologists to achieve this honor. Simons also occupied an en-dowed professorship at Duke University, where he was in charge of the university's primate center—a world-renowned facility dedicated to the biology of living prosimians. And, of course, Simons's acknowledged ex-pertise on anthropoid origins gave him fame and recognition within his own discipline. Most introductory textbooks touched on *Aegyptopithe-cus* as a prelude to the topic of human evolution. After such a long and distinguished run, who could fault Simons for viewing anthropoid ori-gins as his own private domain? For whatever reason, Simons and his colleagues wasted little time in challenging the claims that had been made on behalf of *Algeripithecus* and *Eosimias* at the Duke conference. They used two criteria—age and anatomy—to make their case.

Godinot's significant head start describing *Algeripithecus* placed his fos-sils under scrutiny first. The distinctly anthropoidlike anatomy of its cheek teeth prevented *Algeripithecus* from being dismissed as an anthropoid out-right. Instead, its age came under fire. As we noted earlier, the Glib Zeg-dou site that yielded *Algeripithecus* lacks the widespread ash layers that bracket so many fossil hominid sites in the East African Rift Valley. In the absence of volcanic rocks suitable for direct radiometric dating, estimat-ing the age of the Glib Zegdou site depends on widespread and abundant organisms that can be used as "guide fossils." Among the mammals that lived alongside *Algeripithecus* at Glib Zegdou, only the hyraxes and ro-dents are cosmopolitan enough to be useful in this way. Recall that hyraxes, which are small and inconspicuous components of African ecosystems to-day, occupied a much broader range of ecological niches during Eocene and Oligocene time. In the Fayum ecosystem, hyraxes evolved into piglike, antelopelike, and rhinolike forms that differed substantially from their modern brethren. After making a few cursory comparisons between the fossil hyraxes of Glib Zegdou and those from the Fayum, Simons and his team announced that the hyraxes from both sites were so similar that they must be nearly the same age. As a result, *Algeripithecus* could not be older than the primates from the Fayum's L-41 site.[8] Their revised estimate for the age of Glib Zegdou conflicted with Godinot's original assessment by roughly ten million years.

Such an immense chronological discrepancy was necessary to bring *Algeripithecus* back into the fold of traditional theories of anthropoid origins. But even by geological standards, ten million years is a substantial span of time. To place this number in perspective, ten million years ago, during the late Miocene, three-toed horses known as *Hipparion* roamed across much of the globe. The geographic range of *Hipparion* included parts of Africa, where these early horses would have encountered various species of apes, one of which—millions of years later—gave rise to the three lineages leading to modern gorillas, chimps, and humans.

Godinot did not allow the sudden disparity in dating Glib Zegdou to remain unanswered. Instead, he reassessed the evidence from fossil hyracoids and rodents and determined that his earlier age estimates for Glib Zegdou still made sense. In every case, the hyrax species from Glib Zegdou appeared to be more primitive (and therefore older) than their counterparts at L-41. For example, the upper molars of *Megalohyrax gevini* from Algeria differ from those of *Megalohyrax eocaenus* from the Fayum in ways that suggest that the Egyptian species might be descended from its Algerian relative.[9] Likewise, *Titanohyrax mongereaui* from Glib Zegdou has lower-crowned cheek teeth than *Titanohyrax ultimus* from the Fayum. Because all early hyracoids possess low-crowned teeth, the Algerian species of *Titanohyrax* is more primitive than its Egyptian relative, at least with respect to this single characteristic. Yet a third Algerian hyracoid, an unusually small form known as *Microhyrax lavocati,* simply lacks any comparable species in the extensively documented Fayum fauna.

If fossil hyracoids hinted that Glib Zegdou was older than the Fayum, the evidence from fossil rodents amounted to a smoking gun. Rodents are the dominant small mammals in most modern ecosystems, and nine species of rodents have been described from the Fayum. Their abundance and diversity make rodents especially useful as guide fossils. All of the Fayum rodents belong to a single specialized family known as the Phiomyidae. Living relatives of these Fayum rodents survive in Africa today in the form of the widespread and plentiful cane rats (genus *Thryonomys*) and the less common dassie rats (genus *Petromus*). The earliest and most primitive African phiomyids come from the late Eocene Bir el Ater locality in northeastern Algeria, the same site that yielded the early anthropoid *Biretia* (see chapter 7).[10] Although several species of rodents have been discovered at Glib Zegdou and sites of similar age in Algeria and Tunisia, none of them is closely related to the phiomyids that so utterly dominated the Fayum ecosystem.[11] Such a dramatic change in the African rodent fauna can only be explained by evolutionary turnover through time. Apparently, ancestral phiomyids were not among the earliest African rodents. Instead, the group must have dispersed to Africa near the end of the Eocene, after the interval represented by Glib Zegdou but before the time documented by Bir el Ater and the Fayum's L-41 site. Currently, this phiomyid datum is the strongest evidence supporting Godinot's original contention that *Algeripithecus* is older than any of the Fayum anthropoids, although by exactly how much remains debatable.

While efforts to bring *Algeripithecus* in line with traditional theories of anthropoid origins focused on its age, *Eosimias* posed a different problem entirely. No one could seriously protest the fact that *Eosimias* dates to roughly ten million years before the Fayum's L-41 site. Too many middle Eocene guide fossils had been found alongside *Eosimias* in the Shanghuang fissure-fillings for its age to be challenged successfully. Instead, as soon as my colleagues and I published our initial description and interpretation of *Eosimias* in *Nature* in 1994,[12] Elwyn Simons and others began to question whether its anatomy supported its anthropoid status. They did so on two separate fronts.

The more conventional of these entailed highlighting the many primitive features found in *Eosimias* and then claiming that the Chinese fossil was simply not advanced enough to be an anthropoid. Marc Godinot, who was busy fighting his own battles on behalf of *Algeripithecus,* took this position. From our private summit at the Carnegie Museum, I knew that Marc and I disagreed about whether *Eosimias* was a basal anthropoid. Soon enough, Marc published his dissenting viewpoint. Citing the presence of enlarged paraconid cusps on its lower molars as well as other aspects of its dentition, Marc suggested that *Eosimias* might be related to tarsiers. At the same time, Marc left open the possibility that *Eosimias* might not be a primate at all. Like Gingerich, Marc felt that *Eosimias* might conceivably be an insectivore, although Marc was far less strident about this possibility. In any case, Marc was convinced that *Eosimias* should not be accepted as an early anthropoid.[13]

Simons and his colleagues employed shrewder tactics to assail the anthropoid status of *Eosimias.* After spending several decades describing Fayum anthropoids on the basis of their fossilized teeth and jaws, Simons suddenly raised the bar for telling early anthropoids apart from their prosimian brethren. According to Simons, it was no longer possible to discriminate between anthropoids and prosimians from their teeth and jaws alone. To be certain that a fossil primate was an anthropoid, you now had to have a skull.[14] Any fossils that fell short of this new and much higher standard—like *Eosimias* and *Algeripithecus*—were immediately suspect and could not be used to substantiate any theory of when, where, and how early anthropoids evolved.

It is impossible to argue against the need for superior specimens in paleontology. Relatively complete fossils obviously inspire greater confidence than do bits and scraps. Still, Simons's rapid conversion to making skulls the new benchmark for anthropoid origins was no accident. Only a few years earlier, Simons and his team had achieved a substantial breakthrough when they uncovered a crushed skull of *Catopithecus browni* at the L-41 locality in the Fayum.[15] Previously, *Catopithecus* had been documented only by its teeth and jaws. These corresponded so precisely with teeth and jaws of the slightly younger Fayum primate *Oligopithecus* that *Catopithecus* could be grouped with confidence in the family Oligopithecidae (see chapter 7). But this only solved part of the evolutionary puzzle surrounding *Catopithecus*. The broader relationships of oligopithecids were still mired in controversy. For example, Philip Gingerich had shown that *Oligopithecus* shares many aspects of dental anatomy with the lemurlike adapiforms. While Gingerich interpreted *Oligopithecus* as an important transitional fossil linking anthropoids with adapiforms, others maintained that *Oligopithecus* was simply one of many Fayum anthropoids. Alternatively, a few prominent researchers regarded *Oligopithecus* as an unlikely candidate for anthropoid status.[16] So long as teeth and jaws were all that documented *Catopithecus* and *Oligopithecus,* their position on the family tree would remain murky.

The crushed skull of *Catopithecus* from L-41 resolved all of these issues at once. Simons used it to show that *Catopithecus* differs from living and fossil prosimians—and resembles anthropoids—in several critical features of cranial anatomy. Like modern anthropoids, *Catopithecus* possesses a bony septum behind its eye sockets. On the side of its skull, a bone known as the ectotympanic frames the external opening that leads to the soft, fleshy part of the ear. Just above the eye sockets, the paired frontal bones of *Catopithecus* fused together seamlessly at an early age, leaving no hint of the metopic suture that marks the midline of the skull in prosimians. Citing these and other skull characters, Simons demon-

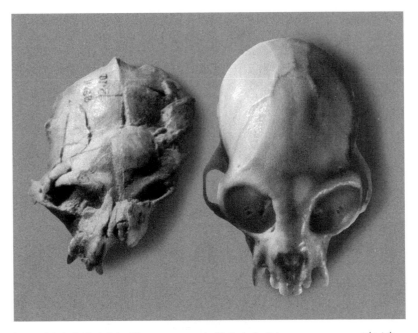

Figure 31. A skull of *Catopithecus* compared with that of a living pygmy marmoset (right). Photograph by R. L. Usery, courtesy of and copyright by Elwyn Simons.

strated once and for all that *Catopithecus* was an anthropoid, even though its teeth and jaws could be interpreted very differently.

Although at first the crushed skull of *Catopithecus* simply clinched the anthropoid status of oligopithecids, it soon assumed far greater significance. Simons feared that *Algeripithecus* and *Eosimias* would render *Catopithecus* and its Fayum contemporaries redundant in the all-important sense of time. Moreover, he continued to doubt that a primate as primitive as *Eosimias* could have any relevance for anthropoid origins. If the skull of *Catopithecus* was necessary to verify its rightful place in anthropoid evolution, surely nothing less was required for fossils like *Eosimias* and *Algeripithecus* that were millions of years older—and proportionately more primitive. By insisting that early anthropoids could only be distinguished from prosimians on the basis of their skulls, Simons was merely adopting the stance that any cautious scientist might take. Of course, in moving the goalposts this way, Simons might also extend his lock on the earliest undoubted anthropoids, since *Eosimias* and *Algeripithecus* lacked skulls to back up their claims to anthropoid status.

Coming from the person who had boldly anointed *Ramapithecus* as the world's oldest hominid on the basis of a fragmentary upper jaw, I found Simons's sudden enthusiasm for complete skulls to be disingenuous at best. After all, he and his predecessors had failed to find skulls for the vast majority of the Fayum anthropoids that they had named. Did this mean that we now had to reject the anthropoid status of *Propliopithecus, Arsinoea,* and *Qatrania?* No one, myself included, wanted to pursue Simons's stringent new guidelines to their logical conclusion. Still, the idea that skulls might hold the key to unlocking the mystery of anthropoid origins deserved deeper reflection. I was reminded of an earlier debate in paleoanthropology, whose resolution might provide some guidance now.

Early in the twentieth century, before the hominid fossil record had accrued much beyond the discovery of Neanderthals and fossils that became widely known as "Java Man," the biggest dispute in paleoanthropology focused on the sequence in which modern human attributes evolved from our more apelike ancestors. Did modern human characteristics arise as a package, or did certain features antedate others? If the latter were the case, did humans acquire big brains before they began to walk upright on two legs, or vice versa? And did these physical changes occur before or after early hominids came down from the trees to live mainly on the ground? The prominent British anthropologist Sir Grafton Elliot Smith (whose contribution to the "arboreal theory" of primate and human evolution is discussed in chapter 5) favored the view that big brains evolved first. Other experts—like Henry Fairfield Osborn of the American Museum of Natural History—believed that upright bipedalism was the true hallmark of humanity. Depending on which part of the human body—head or legs—you believed was more precocious in an evolutionary sense, you would naturally tend to favor skulls or knees as the more important element for distinguishing early hominids from contemporary apes.

These were topics of heated deliberation during the first decades of the twentieth century, but from our perspective nearly a century later they appear rather quaint. Thanks to a greatly enriched hominid fossil record, we now know that early hominids became habitual bipeds millions of years before their brains enlarged much beyond those of chimpanzees. Hence, no modern paleoanthropologist would suggest that skulls are required to tell an early member of the human lineage apart from an ape.

A recent discovery illustrates this point in a particularly dramatic way. In a remote and inhospitable desert region of Chad, a team of scientists led by the French paleoanthropologist Michel Brunet unearthed a nearly complete hominoid skull in 2001. The specimen, which Brunet and his colleagues soon named *Sahelanthropus tchadensis,* sparked an immediate and ongoing debate. It is not hard to understand why. Hailing from the vast and arid belt that separates the Sahara Desert from the forests and savannas of equatorial Africa, Brunet's fossil shows that early ape and human relatives lived far beyond the narrow East African corridor stretching from Ethiopia to South Africa. While its geography alone poses many interesting questions, Brunet believes that these pale in comparison to how *Sahelanthropus* affects the human family tree. At nearly seven million years in age, *Sahelanthropus* lies at or near the juncture when ancestral chimpanzees and humans are thought to have embarked on their separate evolutionary paths. Despite its antiquity, the vertical face and small canine teeth of *Sahelanthropus* look surprisingly human. Citing these and other anatomical features, Brunet and his team regard *Sahelanthropus* as the oldest known hominid. Such assertions rarely go unchallenged in paleoanthropology. Predictably, a rival group maintains that *Sahelanthropus* bears no direct relation to human origins, and that it lies on the gorilla branch of the primate family tree instead. However this debate ultimately gets resolved, the controversy surrounding *Sahelanthropus* shows that, even when a relatively complete skull is available, bitter disagreements can persist.[17]

Today, the quality of the fossil record of anthropoid origins approximates that which illuminated human origins during the time of Elliot Smith, Osborn, and their peers. We simply do not know which of the many characteristics that distinguish living anthropoids from their prosimian relatives evolved first. *Catopithecus,* with its long list of diagnostic anthropoid traits in the bony anatomy of its eyes, ears, and forehead, can be readily discerned from Eocene prosimians. Possibly, this means that Elwyn Simons is correct—skulls can segregate early anthropoids from prosimians even when teeth and jaws alone cannot. Alternatively, *Catopithecus* may be so far removed from the origin of anthropoids that it is trivial to identify it once adequate fossils become available. Which of these options is more likely, and what ramifications arise for evaluating anthropoid origins?

Curiously, Simons himself has always placed *Catopithecus* and *Oligopithecus* high up on the anthropoid family tree, near *Aegyptopithecus* and *Propliopithecus.*[18] If Simons's version of early anthropoid evolution is correct, it should come as no surprise that *Catopithecus* possesses so many key anthropoid features. Indeed, more recent discoveries of *Catopithecus* bear out this view. Much of the new evidence comes from limb and ankle bones, previously an unknown aspect of oligopithecid anatomy. These new elements show that *Catopithecus* closely resembles *Aegyptopithecus*, especially in terms of its elbow and ankle morphology.[19] Other early anthropoids from the Fayum, including both parapithecids and proteopithecids, have elbows and ankles that are decidedly more primitive. Had these postcranial bones of *Catopithecus* been discovered before its skull was unearthed, Simons might well have decided that ankle bones were necessary to distinguish early anthropoids from their prosimian relatives. A more defensible—although rhetorically less effective—conclusion would depict *Catopithecus* as a fairly advanced anthropoid that offers only modest insight into anthropoid origins. To reiterate a point made earlier, using *Catopithecus* to untangle anthropoid origins is akin to reconstructing human origins with Neanderthals as our only guide.

The early debates among Elliot Smith, Osborn, and others on the sequence in which key hominid features evolved make one thing abundantly clear. It is futile to predict how any group of animals evolved until the fossil record becomes sufficient to sustain the effort. As such, Elwyn Simons's contention that skulls are required to distinguish early anthropoids from their prosimian cousins places the cart squarely before the horse. Just as significant expansion of the brain happened only late in hominid evolution, major changes in the anthropoid skull may have been deferred until long after the anthropoid lineage was established. Even if I were lucky enough to find a skull of *Eosimias,* it might lack the anthropoid characteristics found in *Catopithecus* and nonetheless still be an anthropoid!

In some ways, the rancor over how to distinguish early anthropoids from other primates brought everything full circle for me. A few months after my study of *Shoshonius* christened the ghost lineage theory of anthropoid origins, I posed much the same question during a seminar at Stony Brook University. Even if we were lucky enough to discover early fossils of very primitive anthropoids, would we be able to recognize them for what they are? The fragmentary jaws of *Eosimias* from Shanghuang convinced me that we could now answer this question with a resounding yes. But the lingering skepticism of other experts on primate evolution meant that we sorely needed to uncover additional evidence. Besides, by my own calculation, *Eosimias* was too young to document the very first stages of anthropoid evolution. For that, we would need to find fossils that were at least as old as *Shoshonius*. According to my best estimates, fossils documenting such a truly primordial stage of anthropoid evolution would have to be older than *Eosimias* by at least five million years—an interval equal to most of the span of hominid existence. And doubling that vast expanse of time was hardly beyond the realm of possibility.

Progress in science often resembles the ascent of a mountain climber struggling to reach a difficult summit. To advance, the mountaineer may have to take two steps forward to overcome the backsliding caused by the shifting rock underfoot. The discovery and technical description of *Eosimias* propelled my team several steps toward our destination, but our upward trajectory was stalled by the clamor of the ghost busters. To maneuver past this obstacle, we needed to unearth more compelling fossils of *Eosimias*—specimens that would demonstrate beyond any reasonable doubt that *Eosimias* occupies a pivotal position on the anthropoid family tree. At the same time, we needed to learn as much as possible about the paleobiology of these intriguing primates in order to identify the factors that contributed to their evolutionary success. For me, the only path leading to these lofty scientific objectives ran squarely through China.

9

Resurrecting the Ghost

D usk still alters the scope and pace of daily activities in rural China, just as it affected countless human generations before the era of rural electrification projects. As I stroll the main avenue of Yuanqu, the county seat of this part of Shanxi Province, its nonfunctional street lamps actually work to my advantage. The hour is still early, and many of the locals are outdoors, promenading with their family and friends after the evening meal. For a *wei guo ren* (foreigner) like me, this is the best time to get out and see the sights. Even along the town's busiest thoroughfare, the ambient lighting is so dim that faces can only be recognized from a distance of two or three yards. As a result, I saunter along incognito where, during daytime, my presence would attract a small crowd curious to see their first live Caucasian.

For my field crew and me, the shortage of electric power in this part of central China definitely has its pluses and minuses. Aside from our brief forays about town after dinner, there isn't much to do after dark. In the dormlike hotel that serves as our base of operations, hot water flows for only an hour or so each day, which strongly constrains our schedule if personal hygiene is any priority. Moreover, the paucity of refrigeration obliges us to drink the local beer at room temperature. On the positive side, the dearth of television promotes lively discussion. Our rambling conversations, which range from Chinese history to academic politics back home, foster a sense of camaraderie and group cohesion. But the biggest effect of the local power shortage is more fundamental. Simply put, none of us would be here were it not for the daily blackouts.

To feed the country's starving grid of power lines, China's leaders had recently embarked on an ambitious plan to increase the nation's capacity for generating electricity. Achieving this goal meant building new hydroelectric dams, along with other initiatives. Compared to many of its alternatives, hydroelectric power is safe, clean, and renewable. Still, imposing a dam on an untamed river raises its own set of problems. In a densely populated country like China, these challenges can be greatly exacerbated. Multiple villages and thousands of people must be permanently displaced. Unique cultural heritage sites and endangered flora and fauna may likewise stand in the way of progress, each demanding mitigation of one sort or another. Against this backdrop, the Chinese government had invited scores of foreign specialists and international agencies to provide assistance. One such request landed on my desk at the Carnegie Museum. It sought to establish a new collaborative project between my American team of geologists and paleontologists and some of our colleagues at the Institute of Vertebrate Paleontology and Paleoanthropology in Beijing. Our task was to salvage as much as we could from the fossil record around Yuanqu before the construction of a new dam downstream drowned its most important sites. My colleagues and I leapt at the chance to get involved, and by May 1994, we had secured preliminary funding to begin our work there.

Yuanqu County lies in a remote, mountainous region of central China. To get there, you take an overnight train from Beijing to the nearest railway station, then clamber aboard a four-wheel drive vehicle for the last thirty miles (fifty kilometers) or so. The surrounding terrain is rocky and surprisingly dry, despite the agricultural focus of the local economy. Such a marginal environment dictates a hardscrabble existence for the local population. The combination of difficult circumstances and deeply ingrained cultural norms made the region fertile ground for Mao Zedong and his Communists during their long civil war against the Guomindang (Nationalists) of Chiang Kai-shek. Some of the older peasants in the region can even recount childhood memories of a pitched battle between Mao's Communists and Guomindang troops under Chiang's command. It took place in an otherwise nondescript wheat field, near the southern border of the county. Just a stone's throw from the neglected battlefield, the mighty Huang He (or Yellow

River) rushes through steep, narrow gorges on its headlong quest to reach the East China Sea.

Most historians and archeologists agree that the origins of Chinese civilization lie in the Yellow River valley. Archeological excavations along its banks have uncovered early evidence of agriculture, as well as such important cultural innovations as the famous Shang oracle bones—the earliest examples of Chinese writing. As I watch its variegated hues of green, brown, and blue flow by, I am reminded of the central role the Yellow River continues to play in the commercial and cultural life of China. Once the river's untapped potential can be harnessed, its currents will help solve the local power shortage. A few miles downstream, an army of construction workers labors to make that vision a reality. But the reason why we so eagerly agreed to come here has nothing to do with history or economics. Our enthusiasm, like the course of the river itself, is dictated by the local geology.

About forty million years ago, a more sluggish forerunner of the Yellow River flowed across this same landscape. Periodic floods caused the meandering river to break through the natural system of levees that defined its channel. Afterward, the ancient river often settled into a different and more efficient course. The abandoned river channel survived, however, as an oxbow lake lying adjacent to the new and more streamlined course of the river. Its quiet waters provided ideal habitat for wallowing herds of anthracotheres known as *Anthracokeryx,* cow-sized progenitors of modern African hippos. Without any winnowing current of its own, the oxbow lake also functioned as a natural trap for the accumulation of all manner of debris, including the carcasses of animals inhabiting the surrounding terrain. Over time, through multiple iterations of flooding, the oxbow lake became so filled with mud and animal bones that it disappeared from the local landscape entirely. Eventually, the constantly migrating river retraced its former path, settling once again into a channel above the site where the oxbow lake had once been. The river's undulating currents draped large volumes of sand over the site, sealing the contents of the former oxbow lake as effectively as the lid of some giant Egyptian sarcophagus. The resulting time capsule would not be opened until the early part of the twentieth century.

In May 1916, a Swedish geologist by the name of Johan Gunnar Andersson happened to be traveling through Yuanqu County in search of commercially viable copper deposits.[1] Only two years earlier, Andersson had relinquished his position as director of the Geological Survey of Sweden in order to serve as a mining advisor to the Chinese government.

Figure 32. Johan Gunnar Andersson, the Swedish geologist who discovered Locality 1 in the Yuanqu Basin and unearthed the first Chinese Eocene primate fossils.

Having completed his copper survey in the Chinese hinterland, Andersson was anxious to return to his base of operations in Beijing. The fastest way to do this was to cross the Yellow River into adjacent Henan Province, where a major rail line provided a relatively fast and efficient means of returning to the capital. Then as now, the only means of traversing the Yellow River in this region was by boat. Because of the rugged terrain and long stretches of whitewater on the river, the few reliable sites for navigating it had long since been established. Andersson chose to cross the Yellow River about two and a half miles (four kilometers) upstream from the modern village of Gucheng, purely for logistical convenience. His choice proved to be extremely fortuitous.

As his boat progressed southward across the Yellow River, Andersson was rewarded with a panoramic view of the rock strata exposed in the steep riverbank that he had just left. Much to his surprise, he noticed several brightly colored layers of rock cropping out beneath the thick, drab beds of wind-blown sediment known as loess that prevail throughout central China. Andersson's cursory geological survey of the region had failed to register strata of this type. Yielding to his own innate curiosity, Andersson decided that the colorful rocks at the base of the Yellow River's northern bluffs deserved a closer look. He ordered the boat to be turned around so that he could inspect the geological section more thoroughly.

As soon as his boat ran aground on the river's opposite bank, Andersson leapt ashore and made his way upstream. Clambering over rockslides and river cobbles, he soon reached a deep, narrow gully where a trickle of water drained into the Yellow River from the north. In the bottom of the gully, Andersson found the intriguing beds of mudstone and limestone that he had initially seen from the boat. After a few minutes of searching, Andersson discovered that some of the multihued strata contained the fossilized remains of freshwater snails. He hastily collected a few samples to take back to Beijing. With luck, the fossils might give some clue as to the age of the mudstone and limestone layers at the bottom of the small gully. This, in turn, would provide an upper limit on the antiquity of the overlying loess.

Soon after Andersson returned to the Chinese capital, he shipped the fossil-bearing rock samples to Nils Odhner, a specialist on snails based at the Swedish Museum of Natural History in Stockholm. Andersson fully expected Odhner to confirm his suspicions about the age of the snails. The fossils were entombed in beds that were immediately beneath the loess, and Andersson suspected that they were only slightly older than the loess itself. The geological evidence available to Andersson suggested that the loess dated to the middle Pleistocene. Hence, the fossil snails might date to the early Pleistocene (about two million years ago), but they could hardly be much older.

The response he received from Odhner a short time later stunned him. Most of the fossil snails Andersson had collected along the north bank of the Yellow River were identical to Eocene species from France and Germany! This made finding the multicolored beds of mudstone and limestone a significant geological breakthrough. For the very first time, strata dating back to the Eocene had been discovered in China. Odhner had also identified a tiny fragment of bone embedded within the rock samples Andersson had sent him, demonstrating the potential for even more exciting scientific discoveries. With only a few strategic whacks from his rock hammer, Andersson had pried open the lid of a forty-million-year-old time capsule. In time, its contents would offer a rare glimpse of the animals inhabiting a lost world, including two very different species of early primates.[2]

Unfortunately, Andersson himself never realized the full potential of his most important contribution to Chinese paleontology. A major barrier was the highly unstable political environment that Andersson had to navigate every time he left the vicinity of Beijing. Within two years of Andersson's arrival in China, the legitimacy of the central government collapsed as a result of a series of military coups and failed attempts to reinstate imperial authority. Meanwhile, a motley assemblage of regional warlords declared their independence from Beijing, and much of rural China descended into chaos. Yan Xishan, the warlord who ran Shanxi Province, personified the unpredictable political landscape. Yan claimed to have established a new and virtually flawless ideology based on the best aspects of "militarism, nationalism, anarchism, democracy, capitalism, communism, individualism, imperialism, universalism, paternalism, and utopianism."[3] Given the daunting prospect of dealing with such a flamboyant leader, Andersson chose to concentrate his fossil-collecting efforts on other parts of China.

As if negotiating with local warlords wasn't difficult enough, Andersson also faced mounting criticism from his superiors at the Geological Survey of China. Originally, Andersson had been hired to assist the Chinese government in developing the country's vast mineral resources. Yet it was apparent to everyone that Andersson had become increasingly distracted from his official duties by the allure of making important discoveries in the realm of paleontology. To relieve some of this pressure, Andersson appealed to personal contacts back in Sweden for funding to support his fossil-collecting campaigns in China. A wealthy Swedish businessman named Axel Lagrelius soon answered Andersson's plea for help. With the understanding that the fossils Andersson collected in China would be deposited in a Swedish museum, Lagrelius established the Swedish China Research Council (or *Kinafond*). Lagrelius's extensive network of social and political connections fostered the international endeavor, and it soon attracted the backing of the crown prince of Sweden among other dignitaries.[4]

The new source of funding left Andersson with only one remaining obstacle. Andersson himself was not trained as a paleontologist, and he desperately needed someone with this type of professional expertise to achieve his goal of putting the Chinese fossil record on the map. Once again Andersson appealed to personal contacts back in Sweden for help. This time, his friend Carl Wiman—a paleontologist based at the University of Uppsala—stepped up to the plate. Wiman persuaded his former student Otto Zdansky to go to China and collaborate with Andersson. Zdansky, an Austrian citizen who had served in the Austro-Hungarian army during World War I, had just completed a doctoral dissertation on the cranial anatomy of fossil turtles. With his professional opportunities in postwar Europe looking bleak, Zdansky departed Sweden in May 1921 on a ship bound for London, South Africa, and eventually China. He arrived in Beijing in June. The terms of Zdansky's employment were hardly lavish. He agreed to work without a salary, although the costs of his subsistence would be covered. In return, Zdansky would be given the privilege of publishing any fossils he might unearth.

While Zdansky prepared for his long voyage, Andersson faced the looming threat of competition in the fossil fields of China. A dapper young American by the name of Roy Chapman Andrews had just announced his intention to lead a high-profile expedition to collect fossils in central Asia. Andersson was fully aware that Andrews possessed the scientific and financial resources to execute his bold plan. After all, Andrews worked for the well-connected Henry Fairfield Osborn, who presided over paleontology at the American Museum of Natural History in New York. Osborn, along with his brilliant protégé William Diller Matthew (whose work on the relationship between climate and evolution is discussed in chapter 5), had become convinced that Asia was the most likely place of origin of humans and many other important groups of mammals.[5] Andrews, who had no prior experience as a paleontologist, gained Osborn's backing by shrewdly offering to test (and, it was hoped, confirm) his pet theory. With a burgeoning bankroll of American dollars, matched by some of the best scientific talent in paleontology, Andrews's expedition jeopardized virtually everything that Andersson hoped to accomplish in China.

Roy Chapman Andrews arrived in Beijing on April 14, 1921. One of his first stops was to visit Andersson and his colleagues at the Geological Survey of China. In part, the visit was a courtesy call. During an earlier trip to Beijing, Andrews had tried to entice Andersson into fully cooperating with his team, offering him the financial backing of the American Museum if he would merely agree to send his fossils to New York rather than Sweden. Andersson interpreted this offer as an affront to his own scientific agenda in China, and he rejected it outright (although not without leveraging the offer into additional funding from his Swedish underwriters). Still, the rival teams reached a gentleman's agreement to avoid any direct competition. The best way to accomplish this was by dividing the region geographically—the American Museum team would work mainly in northern China and Mongolia, while Andersson and his colleagues would continue their efforts in other parts of China.[6]

Notwithstanding his agreement with Andrews to partition China, Andersson remained wary of the upstart Americans and their deep pockets. In hindsight, Andersson had every reason not to trust Andrews to abide by the geographical terms they had just negotiated. From its earliest conception, Andrews's rationale for initiating the Central Asiatic Expeditions included exploiting the most promising fossil sites that Andersson had already located in China. In a letter to Henry Fairfield Osborn dated August 10, 1920, Andrews wrote:

> Dr. [William Diller] Matthew thought that you would consider having Mr. [Walter] Granger go over with us for at least a beginning in the localities which Dr. Anderson [sic] has already discovered. . . . Since Dr. Anderson [sic] has barely touched the fields which he has already discovered, and is not a palaeontologist who is familiar with the fauna which he has unearthed, I am quite sure that Mr. Granger would be able to carry out further investigations with a great deal of profit.[7]

If the tense negotiations with Andrews failed to motivate Andersson to cover his scientific bases, the impending arrival in Beijing of Walter Granger surely must have done so. Osborn knew that Andrews was no paleontologist. If the Central Asiatic Expeditions were to succeed, a paleontologist with international stature and wide-ranging expertise would need to direct the team's fossil-collecting efforts. Walter Granger embodied the best traditions of American vertebrate paleontology, and he possessed all of the scientific and practical skills that both Andrews and Andersson lacked. By 1921, Granger had already led numerous successful fossil-collecting campaigns exploring early Cenozoic basins in the Rocky Mountain West. Outside of North America, Granger had also spearheaded the American Museum's 1907 expedition to the Fayum region of Egypt. At that time, very few people in the world could match Granger's impeccable credentials in paleontology—certainly no one in China possessed a similar level of expertise.

Given the circumstances, it hardly seems coincidental that Andersson finally returned to the banks of the Yellow River—where he had serendipitously discovered those Eocene snails some five years previously—less than two weeks after his meeting with Andrews in Beijing. Andersson spent roughly two weeks in the vicinity of Yuanqu—enough time to reach a basic understanding of the regional Eocene geology. He concentrated his fossil-collecting efforts on the narrow gully north of the Yellow River where he had first inspected those multicolored Eocene strata during his initial visit to the region. Appropriately enough, Andersson designated this site simply as "Locality 1."

During the course of this short field trip, Andersson and his single Chinese assistant unearthed the first Eocene mammals ever found in China. Among the fossils they collected at Locality 1 were two specimens—an isolated upper molar and a fragmentary lower jaw preserving the last two molars—belonging to a primate. Had Andersson himself been trained as a paleontologist, his discovery of these Eocene primate fossils in the Yuanqu Basin—scrappy as they might be—could have been heralded as an important scientific breakthrough. At the time, Eocene primates had never been reported from Asia. Accordingly, no matter where they might fit on the primate family tree, the Yuanqu Basin specimens filled an enormous geographic gap.[8] But Andersson lacked the expertise to describe his primate fossils, and he eventually turned the specimens over to his younger colleague Zdansky.[9]

Even though Zdansky possessed the academic training in paleontology that Andersson lacked, his timid scientific disposition caused him to downplay Andersson's discovery. He could have easily composed a brief article that would have named Andersson's primate, while highlighting the potential for further scientific advances in the Yuanqu Basin. Instead, Zdansky deferred the description of Andersson's primate until 1930. Even then, he failed to emphasize the significance of the fossils, burying them in a long compendium of Eocene mammals collected in the Yuanqu Basin during the course of several field seasons.[10]

When he finally put his pen to paper, Zdansky chose to name China's first Eocene primate *Hoanghonius,* based on the local term for the Yellow River. (In standard Mandarin the Yellow River is known as the Huang He, but in the local dialect of Yuanqu County, the pronunciation more closely approximates "Hoang Ho.") Given the fragmentary nature of the *Hoanghonius* remains available to him, Zdansky was not even fully convinced that *Hoanghonius* was a fossil primate. He dutifully compared it with Eocene adapiforms and omomyids, but he wasted no time speculating on where *Hoanghonius* might fit within the broader scheme of primate evolution.

The same constraints did not dissuade later workers from lifting

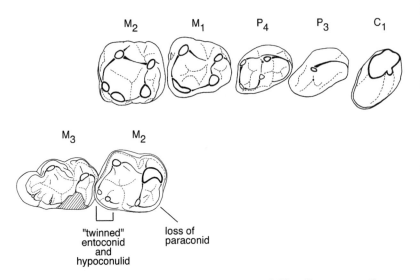

Figure 33. Schematic drawings of the holotype specimens of *Oligopithecus savagei* from the Fayum region of Egypt (top) and *Hoanghonius stehlini* from the Yuanqu Basin of central China (bottom). Note the strong similarities in the anatomy of M_2 (the only tooth that is preserved in both fossils), which led Philip Gingerich and others to postulate that *Hoanghonius* may be a link between Fayum anthropoids and Eocene adapiforms.

Hoanghonius from its state of relative obscurity. By the late 1970s, *Hoanghonius* had become a major figure in the debate over anthropoid origins. Philip Gingerich and other advocates of the adapiform theory of anthropoid origins repeatedly cited *Hoanghonius* as a fossil that linked primitive Fayum anthropoids such as *Oligopithecus* with their putative adapiform ancestors.[11] They emphasized that *Hoanghonius* and *Oligopithecus* share several unusual details of molar anatomy, including the loss of the paraconid cusp and the close approximation or "twinning" between the entoconid and hypoconulid cusps on the rear margin of the lower molars.

By the time my team arrived in the Yuanqu Basin in 1994, the number of opinions about how *Hoanghonius* fits on the family tree exceeded the paltry number of *Hoanghonius* specimens that had been found. Many experts regarded *Hoanghonius* as a run-of-the-mill adapiform, although Gingerich and his colleagues continued to view it as a transitional fossil linking adapiforms with their alleged anthropoid descendants. A different group of scientists held that *Hoanghonius* was an atypical Asian omomyid. Even the possibility that *Hoanghonius* was an early anthropoid could not be dismissed outright. Given this ambiguity, we hoped to locate more nearly complete specimens of *Hoanghonius* that might resolve its evolutionary position once and for all. An even more exciting prospect was the possibility of finding equally complete specimens of *Eosimias*. Although additional fieldwork had been conducted only sporadically in the Yuanqu Basin since the days of Andersson and Zdansky, preliminary efforts to screen-wash some of its classic localities had turned up a few isolated teeth of *Eosimias*. With luck and determination, we might uncover material of *Eosimias* in the Yuanqu Basin that would surpass the fragmentary specimens we had described from the fissure-fillings at Shanghuang, some five hundred miles to the southeast.

Our optimism was well founded. In contrast to the Shanghuang fissure-fillings, the fossil sites in the Yuanqu Basin resulted from a wide variety of geological processes. Some of the Yuanqu Basin sites—notably including Andersson's Locality 1—originated as ancient oxbow lakes. Such low-energy depositional settings frequently preserve extremely complete fossils. Recall that a similarly tranquil setting characterized the Buck Spring Quarries in Wyoming's Wind River Basin (see chapter 6). Our ability to find skulls and skeletal elements of *Shoshonius* there made me hopeful that we might enjoy similar success at Locality 1. Elsewhere in the Yuanqu Basin, specimens had been ushered into the fossil record under significantly different circumstances. At some sites, fossils had slowly become incorporated into ancient soils. At others, sand and mud rapidly buried whatever animal remains happened to be lying nearby when an ancient river spilled its banks.

This diversity of geological settings demanded different methods of retrieving fossils. Locality 1 preserved fossils in sufficient density that they had to be quarried by hand. Other sites could be profitably worked by screen-washing. But the majority of localities in the Yuanqu Basin yielded only a few fragmentary bones or broken jaws, spread across a wide expanse of rock outcrop. The best way to collect fossils at these sites was by prospecting—walking or crawling across the geological exposure, searching intently for any fossils that had recently eroded out of the ground.

The geological heterogeneity of the Yuanqu Basin sites offered some distinct advantages over the Shanghuang fissure-fillings. Our work at Shanghuang convinced me that we could find plenty of fossils there. Unfortunately, the vast majority of the fossils we recovered were highly fragmentary—mainly isolated teeth and broken bones. Many of them bore the telltale signs of having been exposed to the digestive tracts of owls or other raptorial birds. This distinctive taphonomic imprint—chemical etching of tooth enamel and eroded joint surfaces on bones—further reduced the anatomical value of the Shanghuang specimens. To make matters worse, the only way we could extract fossils from the gooey fissure-filling matrix was by screen-washing. The physical agitation of this technique further degraded the Shanghuang specimens beyond the fragmented condition in which they had entered the fossil record. In contrast, the fossilized contents of Locality 1 often came out in relatively pristine condition.

Beyond the possibility of finding well-preserved fossils in the Yuanqu Basin, I hoped that we might also address that most important factor in paleontology—time. Each of the Shanghuang fissures potentially functioned as a separate snapshot of life during the Eocene. Unfortunately, most of those snapshots were taken at roughly the same moment. A fundamentally different situation prevailed in the Yuanqu Basin. According to the early stratigraphic work conducted by Andersson, the Eocene rocks in the Yuanqu Basin spanned at least half a mile (one kilometer) in vertical thickness. No matter how rapidly these rocks were deposited, it must have taken a long time to lay down such a voluminous sequence of sand, mud, limestone, and conglomerate. If we could sample the fossil record across a reasonable fraction of this stratigraphic section, we might be able to track the evolution of *Eosimias* and its relatives through millions of years.

With such high scientific goals in mind, we began our work in the Yuanqu Basin. From the start, the Chinese authorities told us that we could expect to work there for four field seasons—1994 through 1997. After that, the completion of the dam downstream would flood any low-lying sites near the river. Nothing inspires hard work like a hard and fast deadline. Accordingly, we planned an ambitious schedule of activities. Each field season, our team would divide into several smaller groups. Virtually every day, some of us would quarry at Locality 1, while others prospected new or previously established localities. Whenever circumstances demanded it, we would screen-wash sites that could not be exploited by other means. In short, we were prepared to unleash the full arsenal of paleontological collecting techniques on the Eocene strata of the Yuanqu Basin. Each method succeeded in yielding its own unique insights into our deep evolutionary roots.

We made our first two key discoveries by prospecting. Both took place during our 1994 field season in the Yuanqu Basin—a time that I remember best for the fact that virtually each day set new local records for high temperatures. On one particularly miserable day, several of my colleagues and I were prospecting the Eocene strata exposed in a small, ephemeral streambed or arroyo. After spending two-thirds of the day scouting for fossils along the winding path forged by seasonal runoff, the oppressive heat was finally getting to me. I had yet to see the tiniest shard of bone or flake of enamel—not a glimmer of physical evidence that ancient life once inhabited the stark landscape surrounding me. Just as I was about to urge everyone to retreat to the relative comfort of our field vehicles, I heard someone calling my name from around a bend in the narrow drainage. The voice belonged to Alan Tabrum, my preparator and field logistics expert at the Carnegie Museum. From years of collecting fossils with him in China and the Rocky Mountain West, I knew that Alan was immune to unwarranted excitement. Yet from the pitch of his voice, it was obvious that Alan had found something special. I skidded down the steep outcrop I was crawling on and ran up the arroyo to see what Alan had discovered.

To my astonishment, Alan held up a tiny but nearly perfectly preserved fossil jaw. "What do you think it is?" I asked. Alan shrugged and said that he thought it was either a small primate or some type of insectivore. I whisked out my hand lens for a closer look. I could immediately see why Alan was so reticent to commit to a hard and fast taxonomic identification. The left lower jaw held five shiny teeth in place—the last two premolars (P_{3-4}) and all three molars (M_{1-3}). All of the teeth bore sharply defined crests, giving them a vaguely insectivore-like appearance. But the breadth of the lower molar talonids, along with the configuration of the front part of the jaw, made it clear that Alan had found a primate rather than a fossil shrew or hedgehog. In its small size and several details of anatomy, Alan's specimen resembled omomyid primates like *Shoshonius*. Still, I could make out some fundamental differences. Alan's jaw preserved a very large root socket for the canine, while in omomyids this tooth is always much smaller. The anatomy of the premolars and molars in Alan's specimen also differed from omomyids in ways that were oddly familiar. Suddenly, the source of this odd familiarity came to me. The features that, by omomyid standards, seemed peculiar in Alan's specimen were all traits that would be expected in a genuine fossil tarsier. Given that modern tarsiers live nowhere else but southeastern Asia, it made sense that we might find their Eocene relatives in China. Here in my hand was actual evidence that this was indeed the case!

To honor Alan's skill and perseverance, I decided to name his new fossil tarsier *Xanthorhysis tabrumi*.[12] Although it was not the additional ev-

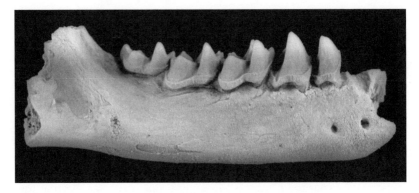

Figure 34. The holotype lower jaw of *Xanthorhysis tabrumi* from the Yuanqu Basin in central China. Photograph courtesy of and copyright by Marc Godinot.

idence of early anthropoids we were seeking, *Xanthorhysis* underscored a point I had been making for years. Ever since my early publications on the cranial anatomy of *Shoshonius*, I had been promoting the existence of a long but poorly documented ghost lineage of early anthropoids. One of the sharpest critiques of this view came from supporters of the tarsier theory of anthropoid origins (see chapters 6 and 8). These experts denied any special evolutionary relationship between *Shoshonius* and modern tarsiers. Accordingly, they doubted the need for an early lineage of anthropoids—especially one that left such an intangible mark on the fossil record. Being a fossil tarsier itself, *Xanthorhysis* altered the established chronology of the tarsier theory of anthropoid origins, leaving little room for disagreement between my view and theirs.

To understand why, let's build on the fork-in-the-road analogy introduced in chapter 6. The relevant context went as follows. You are driving down an unfamiliar road toward Monkeyville, your ultimate destination. Somewhere up ahead, you know that the road forks into a Y-shaped pattern. One branch leads to Tarsiertown, while the other—equal in length to the first—goes to Monkeyville. When you finally reach the critical fork in the road, you are disappointed to see that no sign directs you toward Monkeyville. Instead, a sign points down one of the two forks, bearing the inscription "Fifty miles to Tarsiertown." You conclude that the other fork leads to Monkeyville, and that you must drive an additional fifty miles to get there.

In our initial analogy between the branching sequence of primate evolution and a road trip, *Shoshonius* served as that critical guidepost pointing toward Tarsiertown. Acknowledging *Shoshonius* as a basal twig on the tarsier branch of the family tree allowed us to infer that the anthropoid lineage extends back at least as far as *Shoshonius* (fifty million years), even though fossil anthropoids of that age had never been found. Similar reasoning allowed you to deduce that the fifty-mile stretch of highway heading off in the other direction leads to Monkeyville, even though the road is unmarked. But what happens to our analogy if you were to speed through the fork in road without noticing the helpful sign pointing toward Tarsiertown? This is equivalent to the position adopted by supporters of the tarsier theory of anthropoid origins, who reject the notion that *Shoshonius* is an early relative of tarsiers. In that case, there would be no reason for you to assume that Monkeyville lay fifty miles away. For that matter, you might not even know that you had already zipped past the critical fork in the road between Tarsiertown and Monkeyville. Instead, you might expect this juncture to lie somewhere along the road ahead, with the implication being that the distance between Tarsiertown and Monkeyville is much less than it actually is.

Continuing our analogy shows how *Xanthorhysis* bridges much of the discrepancy between the ghost lineage and tarsier theories of anthropoid origins. Even if you are one of those distracted drivers who ignored the caution sign provided by *Shoshonius,* you might yet reach Monkeyville, especially if additional clues turn up along the way. Under this scenario, *Xanthorhysis* functions as an unmistakable landmark on the great road atlas of primate evolution, because its distinctive anatomy proves that it belongs on the tarsier branch of the family tree, even if the case for *Shoshonius* is not so clear. Driving past the *Xanthorhysis* landmark, you see another helpful sign that reads "Forty miles to Tarsiertown" (*Xanthorhysis* is roughly forty million years old). For the first time, you realize that you have already passed the important Y-shaped intersection, with its separate roads leading to Tarsiertown and Monkeyville. You immediately turn around, knowing that Monkeyville is at least forty miles away.

As the world's oldest reasonably well-preserved fossil tarsier, *Xanthorhysis* requires followers of the tarsier theory of anthropoid origins to admit that anthropoids originated sometime prior to forty million years ago. Accepting *Shoshonius* as an earlier and more primitive member of the tarsier lineage, as my ghost lineage theory of anthropoid origins postulates, pushes the minimum date for anthropoid origins back an additional ten million years.[13] *Xanthorhysis* therefore solidifies the ghost lineage theory of anthropoid origins. But to resurrect the ghost itself, I needed to find additional—and, I hoped, better-preserved—fossils of those elusive early anthropoids.

Slightly off the beaten track, an important clue soon emerged. All of Andersson's Yuanqu Basin sites—and the vast majority of ours—were located north of the Yellow River in southern Shanxi Province. Yet Andersson himself had seen promising outcrops on the south side of the river, in adjacent Henan Province. He urged Zdansky to explore this potential new site, located several miles upstream from his original Locality 1. When Zdansky followed up on Andersson's lead, he found a number of large fossil mammals there, including early relatives of rhinos known as amynodonts. In the sequential numbering scheme set up by Andersson, Zdansky designated the narrow belt of Eocene strata south of the Yellow River as Locality 7.

Being on the opposite side of the Yellow River, Locality 7 posed its own logistical problems for us, although the difficulties were hardly insurmountable. Along this particular stretch of the Yellow River, its waters flow serenely downstream, and boats can easily navigate it. However, steady river traffic never developed here because of the rugged terrain along the river's southern bank. For us to reach Locality 7, we had to persuade one of the operators of a nearby commercial ferry service to move his enterprise several miles upstream. For a surprisingly reasonable price, the captain of a small dinghy agreed to meet us at a village known as Heti the following day. Immediately across the river from Heti village lay the multicolored Eocene strata that form Locality 7.

We reached Heti at the appointed time the next morning, but the boat and its captain were nowhere in sight. With nothing better to do, we hung out on the sandy shoreline and watched the villagers go about their daily chores. Small boys fetched water from the river in buckets dangling from each end of a wooden pole carried over one shoulder. Women washed their laundry on large rock cobbles at the water's edge. A hundred yards upstream, several children skinny-dipped in the cool morning mist. After an hour or so, we could finally make out the distant hum of a boat's engine. Our water taxi was making its way upriver, but its progress was slow because the boat's tiny motor was barely capable of overcoming the opposing current.

Once our means of transportation arrived, it took no time at all to reach the enticing Eocene rocks on the opposite side of the river. As we approached the river's southern bank, I was surprised to see that the Eocene outcrops there were far more extensive than they had appeared from across the river. Even with six veteran fossil collectors, Locality 7 was too large to survey in one day. We decided to split into two groups. Two of our Chinese colleagues from the IVPP, Tong Yongsheng and Wang Jingwen, wanted to show Mary Dawson a place upstream where they had screen-washed some sediment to recover several dozen teeth of small Eocene mammals, mostly rodents. Huang Xueshi—Tong and Wang's colleague at the IVPP—stayed behind with Leonard Krishtalka and me to prospect some of the more promising beds along the riverbank. Krishtalka—whose friends call him Kris—and I had worked together for two field seasons in the Wind River Basin. Our main task there was to excavate the Buck Spring Quarries, but we both preferred to prospect for fossils among the vast Wind River badlands whenever we got the chance.

To be proficient at prospecting, you must first master the fundamentals of the technique. The general idea is to scour the surface of a rock outcrop, using sunlight to illuminate the ground before you. Instead of looking for particular shapes or colors, you are actually searching for subtle changes in texture. Fossilized bone and tooth enamel will glisten in bright sunlight, whereas most rocks will not. I've taught many a student the basic principles of prospecting, and within a week or so they can usually hold their own in the field. But nothing beats experience in separating proficiency from true excellence. Kris has as much experience as anyone when it comes to prospecting for small Eocene mammal fossils. I therefore wasn't surprised when I heard him suddenly yelp for joy as he sat crouched on a small ledge of rock, just a short distance from where I stood.

I needed my hand lens to appreciate the specimen that Kris held up so triumphantly. Kris had found a piece of upper jaw that still held three teeth in place—the last premolar (P^4) and the first two molars (M^{1-2}). The teeth were unmistakably those of a primate. Each molar bore three main cusps that enclosed a relatively large central depression or trigon. A complete shelf of enamel known as a cingulum marked the internal (or lingual) border of each molar. Other details of the molar crown pattern replicated features found in primitive anthropoids from North Africa.[14] Despite the presence of these advanced or anthropoidlike traits, Kris's partial maxilla struck me as that of an extremely primitive fossil primate. It had to belong to some close evolutionary cousin of *Eosimias*, although the specimen was too large to be that of *Eosimias* itself. Later, when our team uncovered the first upper jaw of *Eosimias* at Locality 1, additional differences emerged between it and Kris's new specimen from Locality 7. Because the anatomical divergence was too great to be attributed to different species in the same genus, we named Kris's specimen

Phenacopithecus krishtalkai.[15] We grouped *Phenacopithecus* and *Eosimias* together in the family Eosimiidae, in the same way that *Australopithecus* and *Homo* are distinct, but closely related, genera in the family Hominidae.

When paleontologists find fragmentary jaws of early primates and other mammals, they tend to fixate on the teeth. In part, this is because mammalian teeth are so complicated that most species can be identified from this part of their anatomy alone. As the most durable parts of the mammalian skeleton, teeth also get disproportionately preserved in the fossil record. But as I examined Kris's *Phenacopithecus* jaw from various perspectives, what impressed me most was not its teeth but the structure of its lateral surface. There, where the bone turns vertically away from the tooth row to form the lower part of the face, I saw no hint of the lower rim of the eye socket. If the upper jaw of *Phenacopithecus* had resembled that of an omomyid or tarsier, its greatly enlarged eye sockets would have impinged on the upper tooth row itself. Kris and I had each found enough upper jaws of omomyids in the Wind River Basin and elsewhere to recognize this highly distinctive type of facial anatomy at a glance. The upper jaw of *Phenacopithecus* diverged from this tarsierlike pattern, preserving a significant depth of bone above the upper tooth row. Naturally, the specimen was broken before it reached the lower rim of the eye socket. But the mere fact that so much bone intervened between the upper teeth and the bottom of the eye socket meant that this particular fossil primate could not have possessed hypertrophied eyeballs like those of *Shoshonius* or a tarsier. Instead, it must have had relatively small—and therefore monkeylike—eyes.

Further inspection of the *Phenacopithecus* maxilla turned up another monkeylike feature. About halfway between the upper tooth row and the broken margin of the fossil that approximates the eye socket, I detected a small hole known as the infraorbital foramen. In living primates, the infraorbital foramen serves as a major conduit for nerves and blood vessels supplying the upper lip and nose. Its size varies among primates, depending on the sensory requirements of different species. Living lemurs have extremely sensitive noses and upper lips that bear facial vibrissae (or catlike "whiskers"). As a result, the nose and upper lips of lemurs require rich innervation and must be nourished by a steady and abundant blood supply. Judging by their large infraorbital foramina, fossil adapiforms probably had sensitive lemurlike noses as well. As part of their trend toward relying more heavily on vision than touch and smell, anthropoids have partly desensitized their noses and upper lips.

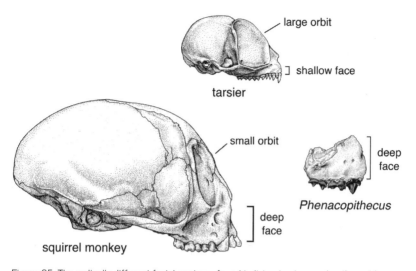

large orbit

shallow face

tarsier

small orbit

deep face

Phenacopithecus

deep face

squirrel monkey

Figure 35. The radically different facial anatomy found in living tarsiers and anthropoids can often be recognized in fragmentary fossils. In this case, the holotype specimen of *Phenacopithecus krishtalkai* possesses a relatively deep face, indicating that this eosimiid species had relatively small (or anthropoidlike) orbits. Original art by Mark Klingler, copyright Carnegie Museum of Natural History.

This region in anthropoids does not require the same level of innervation and blood supply that occurs in lemurs, and anthropoids have relatively smaller infraorbital foramina as a result. In *Phenacopithecus*, the diameter of the infraorbital foramen spans less than one millimeter, about the size of the tip of a standard ink pen. The small size of the infraorbital foramen in *Phenacopithecus* therefore indicates that it also had a monkeylike nose.

I certainly welcomed the new anatomical evidence provided by Kris's upper jaw, but the specimen was equally important for documenting taxonomic diversity among early Chinese anthropoids. In fact, the more we worked in the Yuanqu Basin, the more it became apparent that *Eosimias* was merely the tip of the iceberg of early anthropoid diversity in China. Virtually every fossiliferous locality presented us with a new species of anthropoid. Given the long timespan represented by the Eocene strata in the Yuanqu Basin, we decided to see whether we could find early anthropoids throughout the local stratigraphic section. This would give us some indication of how long *Eosimias* and its relatives had been evolving in central China. It might also give us a better idea of the abundance and diversity of these early anthropoids in the local ecosystem.

We began our quest to push the temporal envelope by searching for younger Eocene localities in the Yuanqu Basin. Based on the physical disposition (technically known as the strike and dip) of the local Eocene strata, we concentrated our search in the eastern part of the basin. Near the village of Nanbaotou, we located a small Eocene outcrop exposed in a dry streambed. We could see tiny bone fragments and snail shells on the surface of the outcrop, so we decided to screen-wash the mudstone sediment to see whether we might retrieve small mammal teeth and jaws. Unfortunately, this meant shoveling hundreds of pounds of mudstone into heavy burlap sacks. Each sack then had to be manually transported to the nearest primitive road, where they were loaded into our field vehicles for the long, bumpy ride to the banks of the Yellow River. The hard work of shoveling and hauling the fossiliferous sediment from Nanbaotou alone took several days. It took even more time to screen-wash the raw matrix in the Yellow River. Huang Xueshi, who had a great deal of experience screen-washing sediment from various fossil sites around China, led the effort to reduce the large volume of Nanbaotou sediment to a more manageable amount of screen-washed concentrate. In the evenings, we picked and sorted through the concentrate under a microscope at our base of operations in the town of Yuanqu.

The fossils that emerged quickly confirmed our suspicion about the relatively young geological age of the Nanbaotou site. Most of the specimens we recovered belonged to small rodents. These were obviously advanced compared to the rodents we had found at other sites in the Yuanqu Basin. The same pattern held for the fossil primate teeth we found there. All of the primate specimens from Nanbaotou were appropriate in size and anatomy to belong to the same species. We found examples of most of the lower teeth (including all of the lower premolars and molars) along with a few examples of upper molars. Like *Phenacopithecus krishtalkai* from Locality 7, the Nanbaotou primate was obviously related to *Eosimias*, although it was much larger than *Eosimias sinensis* from the Shanghuang fissure-fillings. The Nanbaotou species also differed from *Eosimias* in having smaller paraconid cusps on its lower molars and better crest development on its lower premolars. On the other hand, the poorly preserved upper molars we found at Nanbaotou resembled those of *Phenacopithecus*. Although the evidence was thin, it appeared to me that the two new primates from Locality 7 and Nanbaotou were closely related to each other, but more distantly related to *Eosimias*. We therefore named the new Nanbaotou primate *Phenacopithecus xueshii*, in recognition of Huang Xueshi's leadership and hard work in screen-washing the sediment from that fossil site. Because Locality 7 occurs low in the local stratigraphic section, while Nanbaotou is near the top, *Phenacopithecus* must have inhabited the Yuanqu Basin during most of the time that Eocene rocks were being deposited there.

At the same time that we were prospecting and screen-washing various sites in the Yuanqu Basin, most of our time and energy was devoted to quarrying at Andersson's classic Locality 1. Our work there was tedious. At times, it could even be dangerous. In order to access the fossil-bearing strata, we first had to remove the annoying overburden that had so effectively sealed the natural time capsule for the past forty million years or so. Because the entire locality was restricted to the bottom of a deep gully, we also had to contend with the severe constraints imposed by local topography. It was futile to attempt to widen the gully on either side—we simply could not remove the thick beds of sandstone and loess that formed the steep walls of the gully. Instead, we focused our efforts along the length of the gully, where it was merely necessary to strip away the last few feet of rock capping the fossil-bearing strata. Unfortunately, the rock layer immediately above our fossil-bearing unit had the hardness and consistency of cement. With an army of manual laborers, we could have removed it by hand. But with the limited personnel at our disposal we had to resort to other means. We settled on dynamite.

Our first efforts to blast away the pesky overburden at Locality 1 proved to be remarkably effective. Roughly a week before each field season began, a couple of our Chinese colleagues would leave Beijing for the Yuanqu Basin to supervise the demolition in the bottom of the gully. By the time we arrived on the scene, all that remained to be done was to peel back layer after layer of multihued mudstone and limestone, each of which held its own quota of precious fossils. Unfortunately, our luck eventually ran out. When we got off the train to begin the 1996 field season, I noticed that Wang Jingwen—who had left Beijing a few days in advance of us—was nowhere to be found. I asked Huang Xueshi what had happened. Huang reluctantly informed me that Wang Jingwen was being treated in a local hospital. It seems that someone had miscalculated how much dynamite was required to blast away the unwanted overburden that year. The resulting explosion sent huge missiles of rock skyward, which then rained down on our colleagues below. Miraculously, Wang Jingwen was the sole casualty. He suffered a broken arm caused by the flying debris. A few days later when Wang was released from the hospital, I expected him to be upset over this mishap. Instead, Wang smiled at me and beamed: "I'm very lucky! If I had been standing six inches away, the rock would have hit my head instead of my arm. In that case I would be dead." I've never witnessed a more optimistic spin on the old proverb about the glass being half full, as opposed to half empty.

Fortunately, Wang's personal sacrifice was not in vain. Each year our work at Locality 1 took advantage of the newly exposed fossil-bearing strata, with outstanding results. Andersson's pioneering efforts at Locality 1 had already established that fossils of the primate *Hoanghonius* could be unearthed there, but the specimens he collected were so fragmentary that no consensus had emerged as to what type of primate *Hoanghonius* might be. Many scientists regarded *Hoanghonius* as a more likely candidate for anthropoid status than *Eosimias* and its relatives could ever be. We were therefore keen to locate additional specimens of *Hoanghonius,* to see how it compared with *Eosimias* and whether it might alter our views of how early anthropoids evolved.

Our persistence at Locality 1 paid off in the form of many additional examples of *Hoanghonius,* some of them superbly preserved. Our new *Hoanghonius* fossils resolved the long-standing controversy over its potential role in anthropoid origins, by revealing substantial additional information about its anatomy. Several of the better-preserved specimens documented the structure of the front part of the lower jaw in *Hoanghonius* for the first time. Among other things, these fossils show that in *Hoanghonius,* the symphyseal region (equivalent to the human chin) protrudes strongly forward. In this key respect, *Hoanghonius* resembles lemurs, adapiforms, and many omomyids. In anthropoids, the same region is oriented much more vertically. This accounts for the upright disposition of the lower incisors in anthropoids, while in *Hoanghonius* and other primitive primates these teeth angle forward.

In addition to the very primitive nature of the front part of its lower jaw, the lower premolars of *Hoanghonius* contrast markedly with those of *Eosimias* and other early anthropoids. Perhaps the most important distinction lies in the number of roots supporting the front premolar, which in both *Hoanghonius* and *Eosimias* is P_2 (both species lack P_1, a tooth that is retained in the most primitive adapiforms and omomyids). In *Hoanghonius,* P_2 possesses two roots. With two minor exceptions, the only fossil primates that retain this archaic condition are certain adapiforms.[16] A different condition prevails in the large group of primates that includes omomyids, microchoerids, tarsiers, and anthropoids. In order to accommodate the reduced length of their snouts, these animals were forced to compress their lower premolars from front to back. By fusing

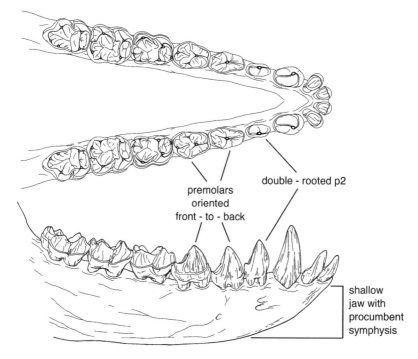

Figure 36. The complete lower dentition of *Hoanghonius stehlini*, based on specimens collected by our expeditions to the Yuanqu Basin. These relatively complete fossils show that *Hoanghonius* cannot be closely related to *Oligopithecus* and *Catopithecus* from the Fayum. Original art by Mark Klingler, copyright Carnegie Museum of Natural History.

the dual roots of P_2 into a single root, the earliest members of this large subdivision of the primate family tree were able to achieve the necessary degree of compaction without sacrificing the lower premolar entirely. The same emphasis on saving space distinguishes the remaining premolars (P_{3-4}) of *Eosimias* from their counterparts in *Hoanghonius*. In *Eosimias*, these teeth are obliquely oriented in the jaw, which is another way to compress the lower premolars from front to back. In contrast, not only are P_{3-4} aligned with the long axis of the lower jaw in *Hoanghonius*, but their crowns are actually elongated in this dimension. All of these aspects of lower premolar anatomy suggest that *Hoanghonius* retained a fairly long, lemurlike snout. Certainly, it lacked the abbreviated muzzle that occurs in such early anthropoids as *Catopithecus*.

If our new evidence from Locality 1 showed that *Hoanghonius* could not be an anthropoid, we still needed to establish exactly where it fits in terms of primate evolution. The double-rooted P_2 of *Hoanghonius* ruled out the possibility that it could be an unusual type of omomyid. This left only two options for placing *Hoanghonius* on the family tree—either it could be an adapiform, or it could belong to some other, equally primitive lineage of early primates. Two lines of evidence rapidly convinced me that *Hoanghonius* is an adapiform. The first of these relates to the unusual anatomy of the rear half (or talonid) of the lower molars in *Hoanghonius*. Unlike most adapiforms, two closely spaced cusps (known as the entoconid and hypoconulid) occur on the internal side of the lower molar talonids in *Hoanghonius*. Ironically, this so-called twinning between the entoconid and hypoconulid cusps is one of the primary reasons why experts like Gingerich had previously argued that *Hoanghonius* is related to early anthropoids, because a very similar condition occurs in Fayum anthropoids such as *Oligopithecus* and *Catopithecus*. However, the same distinctive molar pattern characterizes one of the four major adapiform groups, the Asian sivaladapids. In southern and southeastern Asia, sivaladapids survived much longer than any of their adapiform relatives on other continents (see chapter 2). Instead of revealing the anthropoid affinities of *Hoanghonius*, its unusual molar anatomy actually exposed it as an early member of the sivaladapid lineage.[17] Like so many other fossils that had been nominated as missing links between adapiforms and anthropoids, the discovery of reasonably complete specimens of *Hoanghonius* showed that it could no longer play such a critical evolutionary role.

While our work at Locality 1 succeeded in pruning *Hoanghonius* from the anthropoid family tree, our discovery of a second fossil primate there soon grafted a different species in its place. This second primate—whose body mass of roughly four and a half ounces (130 grams) made it no more than a fifth of the size of *Hoanghonius*—immediately became the focus of our entire expedition. From the very first time I encountered its fossilized remains, its teeth betrayed its identity. Here, along the banks of the Yellow River, we had found exactly what had led us to return to China in the first place—additional fossils of *Eosimias*. The new specimens from Locality 1 departed in a few minor ways from the Shanghuang fossils we had named *Eosimias sinensis*. But these differences were so subtle that they could indicate nothing more than a species-level distinction. Within a few months we named our new species *Eosimias centennicus*, to commemorate the centennial of the Carnegie Museum of Natural History.[18] Along with the two new species of *Phenacopithecus* we had found at Locality 7 and Nanbaotou, our discovery of *Eosimias centennicus* at Locality 1 reiterated that we were sampling a diverse radiation of early Chinese anthropoids. It also revealed important new details about the anatomy and lifestyle of these earliest Asian anthropoids. By far the most important fossils of *Eosimias centennicus* we unearthed at Locality 1 were the delicately preserved lower jaws found by Wen Chaohua, a local villager we had hired as a manual laborer, in 1995 (see chapter 1). Prior to Wen's pivotal discovery, the best material of *Eosimias* consisted of the fragmentary lower jaws of *Eosimias sinensis* from Shanghuang, which had provoked such a controversy at the "Anthropoid Origins" conference in 1992 (see chapter 8). Much of the rancor generated by the Shanghuang specimens could be blamed on their poor preservation. Many of the anthropoidlike features that I had originally attributed to *Eosimias* were based on the size and configuration of empty root sockets rather than the crowns of the teeth that occupied them. In contrast, Wen's exquisite new jaws of *Eosimias centennicus* held every tooth in place. Happily, these anatomically superior specimens confirmed every prediction I had made on the basis of the fragmentary fossils from Shanghuang.

On each side, the lower tooth row of *Eosimias centennicus* consisted of two small upright incisors, succeeded by a stout daggerlike canine, followed in turn by three premolars and three molars. Just as the empty root sockets in the Shanghuang specimens of *Eosimias* had implied, the first premolar (P_2) was small and single-rooted, while the dual roots of P_3 were rotated to match the oblique orientation of P_4. The lower molars in Wen's fossil corresponded closely with those in Qi Tao's jaws from Shanghuang. Yet even in the molars, Wen's star specimen yielded significant new information. For the first time, it showed that the back part of M_3 is highly abbreviated in *Eosimias,* as it is in other early anthropoids. Even the anatomy of the bony part of the jaw itself resembled that of anthropoids. For such a small primate, the lower jaw of *Eosimias* was surprisingly deep—much deeper than that of a similar-sized omomyid or tarsier, for example. Wen's discovery catapulted *Eosimias* from being an interesting yet highly controversial fossil primate to being the earliest anthropoid documented by such complete anatomical material. Rarely has a single fossil had a more decisive impact on the overall picture of primate evolution.

Additional specimens from Locality 1 helped flesh out the anatomy of *Eosimias* even more. In contrast to the Shanghuang fissure-fillings, where we had found a number of small primates in the size range of *Eosimias,* our work at Locality 1 turned up only two primate species—*Hoanghonius stehlini* and *Eosimias centennicus.* Since *Hoanghonius* and

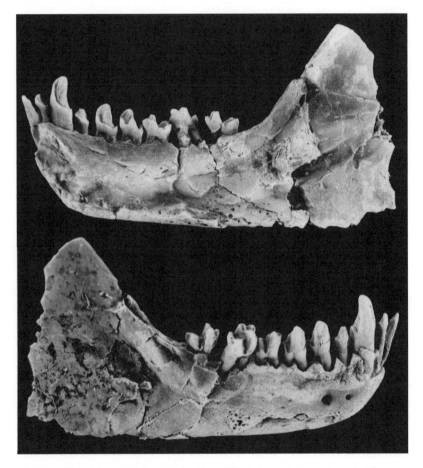

Figure 37. Part of the complete lower dentition of *Eosimias centennicus* found by Wen Chaohua at Locality 1 in the Yuanqu Basin in 1995.

Eosimias differ in size by a factor of five, we could easily allocate isolated teeth and bones to one or the other species. Large primate specimens from Locality 1 belonged to *Hoanghonius,* while small primate fossils from the site had to represent *Eosimias.* This allowed us to identify an upper canine, an upper jaw fragment, and a talus (one of the major ankle bones) as being those of *Eosimias.* At Shanghuang, the same body parts would have been difficult to allocate to *Eosimias,* simply because there were so many other small fossil primate species there to choose from.

The upper jaw fragment and the isolated canine of *Eosimias* from Locality 1 confirmed what I had suspected on the basis of Krishtalka's *Phe-*

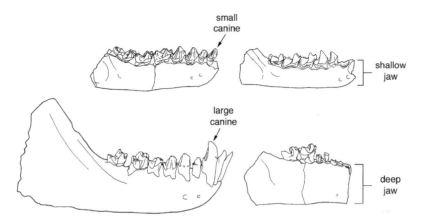

Figure 38. Eosimiid lower jaws (bottom) can easily be distinguished from those of omomyids and tarsiers (top) by their greater depth and their large canines. Original art by Mark Klingler, copyright Carnegie Museum of Natural History.

nacopithecus maxilla from Locality 7. The *Phenacopithecus* specimen preserved enough bone on the lateral side of the face to show that it must have had a small—and therefore monkeylike—eye socket. The upper canine of *Eosimias* from Locality 1 supported a very similar anatomical reconstruction. Its long and voluminous root must have been deeply anchored within the maxilla. If *Eosimias* had huge eye sockets like those of tarsiers or *Shoshonius*, there simply would have been insufficient space between the upper tooth row and the eye to accommodate a canine root with such impressive dimensions. At the same time, detailed comparisons between the teeth preserved in the maxilla of *Phenacopithecus* from Locality 7 and the upper jaw fragment of *Eosimias centennicus* from Locality 1 showed that the two species had to belong to separate but closely related genera.

If the new evidence of the monkeylike facial anatomy of *Eosimias* wasn't reason enough to celebrate, the first information on its postcranial anatomy certainly sufficed. Initially, the evidence consisted of a single ankle bone—a right talus—which was discovered at Locality 1 by my former student Jay Norejko. The talus articulates directly with the two long bones of the lower leg, the tibia and fibula. Together, these three bones form what most people regard as the ankle joint, although the complete anatomical picture is substantially more complicated. In terms of movement, the joint between the talus and the lower leg bones is primarily concerned with simply flexing and extending the foot—that is, increasing and decreasing the angle between the long axis of the foot and that of the lower leg.

Moving from the knee toward the toes, the next ankle bone we encounter is the calcaneus, which in humans forms the bony heel. The talus and calcaneus articulate with one another in a complex fashion, yielding what anatomists refer to as the "lower ankle joint." In primates, motion at this joint allows the foot to assume a variety of postures, including one in which the soles of the feet face inward rather than down. Because most primates utilize a wide range of foot postures as they travel through their arboreal realm, few bones better reflect how primates move than the talus and calcaneus. Unfortunately, we never uncovered a calcaneus of *Eosimias* at Locality 1. For that, we had to turn back to the large unsorted assemblage of bones from the Shanghuang fissure-fillings.

As part of my preliminary work on the Shanghuang fossils in Qi Tao's office in Beijing, I made a first pass at sorting through thousands of miscellaneous specimens searching for primate jaws, teeth, and postcranial elements. I found a reasonable number of primate skeletal elements this way, including a few tali and calcanea. But it rapidly became clear to me that to identify and study all of the fossil primate specimens from Shanghuang was more than any one person could accomplish. I therefore enlisted my friend and colleague Daniel Gebo, one of the world's leading experts on the postcranial anatomy of fossil primates, to spearhead the research on primate limb and ankle bones from Shanghuang. Dan is a veteran of numerous paleoanthropological expeditions to Wyoming, Egypt, Uganda, and other locales. Whenever his academic schedule allowed it, Dan joined me for fieldwork in the Yuanqu Basin and other parts of China. While Dan's experience and skill as a fossil collector contributed in many ways to our fieldwork, he had to defer advancing his own scientific agenda until the end of each field season, when he got to rummage through the unsorted Shanghuang collections back in Beijing. Dan spent days combing through Qi Tao's ever-growing sample of bones from the Shanghuang fissure-fillings. His efforts soon paid off handsomely.

One day in March 1995, while Dan and I were both sorting through new Shanghuang material at the IVPP, Dan rushed into the office where I was working, flushed with excitement. "I didn't know whether to believe your arguments about *Eosimias* being an anthropoid until now," he said. Being an expert on postcranial anatomy, Dan had always remained noncommittal about evolutionary reconstructions that were based entirely on the anatomy of jaws and teeth. "So what has finally changed your mind?" I asked. Dan held out a small primate calcaneus, which he had just segregated from a pile of unidentified bones and teeth from Shanghuang. "This obviously belongs to a small anthropoid," he said. "It's remarkably similar to the calcaneus of a South American squirrel monkey. If this bone belongs to *Eosimias,* it seals your case."

To establish whether the monkeylike calcaneus belonged to *Eosimias,* we had to forge a logical connection between the calcaneus and our available sample of *Eosimias* fossils—the vast majority of which were teeth and jaws. On the basis of its compatible size, it was certainly plausible that the monkeylike calcaneus belonged to *Eosimias.* But a more definitive determination required more compelling evidence. Fortunately, our discovery of the talus of *Eosimias centennicus* at Locality 1 provided a possible way forward. Using that specimen as a guide, Dan located additional examples of this bone from the Shanghuang fissure-fillings. Aside from being slightly smaller than the *Eosimias* talus from Locality 1, the specimens from Shanghuang were identical to it. This difference in size corresponded precisely with what we expected on the basis of jaws and teeth—in other words, postcranial remains of *Eosimias sinensis* from Shanghuang should have been slightly smaller than their counterparts in *Eosimias centennicus* from the Yuanqu Basin. When we articulated the Shanghuang tali of *Eosimias* with Dan's new calcaneus, several of them fit together perfectly. This close anatomical correspondence confirmed that *Eosimias* was the rightful owner of the monkeylike calcaneus. The only way that we could ever make a more convincing case would be to find a complete articulated *Eosimias* skeleton, with every bone in place.

Armed with the two most critical ankle bones of *Eosimias,* we could finally begin to reconstruct how this animal moved through the middle Eocene forests of central and eastern China. Our most important evidence came from the calcaneus. Most small prosimian primates overcome the difficulty of traversing gaps in the forest canopy by leaping. To accomplish this, they leverage their elongated feet to propel themselves many times the length of their own bodies. In prosimians, most of the elongation of the foot resides in the calcaneus, which is long and slender in species such as tarsiers, bushbabies, and most omomyids. Anthropoids leap less frequently, and their calcanea are shorter and wider as a result. The calcaneus of *Eosimias* shows only a modest amount of elongation, similar to that in many South American monkeys. From this we concluded that *Eosimias* was not a specialized leaper. Other details of its ankle anatomy suggest that *Eosimias* preferred a monkeylike foot posture, in which the soles of the feet face downward rather than internally.[19] Small

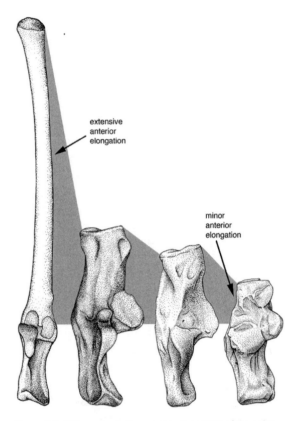

Figure 39. Differences in the anterior elongation of the calcaneus (or heel bone) reflect how various species of primates prefer to move. In tarsiers (left) the calcaneus is greatly elongated, an adaptation for leaping in this small primate. In contrast, baboons (right) show very little calcaneal elongation. Intermediate conditions exist in *Shoshonius* (second from left) and in eosimiids (second from right). Original art by Mark Klingler, copyright Carnegie Museum of Natural History.

prosimians, which actively grasp vertical supports more often than anthropoids do, favor a habitual foot posture in which the feet are inverted. Despite its small size, *Eosimias* apparently moved through the trees primarily by climbing and walking on all fours along the tops of branches. This very monkeylike mode of locomotion contrasts with the typical prosimian pattern, which emphasizes leaping and clinging over quadrupedal walking and running.

Looking back on those four seasons of intensive fieldwork in the

Yuanqu Basin, I feel deeply gratified. Our knowledge of *Eosimias*—an animal that I had only recently ushered onto the scientific stage—had improved rapidly and immensely. *Eosimias* had been introduced to the paleoanthropological community as a humble waif of a fossil whose claim to anthropoid status dangled by the thread of two scrappy jaws. Now, its place near the base of the great anthropoid branch of the primate family tree rested on a firm anatomical foundation. Complete lower jaws of *Eosimias* reveal that it possessed a primitive yet unmistakably anthropoidlike dentition. Evidence from its upper canine and maxilla show that *Eosimias* had a monkeylike face with relatively small eye sockets. Its ankle anatomy indicates that *Eosimias* preferred to amble along the tops of branches in a distinctly monkeylike fashion. No other fossil bearing on the very root of the anthropoid family tree can marshal such an extensive litany of anatomical features to support its pivotal evolutionary position.

But simply solidifying the anthropoid credentials of *Eosimias* no longer sufficed. *Eosimias* was christened in the midst of an acrimonious debate about how, when, and where our distant anthropoid ancestors evolved. What may have begun as a quixotic attempt to throw a gauntlet down at the feet of the paleoanthropological establishment now demanded resolution. I owed that much to my Chinese and American colleagues. I also felt indebted, historically and intellectually, to Johan Gunnar Andersson, whose pioneering work at Locality 1 had come so tantalizingly close to unveiling *Eosimias* decades before I was born. I couldn't help thinking how differently the search for anthropoid origins would have evolved if Andersson, rather than my colleagues and I, had been the first to discover *Eosimias*. Now it was time to complete the scientific upheaval that Andersson had set in motion so many years ago.

10
Into the African Melting Pot

According to the renowned philosopher of science Thomas Kuhn, science does not progress by marching slowly and steadily toward enlightenment. Rather, scientific advances occur in fits and starts. Most of the time, scientists go about their business in workmanlike fashion. They labor to reconcile assorted classes of data and observations with the organizing principles that dictate how their field of science operates. Occasionally, these long interludes of "normal science" are punctuated by dramatic intellectual transformations, which Kuhn called "paradigm shifts." Whenever one of these paradigm shifts occurs, the old way of thinking gets discarded, often over the objections of scientists who have worked long and hard to validate it.[1]

Measured against the textbook examples of scientific paradigm shifts—plate tectonics versus a stable Earth in geology, Einstein's relativity as opposed to Newtonian physics, and so forth—the impact of *Eosimias* on paleoanthropology hardly seems to qualify. Yet the discovery of *Eosimias* has clearly overturned what was previously accepted as conventional wisdom in the field. Earlier conceptions of when, where, and how our earliest anthropoid ancestors evolved can no longer be sustained. In their place, a new account of anthropoid origins is emerging, even if certain details remain to be worked out. Many of the most pressing questions that surround anthropoid origins today would not even have been entertained a few years ago, because our theoretical outlook is now so fundamentally different. To illustrate how our perspective has changed, let's briefly review the earlier consensus.

Previous models of anthropoid origins posited that the first anthropoids evolved sometime near the Eocene-Oligocene boundary, roughly thirty-four million years ago. A number of observations and widely accepted interpretations supported this view. For example, Le Gros Clark's ladderlike evolutionary progression from tree shrew to lemur to tarsier to monkey to ape to human implied that, because anthropoids are "more advanced" than their prosimian relatives, they must have taken longer to evolve. At least in general terms, the fossil record agreed. In the same way that the Mesozoic was the age of dinosaurs, the Eocene could be construed as the age of prosimians. By the succeeding Oligocene, the vast majority of these Eocene prosimians were extinct, having been supplanted by early anthropoids like those from the Fayum region of Egypt.

For many paleoanthropologists, what seemed like an objective and impartial chronology of anthropoid origins actually carried with it much broader implications. For them, the fact that anthropoids survived the deteriorating climatic conditions that eradicated so many prosimians across the Eocene-Oligocene boundary was hardly a coincidence. Rather, it reflected the inherent biological superiority of early anthropoids. Anthropoids were bigger and brainier than their prosimian ancestors, and their skeletons boasted such new and improved features as complete bony eye sockets and a lower jaw that was fused at the midline. The relatively late appearance of anthropoids and their subsequent evolutionary success were therefore correlated with the deeply entrenched notion of progress in evolution. And since humans stood atop the highest rung on the ladder of evolutionary progress, this theoretical vantage placed the earliest anthropoids on that same evolutionary trajectory—only a few rungs lower down. If the genesis of humans was the climax of the grand evolutionary saga, then the origin of anthropoids certainly qualified as a particularly decisive prelude.

Geography reinforced the theoretical linkage of anthropoid and human origins on what seemed like a long evolutionary march toward biological progress. Because their earliest fossil representatives appeared to be confined to Africa, it had become almost axiomatic that both anthropoids and hominids originated on that continent. In the case of hominids, australopithecine fossils had been recovered from a broad region of sub-Saharan Africa, stretching from South Africa to Ethiopia. Although early anthropoid fossils came mainly from the narrow confines of the Fayum region of Egypt, a surprisingly continuous sequence of fossil monkeys and apes seemed to link these Fayum anthropoids with their hominid brethren much later in time. Anthropoid origins as a whole thus conformed neatly with Le Gros Clark's grand evolutionary progression from tree shrew to human; anthropoids had evolved in the proper chronological sequence, about midway between the first prosimians and the first apes, with more advanced anatomy than their prosimian forebears, and in the right place to give rise to apes and humans.

Eosimias and related lines of evidence cannot be reconciled with this antiquated and rather naïve vision of anthropoid origins. At least in part, this explains why adherents of earlier models of anthropoid origins have objected so vigorously to the novel way of thinking about our deep evolutionary history that the new discoveries require. In terms of simple chronology, fossils such as *Eosimias, Shoshonius, Xanthorhysis,* and *Phenacopithecus* show that anthropoids can no longer be regarded as late arrivals on the stage of primate evolution. Instead, the anthropoid lineage was established during the earliest phases of primate diversification, when the major categories of living primates first came into being. According to this new evolutionary timetable, the anthropoid lineage originated sometime near the Paleocene-Eocene boundary, roughly fifty-five million years ago. It should come as no surprise that, by adding so much time to the beginning of anthropoid history, interesting new possibilities emerge concerning other aspects of anthropoid origins. To begin, let's look at current evidence regarding where anthropoids first evolved.

For years, the fossil record has sufficed to rule out most of Earth's seven continents as potential cradles for anthropoid origins. Aside from our immediate hominid forebears—who recently spread over most of the globe—fossil anthropoids have never been found in North America, Australia, and Antarctica. As a result, no serious scientist would argue that anthropoids originated on any of these three continents. The case against anthropoids arising in either South America or Europe is nearly as persuasive. The earliest fossil anthropoid known from Europe is *Pliopithecus,* which dates back no farther than the early Miocene (about seventeen million years ago).[2] Tellingly, not a single anthropoid occurs among the abundant Eocene primate fossils known from France, Germany, Switzerland, and other parts of Europe. Similarly, although South America harbors a significant fraction of living anthropoid species, the fossil record reveals that anthropoids appeared abruptly there, apparently as immigrants from somewhere else. The earliest South American monkey, *Branisella boliviana,* dates to the late Oligocene (about twenty-six million years ago)—almost ten million years after the earliest Fayum anthropoids.[3] This leaves Africa and Asia as the only remaining contenders to be the ancestral homeland of anthropoids. To evaluate which of these landmasses is more likely to have served as the anthropoid birthplace, we must first review some of what is known about the geology and paleontology of each continent.

Geologically, Africa and Asia parted company in the Jurassic (about 160 million years ago), when the former supercontinent known as Pangea split into northern and southern components, each of which was massive by modern standards. Africa became a central element of the southern landmass, known as Gondwana. Other Gondwanan constituents included South America, Antarctica, Australia, and the Indian subcontinent. Asia contributed its vast surface area to the northern landmass, Laurasia (which also included Europe and North America). For most of the succeeding hundred million years or so, Laurasia and Gondwana experienced separate geological trajectories. Laurasia maintained its core geographic integrity, even as the forces of continental drift caused it to tilt so that its western extremity—which is now North America—moved northward while Asia dipped toward the south. Gondwana, on the other hand, fragmented into progressively smaller units with the passage of time. Geographic isolation spurred the evolution of very different types of plants and animals on the Laurasian and Gondwanan landmasses, although limited biotic interchange must have occurred sporadically. By the time anthropoids diverged from their nearest primate relatives during the early part of the Cenozoic, very different groups of mammals inhabited Africa and Asia. Recent advances in paleontology and molecular biology have gone far toward delineating these distinctions.

Despite the teeming diversity of mammals that thrives in the jungles and savannas of Africa today, only a few mammalian orders can reliably trace their evolutionary origins to that continent. Some of these native African mammals—such as elephants—are familiar to anyone who has ever visited a zoo or attended a circus performance. Most are well known only to zoologists and conservation biologists. These less recognizable groups of native African mammals include such oddities as aardvarks, hyraxes, elephant shrews, tenrecs, golden moles, dugongs, and manatees.[4] In a crude but effective way, this measures the ecological insignificance of native African mammal groups—even in modern African ecosystems.

Elsewhere, these native African mammals have made even fewer inroads. Elephants and their extinct relatives (mammoths, mastodons, and such) once roamed over large parts of the globe, and one species of elephant continues to inhabit forests of southern and southeastern Asia today. Similarly, the geographic range of one species of hyrax extends slightly beyond northeastern Africa to include adjacent parts of Israel, Jordan, Lebanon, and the Arabian Peninsula. Otherwise, only the aquatic dugongs and manatees occur naturally outside of Africa.

Most tourists on safari would probably be astonished to learn that the dominant groups of mammals they observe on Africa's vast savannas actually evolved somewhere else. Yet recent scientific advances prove this to be the case. The long list of evolutionary aliens currently residing in Africa includes even-toed ungulates or artiodactyls (hippos, warthogs, gazelles, eland, impala, wildebeest, Cape buffalo, giraffes, and such), odd-toed ungulates or perissodactyls (rhinos, zebras, and asses), carnivores (lions, cheetahs, leopards, jackals, hyenas, mongooses, civets, and so forth), rodents (springhares, porcupines, rats, gerbils, and squirrels), and lagomorphs (rabbits and hares). The earliest members of all of these mammalian orders first evolved in Laurasia, and only later (often much later) made their way to Africa. This simple biogeographic pattern leads to a rather jolting evolutionary conclusion. Despite the substantial evolutionary head start that the vagaries of continental drift bestowed upon native groups of African mammals, these animals have been ecologically swamped by newcomers arriving from the north. For modern mammals as a whole, Africa has functioned more like a melting pot than a fountain of evolutionary innovation. Any conclusion as wide-ranging as this inevitably inspires further questions. Accordingly, how reliable are the data that support the African melting pot? And what does this pattern imply for anthropoid origins?

Two main lines of evidence uphold the African melting pot theory. The first of these comes from the burgeoning field of molecular evolutionary biology. Several different labs have independently analyzed long sequences of DNA from species representing all living orders of placental mammals (that is, all living mammals other than monotremes, such as the duck-billed platypus of Australia, and marsupials, such as the familiar Virginia opossum). The results of these molecular analyses sometimes conflict with more traditional reconstructions of mammal evolution based on comparative anatomy. Still, the molecular results—often derived from separate labs working on unrelated genes—correspond remarkably well with one another. From an anatomical perspective, the most startling conclusion from these molecular analyses is that the various groups of native African mammals appear to have descended from a single common ancestor.[5] In other words, African mammals with radically different body plans, and ranging in size from elephants to golden moles, share a unique evolutionary heritage rooted in Africa's deep and mysterious past—perhaps as early as a hundred million years ago, during the heart of the Cretaceous.[6] Molecular evolutionary biologists have coined the name Afrotheria (literally "African beasts") to refer to this large and diverse group of native African mammals. Hence, if long sequences of DNA can be trusted to reconstruct family trees, the original mammalian inhabitants of Africa must have consisted mainly of afrotheres.

The formidable amount of raw data enshrined in the DNA of every living organism requires us to take the concept of Afrotheria seriously. But the DNA of living animals alone does not allow us to reconstruct when, where, and how their ancestors lived. For that, we must turn to the fossil record. Unfortunately, the early Cenozoic fossil record of Africa is poorly sampled, especially compared to our knowledge of the contemporary mammals of North America, Europe, and Asia. Still, enough fossils have been unearthed recently in Africa to sketch the basic outlines of its mammalian fauna during that critical interval. Just as molecular evidence from living mammals would predict, the African fossil record documents the presence of several afrotherian groups by early or middle Eocene time. These include early relatives of elephants, elephant shrews, and hyraxes.[7] In each case, the African fossils are sufficiently old and primitive to confirm that each of these groups of mammals must have originated on that continent. The fossil record of the two major groups of insectivorous afrotheres—golden moles and tenrecs—only extends back to the Miocene. However, their fossils have never been found anywhere other than Africa, so an African origin for tenrecs and golden moles also seems secure.

Paleontological evidence is more ambiguous for the two remaining groups of living afrotheres—aardvarks and sirenians (living sirenians include dugongs and manatees)—because early fossils pertaining to these mammals have been unearthed outside of Africa. Still, an African origin for these groups remains extremely likely. The fossil record for aardvarks is poor, but two of the three extinct genera were exclusively African. As aquatic mammals, dugongs and manatees were able to disperse rapidly from their ancestral homeland. These early migrations led to a broad geographic distribution for early sirenians, effectively obscuring an African origin for the group. Evidence in support of this interpretation comes from both anatomical and molecular analyses of their evolutionary position, which strongly favor a close relationship between sirenians, elephants, and hyraxes. Because elephants and hyraxes are so firmly rooted in Africa, any close evolutionary connection between sirenians and these undoubted afrotheres would establish African roots for dugongs and manatees as well.

While the fossil record confirms that most, if not all, of the living afrotheres did indeed originate in Africa, even more enlightening is the information it provides on the mammals that were formerly absent from the continent. Rodents—by far the most diverse and abundant of all living groups of mammals—are conspicuously missing from the earliest Cenozoic fossil sites of Africa, despite the fact that hundreds of small mammal specimens in the size range of rodents have been recovered there.[8] Artiodactyls—the ecologically dominant group of living ungulates—do not appear in the African fossil record until late in the Eocene. By then, an imposing diversity of artiodactyls inhabited the Laurasian continents of Asia, Europe, and North America. Yet only a single group of artiodactyls—the anthracotheres, likely relatives of living hippos—roamed across Africa at that time. The ancestors of other living groups of African artiodactyls—including pigs, giraffes, buffalo, and antelopes—did not arrive in Africa until millions of years later. Likewise, perissodactyls, carnivores, and lagomorphs do not enter the African fossil record until the Miocene. Hence, rhinos, horses, cats, civets, and rabbits are missing from the Oligocene Fayum ecosystem of Egypt, despite the fact that all of these groups were well established on Laurasian continents by that time.

Integrating the separate lines of evidence from molecular evolutionary biology and the fossil record bolsters the image of Africa as a biogeographic melting pot. Modern African mammal faunas arose as successive waves of mammalian immigrants from Laurasia intermingled with a small core of afrotheres—the only mammals that can rightfully claim a deep African heritage. Competition and predation at the hands of these northern invaders impeded the evolutionary success of afrotheres in much the same way that Native American cultures have dwindled in the face of repeated colonization of the New World by human migrants from other parts of the world. Various kinds of afrotheres—including an enigmatic, vaguely otterlike group called ptolemaiidans and the dual-horned, elephantine arsinoitheres—have long since vanished entirely.[9] Other afrotheres, notably including the hyracoids, now play a greatly diminished ecological role that contrasts sharply with their former prominence (see chapter 4). As we shall see, the evolutionary roots of most (and possibly all) of the invading mammals that contributed to the afrotheres' decline can be traced to Asia.

Assuming that it accurately reflects the history of mammal evolution, what does the African melting pot model tell us about the geography of anthropoid origins? Molecular data bearing on this issue are unanimous. Evidence from the DNA of living primates places the group as a whole— and its large anthropoid subdivision—far away from afrotheres on the mammalian family tree. In fact, both molecular studies of living mammals and traditional analyses of comparative anatomy and fossils indicate that the nearest living relatives of primates are flying lemurs and tree shrews. Although each is primatelike in various ways, flying lemurs and tree shrews belong to separate orders of mammals. The two living species of flying lemurs—which, as George Gaylord Simpson and others have noted, are not lemurs and cannot fly—are nocturnal leaf-eaters with remarkable anatomical specializations for gliding. Tree shrews are vaguely squirrellike omnivores with relatively long snouts. Molecular evolutionary biologists have coined the name Euarchonta for the branch of the mammalian family tree that includes primates, tree shrews, and flying lemurs.[10] As far as we know from the fossil record and their modern geographic distributions, both flying lemurs and tree shrews are strictly Asian mammals. Similarly, fossils of the extinct plesiadapiforms—a group commonly regarded as "archaic primates"—have been recovered on all three Laurasian continents, but remains of these animals have never been unearthed in Africa. If we cast the molecular net more broadly, the nearest living relatives of Euarchonta appear to be rodents and lagomorphs.[11] The fossil record for both of these groups of living mammals shows that they too originated in Asia. Therefore, primates form one twig on what appears to be an exclusively Asian branch of the mammalian family tree.

If primates as a group originated in Asia, what further evidence can be marshaled to establish where the first anthropoids evolved? The data bearing on this issue are similar to those that allow us to reconstruct where primates, afrotheres, and other mammalian groups originated. A wide range of biological evidence from living primates suggests that tarsiers are the nearest living relatives of anthropoids (see chapter 1). Tarsiers and anthropoids share such fundamental (and evolutionarily advanced) similarities as having lost the reflective tapetum lucidum layer in the eyeball, lacking a lemurlike external nose that is naked and moist, and being incapable of synthesizing vitamin C (which tarsiers and anthropoids must therefore ingest in order to meet their basic nutritional requirements). Molecular evidence also supports a close evolutionary relationship between tarsiers and anthropoids, although the weak signal from DNA suggests that the two lineages must have separated early in the Cenozoic.[12] Today, tarsiers occur naturally on Borneo, Sumatra, Sulawesi, Mindanao, and nearby Southeast Asian islands. Likewise, the fossil record of tarsiers is strictly Asian.[13] If we expand our scope to include fossils that are more distantly related to tarsiers—such as omomyids and microchoerids—we find that these animals once ranged over all three Laurasian continents, but there is no convincing evidence that they ever lived in Africa.

Based on the evidence reviewed so far, primates appear to conform to the same large-scale biogeographic pattern that has so strongly influenced the composition of modern African mammal faunas. Like most of the ecologically dominant mammals inhabiting Africa today, primates joined the African melting pot only after the initial chapter of their evolutionary history had been written elsewhere. In the revised chronology of anthropoid origins suggested by *Shoshonius, Eosimias,* and other key fossils, very little time elapsed between the origin of primates as a group and the diversification of primates into their primary living lineages, including anthropoids. Assuming that these evolutionary events occurred in fairly rapid succession, there would have been little chance for basal primates to make their way from Asia to Africa in time for anthropoids to originate there. But aside from the inherent difficulties imposed by timing, how confident can we be that anthropoids—like their basal primate ancestors—originated in Asia rather than Africa?

A common method of choosing between competing scientific theories is to adhere to the principle known as Occam's razor. This criterion holds that the simplest explanation is always preferred over its more complex alternatives. In the case at hand, the hypothesis that anthropoids originated in Asia turns out to be much simpler than the traditional "Out of Africa" model. To illustrate, let's consider the biogeographic complexity of both hypotheses. Remember that we have already established that primates originated in Asia, and that living and fossil tarsiers have never been found in Africa. Accordingly, before we can postulate that anthropoids originated in Africa, we must first propose that the common ancestors of anthropoids and tarsiers made their way from Asia to Africa in time for that lineage to split into its two main components. This early migration event is not required if anthropoids originated in Asia, given that ancestral primates already lived there. Because it entails an extra biogeographic step, the "Out of Africa" model is already more complicated than its Asian alternative, even before the first anthropoids have evolved.

Yet the disparity only gets worse. By definition, the tarsier and anthropoid lineages were born when a single ancestral lineage bifurcated into two daughter species. Because speciation is local rather than global, tarsiers and anthropoids must have originated on the same continent. Therefore, by positing that the anthropoid lineage originated in Africa, we are forced to assert that the tarsier lineage originated there too. This poses additional complications for the "Out of Africa" model. First we must claim that, after having arisen in Africa, ancestral tarsiers returned to Asia sometime prior to the middle Eocene, when the earliest fossil tarsiers appear in central and eastern China. Second, we must explain how the African tarsiers that were left behind became extinct without leaving any fossil record of their former presence. Finally, we must acknowledge that early anthropoids such as *Eosimias* also returned to Asia sometime prior to the middle Eocene, when they appear alongside early tarsiers at fossil sites in China. Overall, the "Out of Africa" model requires no fewer than three episodes of intercontinental dispersal between Africa and Asia, in addition to one episode of continental-scale extinction (of early African tarsiers) for which there is no fossil evidence. On the other hand, if anthropoids originated in Asia, the only biogeographic complication that must be explained is how and when early anthropoids dispersed from Asia to Africa. Given what we know about the architecture of the primate family tree and the geographic distribution of its living and fossil members, Occam's razor clearly dictates that we prefer an Asian origin for anthropoids.

Recent discoveries of early anthropoid fossils in Asian countries other than China substantiate the logic of Occam's razor. From my perspective, the most exciting of these discoveries demonstrates that close relatives of *Eosimias* once lived outside of China. The relevant specimens were found in Myanmar by an international team of French, Burmese, and Thai scientists under the leadership of Jean-Jacques Jaeger, whose pioneering work in Algeria during the 1980s led to the discovery of the early anthropoid *Biretia* in that North African country (see chapter 7). After decades of neglect caused by the geopolitical isolation of Myanmar, Jaeger's team returned to the same region where British and American expeditions had discovered the controversial fossil primates *Pondaungia* and *Amphipithecus* in the early part of the twentieth century (see chapter 1). In addition to finding further material of *Pondaungia* and *Amphipithecus*, Jaeger and his colleagues unearthed the first eosimiid fos-

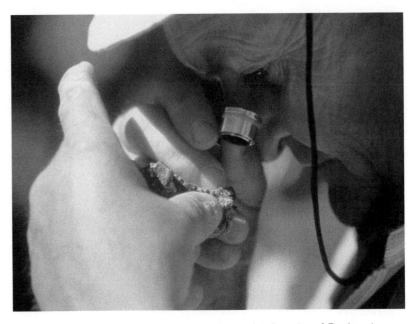

Figure 40. Jean-Jacques Jaeger inspects a nearly complete lower jaw of *Pondaungia cotteri* in the field in Myanmar. Photograph courtesy of and copyright by Patrick Aventurier/ Gamma Agency.

sils ever found outside of China. These specimens, which consist of upper and lower jaw fragments of a single individual, document a previously unknown primate that is closely related to *Eosimias*. Jaeger and his colleagues chose to name their new primate *Bahinia pondaungensis*.[14]

Dating to roughly thirty-seven million years ago, the Pondaung Formation strata that yielded *Bahinia* are from three to eight million years younger than the Chinese sites where we found fossils of *Eosimias*.[15] In agreement with its younger age, *Bahinia* differs from *Eosimias* in ways that reveal it to be a slightly more advanced primate. Tipping the scales at roughly fourteen ounces (four hundred grams), *Bahinia* would have weighed almost three times as much as *Eosimias* in life. Its upper molars differ from those of *Eosimias* in having higher, more trenchant crests and a stronger shelf of enamel (known as the lingual cingulum) lining the internal margin of each tooth. Both of these features are more derived than their counterparts in *Eosimias*. Still, the close anatomical correspondence between *Bahinia* and *Eosimias* indicates that they belong to the same major group of early anthropoids, the family Eosimiidae. Like *Eosimias*, *Bahinia* possesses vertically oriented incisors rooted in a deep mandibular symphysis, stout upper and lower canines with a circular cross-section, obliquely oriented lower premolars, and a deep facial region bearing an abbreviated muzzle. Based on its sharply crested teeth and its relatively small size, *Bahinia* probably ate insects and other small animal prey as well as fruits.

In terms of fleshing out the picture of early anthropoid evolution in Asia, the geographic and temporal contributions of *Bahinia* are more notable than its anatomy. *Bahinia* shows that eosimiid anthropoids once inhabited a vast region of Asia, from as far north and east as the Yellow River valley of central China and the East China Sea all the way to the Bay of Bengal in the southwest. Given the spotty nature of the fossil record, the currently documented range for Asian eosimiids can only be viewed as a minimum estimate of their actual geographic distribution in the Eocene. *Bahinia* also reveals that eosimiids were not evolutionary flashes in the pan. They persisted in Asia for at least eight million years, spanning the interval from the middle Eocene (about forty-five million years ago, which is the age of the oldest Shanghuang fissures) to the late Eocene (about thirty-seven million years ago, which is the age of the Pondaung Formation). Once again, the vagaries of the fossil record force us to consider this as the minimum duration of the eosimiid lineage. Both older and younger eosimiid fossils will likely be discovered in the future.

Jaeger's discovery of *Bahinia* in Myanmar also showed that eosimiids lived side by side with a second group of early Asian anthropoids, the amphipithecids. Although the earliest fragmentary specimens of *Pondaungia* and *Amphipithecus* suggested that these two Burmese primates might not be closely related, the recovery of further and more nearly complete specimens has erased many of the apparent distinctions between them. Both are now classified in the family Amphipithecidae, along with two other Eocene primates that have been discovered in southeastern Asia more recently. Jaeger's team of French and Thai scientists collected the first of these additional amphipithecids, which they described as *Siamopithecus eocaenus*.[16] As its name suggests, *Siamopithecus* was found in southern Thailand (formerly known as Siam) in late Eocene strata that were exposed by a large-scale strip-mining operation. A fourth amphipithecid, *Myanmarpithecus yarshensis,* was found even more recently by a team of Japanese and Burmese scientists working in the Pondaung region of Myanmar.[17]

Amphipithecids differ from eosimiids in being larger and in having cheek teeth that are adapted for eating nuts, seeds, and other hard objects. *Siamopithecus, Pondaungia,* and *Amphipithecus* all weighed

Figure 41. A virtually complete lower jaw of *Siamopithecus eocaenus* from southern Thailand. From Chaimanee et al. 2000, copyright Académie des Sciences. Reproduced by permission.

from fourteen to twenty pounds (six and a half to nine kilograms), making them roughly the size of a South American howler monkey. At four and a half pounds (two kilograms), *Myanmarpithecus* was substantially smaller than other amphipithecids, but still larger than *Bahinia*, the largest known eosimiid. In keeping with their comparatively large size, amphipithecids evolved the dental and jaw anatomy that is required to masticate the tougher, more fibrous plant resources preferred by primates of this body mass. While the molars of eosimiids bear high cusps and sharp crests indicating a diet based on insects and fruit, the molars of amphipithecids show comparatively little topographic relief and have greatly diminished crests. Similarly, the lower jaws of amphipithecids are remarkably deep and robust, especially in the symphyseal region near the chin. In all of these respects, amphipithecids are more derived than eosimiids, even though one eosimiid—*Bahinia*—lived alongside several of them in Myanmar.

Given their large size and relatively advanced tooth structure, amphipithecids have diverged significantly from the anatomical pattern shown by eosimiids, the most primitive anthropoids currently known. There are several ways to interpret the evolutionary signal provided by these anatomical differences. One option, based on similarities in dental anatomy, places amphipithecids near *Aegyptopithecus* and closely related anthropoids from the Fayum.[18] In that case, amphipithecids should possess most, if not all, of the anatomical features characteristic of living monkeys, given that they would occupy an evolutionary position intermediate between New World monkeys on the one hand and Old World monkeys, apes, and humans on the other. Two, or possibly three, lines of evidence suggest that this is unlikely.

The first of these consists of primate postcranial bones collected from the Pondaung Formation in December 1997 by a group of American and Burmese scientists under the leadership of Russell L. Ciochon.[19] Ciochon's team recovered a virtually complete left humerus (the bone that forms the upper arm between the shoulder and elbow) and most of a left calcaneus (the bone that lies in the heel of the foot), alongside other less diagnostic fragments. Based on their size, these postcranial elements probably belong to *Pondaungia*, the largest and most abundant primate currently known from the Pondaung Formation. The Pondaung postcranial bones, especially the humerus, differ in several respects from those of *Aegyptopithecus*. For example, the anatomy of the shoulder and elbow joints in *Pondaungia* resembles that of Eocene primates such as *Notharctus* and *Shoshonius* more than the counterpart in *Aegyptopithecus*. Assuming that the Pondaung limb bones actually pertain to *Pondaungia* or some other amphipithecid, such primitive anatomy would not be expected in a close relative of *Aegyptopithecus*.

Details of lower premolar anatomy likewise indicate that amphipithecids are only distantly related to *Aegyptopithecus* and other advanced anthropoids from the Fayum. For example, amphipithecids retain three lower premolars on each side, as do eosimiids, parapithecids, proteopithecids, and living and fossil South American monkeys. In contrast, more advanced Fayum anthropoids, such as *Aegyptopithecus, Oligopithecus,* and their close relatives, resemble living and fossil Old World monkeys, apes, and humans in having only two premolars bilaterally. Beyond this basic difference in counting premolars lie important distinctions in the anatomy of their crowns. In all living anthropoids, the two main cusps (technically known as the protoconid and the metaconid) on the last lower premolar, or P_4, are transversely aligned, and their summits are connected by a strong crest. The same condition exists in relatively advanced fossil anthropoids, including several of the main groups known from the Fayum (proteopithecids, oligopithecids, and propliopithecids). Amphipithecids retain a more primitive type of lower premolar anatomy, in that the two main cusps of P_4 are not transversely aligned, and no major crest unites them. Premolars like those of amphipithecids are characteristic of primates that are thought to lie near the base of the anthropoid family tree, including eosimiids and such primitive Fayum anthropoids as parapithecids and *Arsinoea*.

The final and most tentative line of evidence suggesting that amphipithecids are primitive anthropoids comes from fragmentary fossils that offer hints about their facial anatomy. A recently discovered upper jaw of *Pondaungia* from Myanmar shows that a thick region of bone intervenes between the upper tooth row and the eye socket in amphipithecids, as it does in eosimiids (see chapter 9).[20] This means that amphipithecids had relatively small eye sockets like those of anthropoids, rather than the grotesquely enlarged orbits found in tarsiers and *Shoshonius*. Unfortunately, this specimen is too fragmentary to show whether the eye sockets of *Pondaungia* were completely enclosed by bone, as they are in living anthropoids and fossils such as *Aegyptopithecus* from the Fayum. Another new specimen from the Pondaung Formation potentially sheds more light on this aspect of amphipithecid anatomy. This fossil—thought to be a fragmentary piece of the frontal bone of the skull—was found near an upper jaw of *Amphipithecus*, and it possibly represents the same individual.[21] If this association is correct, the new Pondaung fossil preserves just enough anatomy to show that the eye sockets of *Amphipithecus* were not completely enclosed by bone. However, we cannot be certain that this specimen actually belongs to *Amphipithecus* (or even to a primate), because there is no bony contact between it and the upper jaw of *Amphipithecus* that was found nearby. If additional specimens confirm that amphipithecids possessed such primitive eye sockets, then this feature alone would demonstrate that amphipithecids are only distantly related to *Aegyptopithecus* and other advanced anthropoids. Still, the absence of monkeylike eye sockets is insufficient to prove that amphipithecids are not anthropoids, for the same reason that a small, chimpanzee-sized brain does not necessarily render a fossilized skull that of an ape rather than a hominid (see chapter 8).

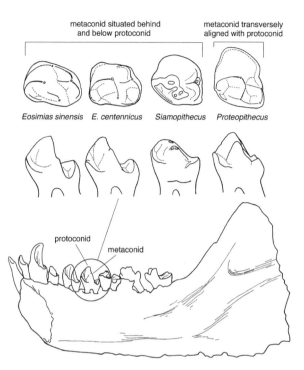

Figure 42. Differences in the anatomy of P₄ distinguish primitive Asian anthropoids such as *Eosimias* and *Siamopithecus* from more advanced anthropoids such as *Proteopithecus* from the Fayum region of Egypt. Original art by Mark Klingler, copyright Carnegie Museum of Natural History.

Although additional fossils might lead to different conclusions, everything that we currently know about the Eocene anthropoids of Asia—including the eosimiids of China and Myanmar and the amphipithecids of Myanmar and Thailand—indicates that both groups belong near the very base of the anthropoid family tree. This pivotal evolutionary position makes Asian Eocene anthropoids more important than their African counterparts for reconstructing anthropoid origins. Not only do eosimiids and amphipithecids conflict with the traditional view that anthropoids originated in Africa; they also overturn previous theories that sought to explain how and why anthropoids evolved at all.

For years, the most influential idea purporting to explain anthropoid origins from a functional or ecological perspective viewed it as a response to the deteriorating climatic conditions of the Eocene-Oligocene boundary (see chapter 5).[22] Let's refer to this idea as the "global cooling hypothesis" of anthropoid origins. It holds that, as environmental conditions became cooler and drier across the Eocene-Oligocene boundary, early anthropoids sought energetic efficiency by evolving larger bodies. This allowed anthropoids to retain more of their precious body heat by decreasing the key ratio of bodily surface area to volume. Being bigger forced early anthropoids to shift their diet from insects to fruits, prompting further changes in tooth, jaw, and skull anatomy. Anthropoid molars became flattened and lower-crowned, since they no longer required the sharp crests that their prosimian ancestors had used to slice through the rigid exoskeletons of insects. The switch to a more vegetarian diet also led to deeper and more robust lower jaws that eventually fused into a single bony element at the mandibular symphysis. Bony plates developed behind the eye sockets as an additional place to anchor the large temporalis muscles that were needed to power long bouts of chewing. In order to distinguish ripe from unripe fruits, early anthropoids began foraging during daytime rather than at night. As a result, they lost the reflective tapetum lucidum layer in their eyes, while their ability to perceive multiple colors became enhanced. As early anthropoids increasingly relied on their sense of sight, their brains enlarged to accommodate the newly urgent need for complex neural processing of visual information.

To its credit, the global cooling hypothesis provides a plausible account for how, when, and why most of the features that distinguish anthropoids from their prosimian brethren evolved. Like many good ideas in science, it is also remarkably simple. The cooler and more seasonal climatic conditions of the Eocene-Oligocene boundary sparked a chain reaction that reverberated throughout the biology of early anthropoids, starting with an increase in body size. But for many paleoanthropologists, the global cooling theory of anthropoid origins fulfilled another, more philosophical role. It brought anthropoid origins under the triumphant umbrella of evolutionary progress that had dominated the field since Le Gros Clark's proclamation that primates had evolved in a ladderlike sequence of steps from tree shrew to lemur to tarsier to monkey to ape to human. According to the global cooling hypothesis, the evolutionary success of anthropoids was no accident. Anthropoids successfully coped with the cooler Oligocene climate and prospered, while most of their prosimian relatives failed to adapt and paid the ultimate evolutionary price for having done so. Natural selection might be cruel, but it is also impartial—the trade-off that biology demands for evolutionary progress.

The only thing wrong with this neat and tidy account of anthropoid origins is that it cannot possibly be true. For one thing, its timing is off. Fossils like *Eosimias* and *Algeripithecus* effectively torpedo the global cooling hypothesis by showing that anthropoids had been evolving for more than ten million years by the time that Earth cooled down near the Eocene-Oligocene boundary. Yet the damage that these early anthropoid fossils inflict on the global cooling hypothesis goes much deeper than that. After all, most of its key elements might remain intact, by simply conceding that Eocene-Oligocene cooling was not the ultimate catalyst for the chain of events culminating in anthropoid origins. Perhaps early anthropoids initially got bigger for some reason other than climate change. In that case, the acquisition of large body size itself might trigger the same evolutionary cascade that the global cooling hypothesis pulled together so well.

Even this pared-down version of the global cooling hypothesis disagrees with the rapidly expanding fossil record of early anthropoids in Asia, however. Eosimiids and amphipithecids conflict with it on two levels. First, because amphipithecids had already achieved the body size of a large monkey or small ape, they should by all rights display most of the diagnostic anthropoid traits found in living monkeys. Yet as we've seen, the new evidence from the fossil beds of Myanmar and Thailand shows that amphipithecids retained many primitive features that are never found in modern anthropoids. Amphipithecids had yet to fuse the mandibular symphysis; they maintained distinctly primitive lower premolars, shoulders, and elbows; and they may even have lacked the complete bony eye sockets that serve as an anatomical trademark of modern anthropoids. If the evolution of large body size leads inexorably to the entire arsenal of anthropoid anatomy, why don't amphipithecids look more like modern baboons—or at least like *Aegyptopithecus*?

While amphipithecids pose serious problems for the idea that anthropoid origins were linked with increasing body size, eosimiids refute the notion more directly. With its two known species weighing between 3.5 and 4.5 ounces (100–130 grams), *Eosimias* shows that early anthropoids were small, even by prosimian standards. *Eosimias* barely overlaps the bottom of the range in body size found in living anthropoids. The smallest living monkeys are South American pygmy marmosets of the genus *Cebuella*, which tip the scales at about 4.5 ounces (130 grams). However, these animals are widely regarded as evolutionary dwarves, whose ancestors would have been substantially larger.[23] Marmoset anatomy offers numerous hints that they evolved from anthropoids that were more standard in terms of size. For example, marmoset jaws are so foreshortened that they have lost their upper and lower third molars (in human terms, their wisdom teeth) entirely. Marmoset nails have been modified into squirrellike claws, because their hands and feet are so diminutive that it would be hard for them to climb up vertical tree trunks in typical monkeylike fashion. *Eosimias* shows none of these telltale signs of being an evolutionary dwarf. On the contrary, *Eosimias* was actually larger than some of its anthropoid relatives in China during the middle Eocene.

My colleague Dan Gebo uncovered the first compelling evidence that anthropoids of truly diminutive proportions lived alongside *Eosimias*. During his survey of primate ankle bones from the Shanghuang fissure-fillings, Gebo turned up several specimens that are smaller than those of any living primate. The smallest living primates are mouse lemurs from the island of Madagascar belonging to the species *Microcebus myoxinus*.[24] As their name suggests, these tiny primates are roughly mouse-sized, with an adult body weight of about one ounce (thirty grams). Despite the remarkably small size of living mouse lemurs, Gebo found several primate ankle bones from Shanghuang that measure less than half the size of their counterparts in *Microcebus*. Moreover, the Shanghuang specimens once belonged to adult individuals, because their epiphyses (the unfused bony plates that allow bones to lengthen as an individual grows) were already fused to the remainder of the bone.

Regardless of differences in how various species of primates move, their ankle bones must be structurally adequate to support their body weight. As a result, a strong correlation exists between individual measurements that can be taken on ankle bones and overall body size in primates.[25] Applying this technique to the tiny ankle bones from Shanghuang shows that a variety of primates weighing between one-third and two-thirds of an ounce (ten to twenty grams) lived in the region during the middle Eocene.[26] Some of the tiny primate ankle bones from Shanghuang are so primitive that they might belong to several different kinds of early primates. Others are anatomically identical to their counterparts in *Eosimias*, indicating that early anthropoids weighing less than an ounce roamed the coastal forests of China during the middle Eocene.

Demonstrating that early Chinese anthropoids were so small raises several implications about their basic biology. Like shrews, these early "micromonkeys" would have had fast metabolic rates, implying active or even frenetic lifestyles. To meet their basic nutritional requirements, they would have been forced to consume food that is rich in calories— mainly insects and fruits. The small size of the Shanghuang micromonkeys would have subjected them to predation by hawks, owls, and other raptorial birds. Indeed, many of the eosimiid fossils from the Shanghuang fissure-fillings show signs of chemical etching caused by the gastric juices of their avian predators (see chapter 9). Tiny mammals also inhabit fairly restricted home ranges. In an evolutionary context, this tendency to form small, geographically isolated populations promotes speciation, because it inhibits gene flow. Hence, for the same reason that there are so many species of mice and so few species of elephants, the Shanghuang ecosystem was capable of supporting several different species of tiny primates. Some of these tiny primates were anthropoids, while others were prosimians. In order to restrict head-to-head competition among close relatives, each species must have developed its own distinctive ensemble of ecological attributes. This in turn raises the most intriguing issue of all—how did the Shanghuang micromonkeys distinguish themselves from their prosimian peers? Did these tiny primitive anthropoids already possess some critical biological advantage that would ensure their subsequent evolutionary success? And if so, what was it?

Despite many recent improvements in our knowledge of eosimiid anatomy, the structure of their teeth and bones offers no compelling reason why they and their close anthropoid relatives would eventually dominate global primate communities, leaving prosimians to fill only minor ecological roles outside of Madagascar. If anything, the fossil record shows that the tables were turned during the Eocene. At that time, omomyid and adapiform prosimians ruled much of the planet, while anthropoids struggled to hold their ground in a few far-flung corners of the globe. During the middle Eocene in China, eosimiids lived alongside a wide array of prosimians. Omomyids, tarsiers, and adapiforms have all been recovered with eosimiids at Shanghuang. At the geologically younger localities of the Yuanqu Basin, eosimiids continued to share the local ecosystem with tarsiers such as *Xanthorhysis* and adapiforms such as *Hoanghonius*. Hence, it is not at all clear that *Eosimias* and other early anthropoids were "better" in any meaningful sense than their prosimian contemporaries even in their Asian homeland. Early adapiforms like *Hoanghonius* clearly surpassed *Eosimias* in terms of their adaptations for folivory (leaf-eating), while early tarsiers like *Xanthorhysis* must have been at least as skillful at hunting and capturing insects. Despite their anthropoid affinities, in many ways eosimiids resembled the prosimians that lived alongside them rather than their anthropoid successors. The subtle differences between eosimiids and prosimians, together with the fact that various prosimian groups thrived alongside eosimiids for millions of years in China, suggest that the subsequent evolutionary success of anthropoids was hardly preordained.

A second paradox embedded in the early fossil record of anthropoids reiterates why primate evolution fails to conform to any simple gradient of evolutionary progress from tree shrew to human. We have seen how Occam's razor and the geographic distribution of eosimiids and amphipithecids indicate that anthropoids must have originated in Asia rather than Africa. Nevertheless, a major disparity rapidly developed between the Eocene anthropoids of Africa and Asia. Despite having originated there, Asian anthropoids remained persistently primitive, while their African relatives evolved into increasingly advanced species. This discrepancy had already emerged by the middle Eocene, as any comparison between Chinese *Eosimias* and North African *Algeripithecus* quickly reveals. The two species are roughly the same age, but their teeth differ radically in structure. Like those of later and more advanced anthropoids, the lower molars of *Algeripithecus* have already lost the paraconid cusp, while its upper molars already possess a hypocone. Neither of these advanced features occurs in *Eosimias*. If anything, the disparity between Asian and African anthropoids becomes more severe through time. By the late Eocene, when the primitive amphipithecid *Siamopithecus* was still holding out in Thailand, the Fayum region of Egypt was home to *Proteopithecus*, *Catopithecus*, *Abuqatrania*, *Arsinoea*, and other anthropoids, all of which occupy branches closer to the top of the anthropoid family tree.

An obvious ramification of the evolutionary disparity between the Eocene anthropoids of Asia and Africa is that anthropoids must have dispersed between the two continents sometime before the middle Eocene, by which time the disparity had already developed. How and when did this dispersal occur, and what impact did it have on early anthropoid evolution?

No matter how early anthropoids made their way from Asia to Africa, the route must have been treacherous and difficult to cross. During the early part of the Cenozoic, the forces of plate tectonics and continental drift caused Old World geography to deviate substantially from the conditions that are familiar to us today. Africa lacked any direct land connection with Eurasia, because an extensive seaway known as Tethys connected the Indian Ocean with the Mediterranean Sea across what is now the Persian Gulf region. At the same time, because the rifting that created the Red Sea had yet to begin, Africa would have been larger than it is today. It would have included the Arabian Peninsula, so that the Horn of Africa (currently located in Somalia) extended as far north as modern Oman and the United Arab Emirates. Despite the larger size of Africa, any Asian land mammal attempting to migrate directly across the Persian Gulf region would have faced a stern challenge. Africa as a whole lay farther to the south, and the broad Tethys seaway intervened between what is currently the Arabian Peninsula and Iran.

To the north, the geography of Asia also looked unfamiliar by modern standards. India, being one of many fragments of the former southern supercontinent of Gondwana, began to collide with the southern margin of Asia about fifty million years ago, during the latter part of the early Eocene.[27] Prior to that time, India was a large island in the Indian Ocean, drifting steadily northward from its Cretaceous location adjacent to Madagascar. The modern Tibetan Plateau, a vast region raised high above sea level by the subduction of India beneath it, lay at or even below sea level during the early Cenozoic. To the west, another shallow seaway separated central and eastern Asia from Europe near the location of the modern Ural Mountains.

Given the patchy nature of the fossil record, we can only guess how early Asian anthropoids traversed this exotic landscape to invade Africa. Possibly, they used the offshore island of India as a giant stepping-stone to disperse directly from southern or southeastern Asia to Africa. The major obstacles along this route would have been the lingering remnant of the Tethys seaway that still separated the Asian mainland from India and the nascent stretch of the Indian Ocean that intervened between India and Africa (modern Oman) to the west. Either or both of these marine barriers may have been partly bridged by ephemeral island arcs that were created and subsequently destroyed by the constantly shifting forces of continental drift. Alternatively, early Asian anthropoids could have skirted the northern margin of the Tethys seaway to occupy what is now Turkey, which they could have used as a staging area to enter Africa from the north. This northwesterly route would also have required early anthropoids to overcome a formidable series of marine barriers, hopping from island to island across what is now Greece and Sicily before washing ashore in the vicinity of modern Tunisia. For early anthropoids to complete either journey would have entailed a significant element of luck. We can be certain of this not only because of the pitfalls inherent in early Cenozoic Old World geography, but also because other Asian mammals were unable to invade Africa at the same time that early anthropoids did so.

We may never be certain about the exact pathway early Asian anthro-

Figure 43. A cartoon depicting the early dispersal of anthropoids from Asia to Africa, which was dominated by various afrotherian mammals at that time. The exact route followed by these early anthropoids remains unknown. They may have dispersed directly across the Tethys seaway to North Africa, or they may have used the Indian subcontinent as a giant stepping-stone to enter the Arabian Peninsula in the vicinity of modern Oman. Original art by Mark Klingler, copyright Carnegie Museum of Natural History.

poids followed to invade Africa, but we now have a much better handle on its timing. As the African fossil record has improved, the date for the arrival of early anthropoids on that continent has been pushed steadily backward in time. I interpret the African record to indicate that primitive Asian anthropoids had already migrated there by the latest Paleocene, roughly fifty-six million years ago. By several criteria, this is astonishingly early. For example, the other groups of Asian mammals that dispersed to Africa all arrived long after anthropoids had already taken up residence there. Rodents were the first of these additional Asian mammals to succeed in occupying the land of the afrotheres. Yet the oldest African rodents, from the middle Eocene Glib Zegdou locality in Algeria, are about ten million years younger than my estimate of the anthropoid colonization of Africa. Artiodactyls, perissodactyls, carnivores, lagomorphs, and other groups of Asian mammals—many of which currently dominate African ecosystems—didn't arrive there until millions of years later.

My proposal that Asian anthropoids dispersed to Africa by the latest Paleocene also exceeds the boundaries of what many paleoanthropologists would regard as the earliest solid evidence for anthropoids in the fossil record. The earliest widely accepted Asian and African anthropoids—*Eosimias* and *Algeripithecus,* respectively—date only to the middle Eocene, about forty-five million years ago. If we employ ghost lineages to estimate the age of the earliest anthropoids, we can be confident that anthropoids existed sometime prior to fifty million years ago, using *Shoshonius* as a beacon for the earliest evidence of the tarsier lineage in the fossil record (see chapter 6). Pushing the ghost lineage approach even farther—by accepting the earliest known omomyids as evidence of the tarsier lineage—allows us to deduce that the anthropoid lineage was established fifty-five million years ago. Happily, this date coincides with recent estimates based on the molecular clock.[28] However, establishing that anthropoids merely existed is necessary—but not sufficient—to demonstrate that they had already migrated from Asia to Africa. Accordingly, what evidence supports my contention that Asian and African anthropoids parted company at such an early date?

The fossils indicating that Asian anthropoids had already colonized Africa by the end of the Paleocene are highly fragmentary, but they are also quite compelling. The sample consists of ten isolated teeth and a lower jaw fragment bearing the unerupted germ of the second lower molar. All of these specimens, which were described by a team of French scientists in 1990 from the latest Paleocene Adrar Mgorn 1 locality in

Morocco, belong to an enigmatic fossil primate known as *Altiatlasius koulchii.*[29] Despite our poor knowledge of its anatomy, *Altiatlasius* figures prominently in any discussion of primate evolution. Dating to roughly fifty-six million years ago, *Altiatlasius* is the oldest undoubted primate of any kind described to date. Otherwise, the oldest primates documented in the fossil record are primitive adapiforms and omomyids from the earliest Eocene. However, the dental structure of *Altiatlasius* diverges dramatically from the pattern that characterizes the earliest adapiforms and omomyids. Instead, *Altiatlasius* shares numerous features with *Eosimias,* which is the reason I regard *Altiatlasius* as an early anthropoid.

The cheek teeth of the earliest adapiforms and omomyids are extremely similar—so much so that one particularly small and primitive adapiform was originally thought to be an omomyid.[30] Their upper molars usually lack a lingual cingulum, while an extra crest called the postprotocingulum is virtually always present. On the labial side of their upper molars, the two main cusps known as the paracone and metacone closely approximate the margins of each molar crown. The upper molars of *Eosimias* and *Altiatlasius* differ from those of omomyids and adapiforms in having a complete lingual cingulum, in lacking the postprotocingulum entirely, and in having the paracone and metacone situated more internally, away from the labial margin of the tooth. The internal position of the main labial cusps in *Eosimias* and *Altiatlasius* allows enough space for an extra cuspule known as the parastyle to become enlarged. This last structure, though present in early adapiforms and omomyids, is always much smaller than it is in *Eosimias* and *Altiatlasius.*

The structure of the lower molars of omomyids and adapiforms also diverges in important ways from that of *Eosimias* and *Altiatlasius.* The lower molars of *Eosimias* and *Altiatlasius* are unusual in having a greatly enlarged protoconid cusp on the front part of each tooth. In contrast, in adapiforms and omomyids the protoconid is similar in size to an adjacent cusp known as the metaconid. As you move from front to back across the lower molar series (that is, from the first lower molar or M_1 to the third lower molar or M_3) in adapiforms and omomyids, the paraconid and metaconid cusps increasingly approximate one another. In *Eosimias* and *Altiatlasius,* these two molar cusps remain equidistant from each other as you move from front to back in the lower jaw. Finally, the rear margin of the lower molars tends to be squared off in adapiforms and omomyids, while this region bulges backward in *Eosimias* and *Altiatlasius.* Considering how little we know of the anatomy of *Altiatlasius,* the

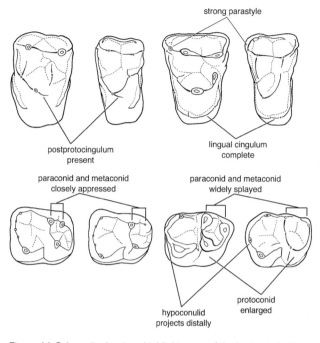

strong parastyle

postprotocingulum
present

lingual cingulum
complete

paraconid and metaconid
closely appressed

paraconid and metaconid
widely splayed

hypoconulid
projects distally

protoconid
enlarged

Figure 44. Schematic drawings highlight some of the basic similarities in upper and lower molar structure shared by adapiforms and omomyids (left pair) on the one hand, and *Altiatlasius* and *Eosimias* (right pair) on the other. From left to right, the fossils depicted are: the adapiform *Cantius*, the omomyid *Teilhardina*, the basal anthropoid *Altiatlasius*, and the basal anthropoid *Eosimias*. Original art by Mark Klingler, copyright Carnegie Museum of Natural History.

number of upper and lower molar traits it shares with *Eosimias* is remarkable. These features indicate that *Altiatlasius* is neither an omomyid nor an adapiform. Instead, they reveal *Altiatlasius* to be an early anthropoid that is anatomically similar in many ways to *Eosimias*. This conclusion affects our understanding of the origin and early evolution of anthropoids in several important ways.

The most obvious implication for anthropoid origins relates to simple chronology. Establishing late Paleocene *Altiatlasius* as an early anthropoid both confirms and extends the ghost lineage model of anthropoid origins that I initially proposed on the basis of *Shoshonius* (see chapter 6). Decades of research by paleoanthropologists who sought to decipher how anthropoids evolved from middle and late Eocene prosimians can now be recognized as being woefully misguided—the academic equivalent of a drunken man stumbling down a blind alley, searching for a way home.

The biogeographic implications of *Altiatlasius* are more nuanced and require further explanation. A literal reading of the fossil record would view the presence of *Altiatlasius* in Morocco at such an early date as confirmation of the old "Out of Africa" theory of anthropoid origins, with the only caveat being that anthropoids arose much earlier than the theory had originally indicated.[31] However, if we place the fossil record of early anthropoids in a broader evolutionary context, we find that the logic of Occam's razor still holds. Anthropoids must have originated in Asia, in spite of the age and geographic location of *Altiatlasius*.

Several factors substantiate this position. So far as we know from the African fossil record and its living mammal fauna, there was never any viable ancestral stock in Africa that might have given rise to early anthropoids there. Primates are entirely unknown in Africa prior to the time of *Altiatlasius*, and the nearest mammalian relatives of primates—such as tree shrews, flying lemurs, and fossil plesiadapiforms—occur in Asia but not in Africa. Likewise, living or fossil tarsiers (along with their extinct relatives the omomyids and microchoerids) have never been found in Africa, although they are becoming increasingly well documented in the fossil record of Asia. Accordingly, *Altiatlasius* does not indicate that anthropoids originated in Africa. Rather, it signals that Asian anthropoids arrived there at a surprisingly early date. Here lies the key to a new and higher level of understanding anthropoid origins.

When the first Asian anthropoids arrived in Africa, they encountered a bountiful continent that was surprisingly free of potential competitors. The vast majority of the resident mammals were afrotheres. This left the ecological niche at which primates excel—that of the arboreal mammalian frugivore and insectivore—largely if not completely vacant. We can only imagine that these earliest African anthropoids were fruitful and that they would have multiplied, following the biblical dictum found in the Book of Genesis. Nothing prevented them from embarking on a rapid and expansive evolutionary radiation. Paleontologists are only beginning to sample this early starburst of anthropoid evolution in Africa, and many of its details remain sketchy. Still, the fossil record offers numerous examples showing that, when new types of animals enter virgin terrain, what often follows is evolutionary experimentation on a grand scale. Two key episodes during primate evolution provide insight in this regard, because each transpired under geographic and ecological conditions similar to those that confronted the earliest anthropoid colonists of Africa. Let's therefore take a closer look at what happened when ancestral lemurs invaded Madagascar, and when early monkeys first made their way to South America.

The fossil record sheds only the dimmest of light on the early colonization of Madagascar by primitive lemurs, because sites of appropriate antiquity remain unknown on the island. However, analyses of the DNA of living and recently extinct lemurs have shown that they all evolved from a single ancestor that invaded Madagascar during the early part of the Cenozoic, about fifty-four million years ago.[32] These early lemurs encountered a depauperate island fauna that offered few competitive obstacles to their evolutionary success. They radiated into a rich array of shapes and sizes, filling much of the ecological vacuum that Madagascar presented. Eventually, the descendants of the first lemur colonists of Madagascar would include species as small as living mouse lemurs and others—such as the recently extinct *Palaeopropithecus* and *Archaeoindris*—that approached the size of large apes. Certain lemurs acquired specializations for eating insects, leaves, bamboo, seeds, fruits, sap, and gum. Various species evolved into slothlike *(Palaeopropithecus)*, monkeylike *(Hadropithecus)*, and even vaguely woodpeckerlike *(Daubentonia)* forms.[33] In contrast, the nearest relatives of lemurs—African bushbabies and African and Asian lorises—span a much smaller range of body sizes and ecological roles. Comparing the diversity of lemurs on Madagascar with the biological monotony of their African and Asian relatives suggests that the invasion of Madagascar itself catalyzed the vast evolutionary experiment that followed. The fact that all of this evolutionary tinkering took place on an island encompassing less than one-fiftieth of the surface area of Africa merely emphasizes this point.

A similar evolutionary radiation occurred after the earliest platyrrhine monkeys colonized South America. Once again, these early primate invaders would have met little ecological resistance, although a variety of tree-dwelling marsupials inhabited South America at the time, some of which may have occupied similar ecological niches as living South American monkeys. In a relatively short timespan, ancestral platyrrhines evolved into the broad range of New World monkeys that live there today. In fact, Miocene fossils from the La Venta region of Colombia show that, by ten to fourteen million years ago, most of the major groups of living South American monkeys were already established. Close fossil relatives of living squirrel monkeys, howler monkeys, owl monkeys, saki monkeys, and marmosets have all been described from the La Venta fossil beds.[34] As in the case of lemurs on Madagascar, the evolutionary radiation of South American monkeys significantly expanded the envelope of primate anatomy and behavior. It produced the world's only nocturnal anthropoid (the living owl monkey, *Aotus*), the smallest living anthropoids (marmosets such as *Cebuella*), and the only primates that are equipped with muscular, prehensile tails (for example, the living spider monkey, *Ateles*). Once again, the factor that almost certainly triggered this explosive bout of anthropoid evolution was the chance dispersal of early monkeys to a large and ecologically appropriate landmass harboring few, if any, competitors.

Although the fossil record remains inadequate to illustrate the sequence of events that transpired after the successful anthropoid colonization of Africa, the evolutionary experimentation that followed must have been truly phenomenal. Like the first lemurs to reach Madagascar and the first playrrhines to invade South America, the earliest anthropoids to colonize Africa faced an enormous frontier offering virtually endless potential for exploitation. They would have diversified rapidly to fill Africa's vast ecological void. Each of the multitude of early anthropoid species that resulted became a new player in the never-ending game of evolution. Some of these new species developed novel anatomical traits, and to the extent that such features proved beneficial, the species that possessed them would have diversified further. Others would have explored new ecological roles by engaging in creative types of behavior, some of which demanded fresh anatomical solutions of their own. The resulting arborescence of African anthropoid evolution spurred the development of innovative anatomical structures and functions in a manner that simply didn't apply to the anthropoids that remained behind in Asia.

By the middle Eocene, African anthropoids had experienced such a long and continuous history of evolutionary ferment that a major anatomical disparity now distinguished *Algeripithecus* and its African relatives from *Eosimias* and its Asian kin. Relatively speaking, *Eosimias* stood frozen in time, still bearing a marked resemblance to older and more primitive anthropoids such as *Altiatlasius* and its unknown Asian contemporaries. Meanwhile, *Algeripithecus* had been markedly transformed, so that it resembled a smaller version of the anthropoids that would dominate the coastal forests of Egypt's Fayum region many millions of years later. By the end of the Eocene, the vast evolutionary experiment that had taken place in Africa had produced a broad range of anthropoids with more or less modern anatomy. These included oligopithecids such as *Catopithecus*, parapithecids such as *Abuqatrania*,

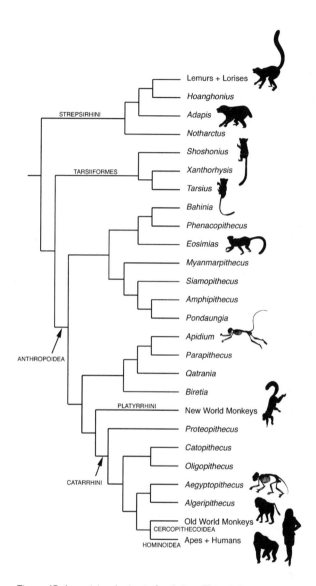

STREPSIRHINI

Lemurs + Lorises
Hoanghonius
Adapis
Notharctus

TARSIIFORMES

Shoshonius
Xanthorhysis
Tarsius

Bahinia
Phenacopithecus
Eosimias
Myanmarpithecus
Siamopithecus
Amphipithecus
Pondaungia

ANTHROPOIDEA

Apidium
Parapithecus
Qatrania
Biretia

PLATYRRHINI

New World Monkeys
Proteopithecus
Catopithecus
Oligopithecus
Aegyptopithecus
Algeripithecus

CATARRHINI

Old World Monkeys
CERCOPITHECOIDEA
Apes + Humans
HOMINOIDEA

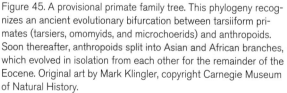

Figure 45. A provisional primate family tree. This phylogeny recognizes an ancient evolutionary bifurcation between tarsiiform primates (tarsiers, omomyids, and microchoerids) and anthropoids. Soon thereafter, anthropoids split into Asian and African branches, which evolved in isolation from each other for the remainder of the Eocene. Original art by Mark Klingler, copyright Carnegie Museum of Natural History.

proteopithecids such as *Proteopithecus* and *Serapia,* and others. Thousands of miles away in southeastern Asia, contemporary anthropoids like the amphipithecid *Siamopithecus* had grown larger, but their anatomy differed little from that of the geologically remote ancestors they shared with their African brethren.

At this point, we can return to some of the interesting questions posed earlier in this chapter. What factors contributed to the ultimate evolutionary success of anthropoids, and when did these factors come into play? We have already established that global cooling near the Eocene-Oligocene boundary and increasing body size (whether as a response to cooling or not) fail to explain the evolutionary success of anthropoids. As a more promising alternative, I nominate the early dispersal of anthropoids to Africa as the mechanism that triggered the events culminating in the evolution of modern anthropoid traits. Anthropoids were lucky enough to be among the first (if not the very first) Asian mammals to enter Africa. The succeeding bursts of evolutionary ingenuity transformed primitive primates such as *Altiatlasius* into creatures equipped with the fundamental features we now associate with monkeys, apes, and humans.

This new understanding of our deep evolutionary roots emphasizes once again the role of historical contingency in evolution. In contrast to older notions of primate and human evolution, nothing about anthropoid origins was preordained. Instead, anthropoids simply won the lottery by getting to Africa early on, thereby preempting the possibility of equally successful evolutionary radiations in other primate groups. Had some other group of Asian primates—adapiforms, omomyids, or even tarsiers—made their way to Africa first, I suspect that primate evolution would have unfolded very differently. In that case, anthropoids would never have enjoyed the chance to radiate in splendid isolation in Africa, and they almost certainly would not have acquired the full range of biological attributes that set them apart from their prosimian relatives today. Following this continuous chain of historical contingency forward in time across million of years, the absence of the basic anthropoid body plan would have precluded the evolution of the earliest apes or hominoids, who inherited numerous features from their immediate anthropoid ancestors. This, in turn, would have prevented early hominids from rising up on their hindlimbs and eventually returning to Asia, thus completing the grand biogeographic cycle initiated by their distant anthropoid ancestors some fifty-five million years earlier.

11

Paleoanthropology and Pithecophobia

When it comes to figuring out where we stand with respect to the rest of Earth's biotic diversity, we humans have always looked at nature through a narcissistic lens. It seems obvious that humans are special. The prevalence of this view comes as no great surprise, given that ours is the only literate, verbally gifted species on the planet. Alternative opinions—if they exist—have never been voiced or written down, at least not by the principals involved.

Prior to the Darwinian revolution, organized religions justified the concept that humanity stood apart from all other life forms. According to the Book of Genesis (1:26–28):

> Then God said, "Let us make man in our image, after our likeness; and let them have dominion over the fish of the sea, and over the birds of the air, and over the cattle, and over all the earth, and over every creeping thing that creeps upon the earth." So God created man in his own image, in the image of God he created him; male and female he created them. And God blessed them, and God said to them, "Be fruitful and multiply, and fill the earth and subdue it; and have dominion over the fish of the sea and over the birds of the air and over every living thing that moves upon the earth."

Generations of early naturalists, aspiring to apply the logic of science to make sense of the world around them, struggled to reconcile the acknowledged superiority and distinctiveness of humanity with an ever-growing body of biological evidence that suggested otherwise. These pre-Darwinian scholars concocted a broad spectrum of taxonomic solutions, arranged on a scale that equated human uniqueness with different ranks in the zoological hierarchy. The most extreme classifications emphasized the mental and spiritual acumen of their proponents by dividing the diversity of life into three separate kingdoms—the Human, the Animal, and the Vegetable. Richard Owen, a British anatomist and paleontologist whose various claims to fame include coining the term *Dinosauria* in 1842, adopted a marginally more realistic stance by isolating humans in one of four subclasses of mammals. Georges Cuvier, who shared Owen's disdain for evolution, segregated humans at a slightly lower taxonomic rank. According to Cuvier's scheme, which was widely embraced by pre-Darwinian zoologists, humans were the sole representatives of the mammalian order Bimana (meaning "two hands") while all other primates belonged in a separate order known as Quadrumana (or "four hands," in recognition of the structure of the primate foot, whose grasping big toe made the entire organ vaguely resemble a human hand). Linnaeus, the father of modern taxonomy, advocated that humans be grouped with apes, monkeys, tarsiers, lemurs, flying lemurs, and bats in the order Primates. However, even the exceptionally inclusive views of Linnaeus did not reflect evolutionary linkages among these species. His classification aimed to organize the natural world on the basis of overall similarity, but for Linnaeus this conveyed nothing in terms of ancestry or descent with modification.

Darwin fully recognized the extent to which his theory of evolution upset the cultural and theological apple cart, even if he remained conflicted about how to express it in terms of a zoological classification. Darwin went out of his way to ridicule the more outrageous schemes seeking to erect high taxonomic barriers between humans and other primates.[1] Darwin and his followers readily conceded that humans differed dramatically from other primates in terms of raw intelligence. Thomas Henry Huxley, Darwin's scientific acolyte and most vocal advocate, recognized this disparity by placing humans in one of three major subdivisions of primates, alongside anthropoids and prosimians. Yet Darwin insisted that such a large taxonomic distinction obscured the close evolutionary affinities of humans and apes.[2] After reviewing the anatomical differences between New World monkeys (or platyrrhines) and their Old World (or catarrhine) relatives, Darwin noted: "And as man from a genealogical point of view belongs to the Catarhine [*sic*] or Old World stock, we must conclude, however much the conclusion may revolt our pride, that our early progenitors would have been properly thus designated."[3]

The popular backlash against Darwin is well documented, and it was hardly restricted to puritanical North America. In Victorian England, upon hearing that Darwin and Huxley supported the notion that humans evolved from apelike ancestors, the wife of the bishop of Worcester is said to have exclaimed, "Descended from apes! My dear, let us hope that it is not true, but if it is let us pray that it will not become generally known."[4] To a lesser extent, this reluctance to acknowledge our apelike—not to mention monkeylike—ancestors extended to well-known paleontologists and anthropologists. Decades after Darwin published *The Descent of Man,* reputable British and American scholars continued to debate exactly how humans fit on the primate family tree. Many of these revisionist scientists, as well as much of the lay public, felt uneasy about where and when the human lineage originated. If it were no longer possible to deny that humans were nestled within the primate evolutionary radiation, a case might at least be made that the human lineage was exceptionally ancient and therefore noble—in other words, as far away from monkeys and apes as science would possibly allow.

In 1918, the British anatomist Frederic Wood Jones launched one of the earliest scientific attacks on Darwin's thesis that humans share a close common ancestry with apes and other anthropoids.[5] Wood Jones, one of the primary architects of the arboreal theory of primate and human evolution, believed that life in the trees inevitably led to a series of evolutionary changes culminating in the upright posture and enlarged brains of modern humans. According to this view, nonhuman primates failed to reach the human condition because their own evolutionary specializations diverted them from the steadily upward path toward humanity. Apes, for example, fell off the bandwagon of progress when they developed specialized forelimbs for swinging beneath the branches of trees, a mode of locomotion known as brachiating. Monkeys deviated by crouching down to walk on all fours, thereby arresting the natural evolutionary trajectory toward upright postures predicted by the arboreal theory. Neither apes nor monkeys qualified as close relatives of humans, because their degenerate evolutionary tendencies moved them increasingly farther away from the human condition. Instead, Wood Jones argued, humans had evolved from an early "tarsioid" stock of primates, whose upright body posture still prevailed in both tarsiers and humans.

Wood Jones found a receptive audience outside the academic establishment, and he avidly promoted his argument that humans lacked monkeylike or apelike ancestors in various public forums. Mainstream British anthropologists proved to be considerably less enthusiastic. When the leading British authorities on primate anatomy gathered to debate the role of tarsiers in human evolution at a special symposium hosted by the Zoological Society of London in 1919, they rejected Wood Jones and his tarsioid theory of human origins one after another. They cited a wide range of biological evidence indicating that humans share a close common ancestry with apes and other anthropoids, while tarsiers occupy a more distant evolutionary position, intermediate between anthropoids and lemurs. Even Sir Grafton Elliot Smith—who, like Wood Jones, was a leading proponent of the arboreal theory of primate and human evolution—repudiated Wood Jones in his presentation on the comparative anatomy of primate brains, showing that humans and other anthropoids share numerous neurological features that are absent in tarsiers and lemurs.[6] Others went further, impugning both Wood Jones's motives and his scientific competence:

> Professor Wood-Jones, to whom we are indebted for originating the interesting discussion, was more reticent to-night than in his addresses to more popular audiences. He attacked Darwinian evolutionists, and Huxley in particular, on the supposition that they believed genealogical trees to be linear, that a higher group took origin from the highest members of a lower group. It was a strange misreading of familiar evidence. Were he to consult Huxley's Essay on "Man's Place in Nature," or any general statement of the case for Evolution, as, for instance, the Article under that heading in the "Encyclopedia Britannica," he would see that he was attacking a bogey that does not exist.[7]

Wood Jones and his tarsioid theory of human origins never recovered from the scholarly drubbing they absorbed that day. Wood Jones exiled himself to Australia, kindled an academic interest in marsupials, and soon abandoned his research on primate evolution altogether.[8]

After Wood Jones's tarsioid debacle, no serious scholar would dare to propose that humans evolved from anything other than an anthropoid. Still, this did not preclude the argument that humans descended from a separate and independent lineage extending far back in geological time. Among those who continued to doubt that humans were closely related to apes was none other than the eminent paleontologist Henry Fairfield Osborn, president of the American Museum of Natural History. Osborn's museum housed many of the most complete and important primate fossils known at the time, including the original skull of *Tetonius* collected by Jacob Wortman in Wyoming in 1881 and several skeletons

Figure 46. The paleontologist Henry Fairfield Osborn, president of the American Museum of Natural History and chief proponent of the "dawn man" theory of human origins.

of *Notharctus*. Osborn himself had described *Apidium phiomense* as the first Fayum anthropoid known to science. Osborn even played a role in the famous Scopes "monkey" trial in Tennessee in 1925, taunting William Jennings Bryan and his fellow creationists with all manner of evidence supporting the scientific validity of evolution. As an avowed evolutionist, Osborn did not doubt that humans and apes shared a common ancestor. Rather, he argued that humans descended from an ancient lineage that bypassed apes and monkeys on the family tree: "The most welcome gift from anthropology to humanity will be the banishment of the myth and bogie of ape-man ancestry and the substitution of a long line of ancestors of our own at the dividing point which separates the terrestrial from the arboreal lines of primates."[9]

As the preceding comment makes clear, Osborn's personal anxieties and philosophical convictions strongly influenced his ideas about human evolution. In this respect, Osborn hardly stands alone. Paleoanthropology has always tolerated interpretations that reach beyond the confines of simple logic and deduction.[10] To his credit, Osborn's unconventional take on human evolution was founded—at least in part—on what seemed like compelling fossils. Unfortunately for Osborn, we now know that the most crucial fossils bolstering his views were either erroneously interpreted or downright fraudulent. At the same time, Osborn's eccentric ideas about the evolutionary process led him to advocate a long and ancient line of human ancestors for purely theoretical reasons. For Osborn, evolution had a purpose, and its results were largely preordained. Through the long span of Earth history, animals became progressively bigger, stronger, smarter, and more complicated. And more than anything else, changing climatic conditions were the driving force behind this long, steady climb up the evolutionary ladder.

The intimate connection between climate and evolution had been explored in detail by William Diller Matthew, Osborn's colleague at the American Museum of Natural History (see chapter 5). Matthew's theory held that life was easy during the greenhouse conditions of the Eocene. With the onset of cooler, more temperate conditions in the Oligocene, the struggle for existence began in earnest, and progressive new types of mammals evolved as a result. These adaptively superior mammals originated primarily in northern regions—especially in central Asia—where the magnitude of the climate shifts was most severe, thereby intensifying natural selection. Those mammals that were unable to adapt to the rigorous new circumstances either went extinct or followed the shrinking forests to the south. According to Osborn, this same evolutionary pattern applied to humans:

> In the succeeding Oligocene time we discover a sharp and world-wide division between plateau-loving and forest-loving types [of mammals]: in the forests remain all the backward conservative types; on the plateaus and uplands are found the alert, progressive, forward-looking types, including all the long-hind-limbed bipedal animals adapted to rapid progression in an open or partly forested country. . . . Is it likely that the primates alone escaped this divorce between backward, forest-loving life and forward, plateau, savanna and upland life, especially as Eocene forest areas in every continent began to contract and upland open plains and plateaus began to expand?[11]

The apparent link between evolutionary progressiveness and open-country living put Osborn directly at odds with Wood Jones's arboreal theory of primate and human evolution. Humans, being primates, must have had tree-dwelling ancestors at some stage, but for Osborn, these ancient human forerunners must have dated back to the Eocene, when forests enshrouded most of the planet. Instead of viewing humans as the logical culmination of life in the trees, Osborn believed that early hominids leapt ahead of their anthropoid relatives precisely because they left the trees to chart a new life on the ground. Osborn characterized these earliest human ancestors—or "dawn men," as he liked to call them—as "ground-living, cursorial, alert, capable of tool-making, and living in a relatively open country on the high plateaus and plains of northern Asia." Apes differed in being "tree-living, brachiating, sluggish, incapable of tool-making, [and] restricted to the forests of south temperate and tropical countries."[12]

A selective reading of the fossil record reinforced Osborn's "dawn man" theory that humans originated in central Asia during the Oligocene. The traditional Darwinian alternative, which Osborn derisively called the "ape man" theory, held that humans had evolved more recently, from apelike ancestors in Africa. Even in Osborn's day, the fossil record hinted that Darwin was right. Raymond Dart published the first African australopithecine—the famous juvenile Taung skull from South Africa— in 1925.[13] Yet Osborn steadfastly ignored the Taung skull in his numerous publications on human evolution. Instead, like many accomplished anthropologists of the day, Osborn relied on the notorious Piltdown forgery—so much so that his "dawn man" theory took its name from *Eoanthropus dawsoni,* the zoological designation for the species that "Piltdown Man" was supposed to represent. A chimera constructed by juxtaposing a modern human skull with the lower jaw of an orangutan, the Piltdown hoax fit perfectly with Osborn's scheme of human evolution.[14] Having been implanted in English gravel beds dating to the Pliocene, Piltdown Man seemingly proved the antiquity of big brains— and increased intelligence—in the human lineage. Osborn relished the way Piltdown Man agreed with his theory of human origins, and he also delighted in citing other evidence of ancient human ancestors, no matter how fragmentary it might be.

Although Osborn's "dawn man" theory predicted that early human ancestors would be found in central Asia, the oldest relevant fossil that he could marshal came from a late Miocene site in the United States. In 1922, a geologist named Harold Cook unearthed a single tooth from the Snake Creek beds of western Nebraska. The highly worn molar preserved few anatomical landmarks. Despite its poor condition, Cook wrote a letter to Osborn stating that the specimen "very closely approaches the human type" and encouraging him to study it. Possibly falling prey to the power of suggestion, Osborn soon agreed that the tooth belonged to an anthro-

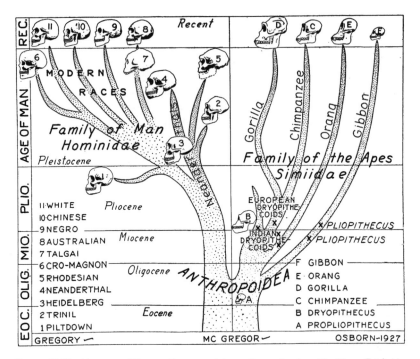

Figure 47. The "dawn man" theory of human origins as it was developed by Henry Fairfield Osborn in the late 1920s. Note that the human lineage, depicted to the left of the family tree, originates in the Oligocene, entirely bypassing the branch to the right leading to modern apes. From Osborn 1927, copyright American Philosophical Society. Reproduced by permission.

poid forerunner of humans, which he promptly named *Hesperopithecus haroldcookii*.[15] In keeping with the geographic aspects of the "dawn man" theory, Osborn interpreted *Hesperopithecus*—colloquially known as "Nebraska Man"—as an immigrant from Asia. The fossil soon took its place near the base of the human family tree in Osborn's publications.[16]

A more timid personality would have hesitated to champion an isolated tooth as crucial evidence in the debate over human origins, but Osborn forged right ahead. Indeed, Osborn highlighted the specimen during his verbal duel with William Jennings Bryan, the creationist prosecutor at the Scopes "monkey" trial—who happened to be a native Nebraskan. With a devilish nod to the biblical passage "Speak to the earth, and it shall teach thee" (Job 12:8), Osborn wrote: "The earth spoke to Bryan, and spoke from his own native state of Nebraska."[17] Ironically, within a few short years, Nebraska Man would fall spectacularly from grace after the solitary tooth of *Hesperopithecus* was exposed as that of an extinct peccary (a group of piglike ungulates that includes living javelinas).[18] Having turned the tables on Osborn, creationists to this day cite Nebraska Man as yet another reason to doubt paleontological accounts of human evolution.

In retrospect, Osborn's exuberance on behalf of Piltdown Man and Nebraska Man was woefully misplaced. At the same time, Osborn downplayed the significance of genuine fossil hominids that could have bolstered the "dawn man" theory. The fact that these two landmark discoveries played less than a pivotal role for Osborn owes partly to chance and partly to geography. The Dutch physician Eugène Dubois unearthed the first substantial Asian hominid remains in the 1890s at a site known as Trinil in Java.[19] These *Homo erectus* fossils, popularly known as "Java Man," did not fully conform to Osborn's "dawn man" theory for two reasons. First, they hailed from the equatorial margin of southeastern Asia, far from the plateaus and steppes that Osborn believed were the birthplace of humankind. Second, the beetle-browed skullcaps of Java Man were so primitive—especially compared with the elegantly modern forehead of Piltdown Man—that Osborn regarded the Indonesian fossils as "a case of arrested development."[20] Instead of citing Java Man as crucial support for the "dawn man" theory, Osborn considered it an irrelevant side branch of the hominid family tree.

The second pivotal discovery of hominid remains in Asia—at Zhoukoudian on the outskirts of Beijing—confirmed Osborn's prediction that fossil humans would be found in central and northern Asia. Despite its promising geographic context, however, "Peking Man" never figured prominently in Osborn's thinking. For one thing, Osborn had already formulated his "dawn man" theory by the time the first isolated hominid teeth were described from Zhoukoudian. Jaws and partial skulls were unearthed slightly later. Yet even if Osborn had known about these specimens, it is debatable whether he would have given them much weight. The Zhoukoudian cranial remains display the same low forehead and strong brow ridges that occur in Java Man. As such, Osborn likely would have regarded them as yet another evolutionary dead end on the human family tree. His "dawn man" theory predicted the discovery of ancient Asian hominids with advanced skulls like that of Piltdown Man. Neither Java Man nor Peking Man satisfied that heady requirement. In fact, the Asian fossil record was remarkably devoid of support for the "dawn man" theory.

If fossils failed to drive Osborn's idiosyncratic ideas about human origins, what else might have done so? Two factors apparently motivated him, both having more to do with Osborn's own ambition and social outlook than with science per se. They illustrate different sides of Osborn's legacy—one benign and the other repugnant.

Osborn began his career as a scientific researcher, but his organizational skills and social connections drew him increasingly into administration. In many ways, Osborn's career trajectory was all but inevitable—his uncle was J. P. Morgan, through whom he maintained strong ties to New York's political and financial elite. By 1908, Osborn was appointed president of the American Museum of Natural History. His tenure in that capacity was long and mostly successful. By leveraging his connections to New York's philanthropic community, Osborn raised the museum's profile as a cultural and educational treasure for the city and the nation. He also spurred the museum's development into a center for groundbreaking scientific research. Osborn's expertise in paleontology played a vital role in this success, because dinosaurs and other prehistoric beasts attracted large crowds to the museum's galleries, bolstering attendance. By mounting major expeditions to search for fossil vertebrates, Osborn believed, the museum accomplished several goals at once. Such expeditions not only enhanced the museum's stature as a research institution; they also allowed it to obtain unique specimens for display. As an added benefit, none of the museum's activities yielded more visible and more positive publicity than its exploration for fossils in remote and exotic locales.

Roy Chapman Andrews emphasized all of these factors when he approached Osborn with the notion that the museum should launch a series of expeditions to the central Asian nation of Mongolia. Osborn enthusiastically supported Andrews's idea, and he avidly participated in the fund-raising and international diplomacy that were required to initiate the project. To potential donors, Andrews and Osborn pitched the whole endeavor as a chance to pinpoint human origins. The tactic proved immensely successful. According to Osborn: "The hunt for fossil man in Asia was the slogan which enabled Roy Chapman Andrews to arouse a nation-wide interest in his great project of Mongolian exploration; no other appeal for financial backing was nearly as strong as this, and it finally enabled him to start his first expedition in the season of 1921."[21]

The basic rationale for Andrews to search for human origins in Mongolia had already been established, because Osborn and Matthew had previously designated central Asia as the most likely birthplace for humans and many other mammals (see chapter 5). Yet Osborn never elaborated his full-fledged "dawn man" theory until after Andrews began his Mongolian expeditions. In a very real sense, the "dawn man" theory needed the Central Asiatic Expeditions as much as the expeditions themselves required a focus on human origins as their raison d'être. Osborn himself claimed that his theory originally came to him in a moment of inspiration on the Mongolian steppes:

> It was at the dramatic moment of the close of the season of 1923 when the discovery of the dinosaur eggs gave our expedition a world-wide fame, that the writer joined Andrews' party in the east central Gobi and began to visualize the life environment of Tertiary time as ideal for the early development of the dawn men or the direct ancestors of the human race. This high plateau country of central Asia was partly open, partly forested, partly well-watered, partly arid and semi-desert. Game was plentiful and plant food scarce. The struggle for existence was severe and evoked all the inventive and resourceful faculties of man and encouraged him to the fashioning and use first of wooden and then of stone weapons for the chase. It compelled the Dawn Men—as we now prefer to call our ancestors of the Dawn Stone Age—to develop strength of limb to make long journeys on foot, strength of lungs for running, and quick vision and stealth for the chase. Their life in the open, exposed to the rigors of a severe climate, prompted the crude beginnings of architecture in their man-made shelters, and the early use of fire for bodily warmth and for the preparation of food.[22]

From a strictly scientific perspective, it hardly matters whether Osborn concocted his "dawn man" theory to drum up financial support for the Central Asiatic Expeditions. The remarkable results of those expeditions—including entire clutches of dinosaur eggs buried in their nests, as well as new types of dinosaurs and fossil mammals—speak for themselves. Sadly, a second issue related to Osborn's development of the "dawn man" theory cannot be so readily dismissed.

By modern standards—and possibly even by the weaker standards of the 1920s and 1930s—Osborn promulgated racist ideas. Not surprisingly, his "dawn man" theory of human origins embraced and incorporated this bias. Osborn argued that the major human races—which he ranked as Caucasian, Mongolian, and Negroid—were distinct biological entities, "which in zoology would be given the rank of species, if not of genera." He also claimed that important "spiritual, intellectual, moral, and physical characters" distinguish these races. According to Osborn, these racial differences arose as early human groups adapted to their divergent environments. Throughout their history, Caucasians and Mongolians were exposed to the invigorating conditions of northern and central Eurasia—the same environmental parameters that fostered the evolution of the earliest "dawn men." The Negroid race, on the other hand, invaded more equatorial regions, where "the quest for food is very easy and requires relatively little intelligence; the environment is not conducive to rapid or varied organic selection; the struggle for mere existence is not very keen; the social and tribal evolution is very slow; intellectual and spiritual development is at a standstill. Here we have the environmental conditions which have kept many branches of the Negroid race in a state of arrested development."[23]

In Osborn's opinion it took a long time for these racial distinctions to evolve, hence the need for the earliest "dawn men" to be so old. At the same time, while Osborn regarded equatorial races of humans as being fairly primitive, by no means did this imply that the earliest "dawn men" evolved in those regions. The warm and luxuriant tropics were the natural habitat of apes and monkeys. To posit that the Negroid race originated in such a decadent environment conflicted with Osborn's most basic premise. Ancestral humans diverged from apes and monkeys by forsaking the safety, the comfort, and the bounty of tropical Eocene forests to embark on a more challenging evolutionary career in the wide-open steppes of central Asia. For Osborn, only after their "dawn man" ancestors adapted fully to life on the ground did the Negroid race return to the tropics. There, plentiful food and easy living conditions hindered their further evolutionary progress.[24]

Today, Osborn's views on race are taken about as seriously as his ideas on human origins. Within a few years of its formulation, Osborn's "dawn man" theory collapsed in the face of overwhelming evidence showing that humans are closely related to chimps and gorillas, as Darwin himself had long maintained. Ironically, it fell to one of Osborn's underlings at the American Museum, the well-known comparative anatomist William King Gregory, to present this case.[25] Osborn himself—through his wide-ranging (and often ill-conceived) political activities—guaranteed that his opinions on race would be rejected. As a leading figure in the American eugenics movement, Osborn lobbied for laws restricting immigration and promoting the involuntary sterilization of individuals who were deemed unfit to reproduce. He even went so far as to endorse the early policies of Hitler during a visit he made to Germany in 1934.[26]

Just as the demise of Wood Jones's tarsioid theory of human origins forced subsequent scientists to admit that humans had evolved from anthropoids, the fall of Osborn's "dawn man" theory made it impossible to disown our apelike hominoid ancestors. Yet paleoanthropologists continued to argue that the divergence between humans and apes was geologically remote, dating back tens of millions of years. As recently as the mid 1960s, two of the world's leading authorities on human origins—Louis Leakey and Elwyn Simons—agreed on little else beyond their shared conviction that the human lineage could be traced back to the Oligocene. As we have seen, only the skimpiest of fossil evidence ever supported such an ancient human pedigree. The reason for this is now obvious: chimps and humans shared a common ancestry until sometime in the late Miocene, probably less than seven or eight million years ago. This raises a rather obvious question. Why did it take nearly a century after Darwin laid out the evidence in *The Descent of Man* for paleoanthropologists—not to mention the public at large—to accept a close (and geologically recent) link between humans and African apes?

I can think of two factors that have contributed to the long delay. The first has been a strong and long-standing incentive to find increasingly ancient human ancestors. The paleoanthropological equivalent of the Holy Grail is a fossil hominid that immediately postdates the ape-human split. For years, the simplest way for a paleoanthropologist to trump his or her rivals has been by claiming to find the oldest known hominid, no matter how thin the underlying fossil evidence might be. Now that truly ancient fossils are no longer in the running, future disagreements about the earliest known hominid will oscillate over specimens spanning an increasingly narrow window of late Miocene time. Appropriately, the quality of the fossils themselves—and the degree of confidence we can have that they actually pertain to the human lineage—has moved to the forefront, while age alone has become less important in determining how we reconstruct the earliest phases of human evolution. Poorly substantiated claims of finding a fossil hominid dating to the early or middle Miocene are simply no longer tenable.

Although the second factor is more subtle, it is also more pervasive and enduring. Many people suffer from a deep-seated anxiety about their origins. No one wants to acknowledge the dirty rotten scoundrel who may be lurking in his or her family tree. By the same token, the bishop of Worcester's wife is hardly alone in being horrified by the prospect of having descended from apelike ancestors. Given the choice, who wouldn't prefer to have been created de novo in the likeness of a supernatural deity? To highlight this dilemma, William King Gregory sardonically claimed to have discovered a new psychological disorder that he called "pithecophobia." According to Gregory's diagnosis, this irrational fear of apes and monkeys as potential ancestors is brought on by greater knowledge of our own evolution.[27] Yet in the same way that we cannot choose to be born into royalty, science dictates that we admit our evolutionary history or risk burying our heads deeply in sand. An ostrich presumably gains some measure of comfort by poking its head in the nearest dune, but the act also imperils the bird unnecessarily. Likewise, certain risks arise from our behavior if humanity cannot overcome its angst about its deep evolutionary roots.

The most immediate threat falls not on us, but on our closest living relatives. Recent surveys of chimps and gorillas in the West African nations of Gabon and the Republic of Congo show that the populations of both species have declined by more than half over the past two decades.[28] If these trends continue, African apes will soon become extinct in the wild. Sadly, human activities have brought about this plight. Throughout most of their natural range, African apes are hunted and killed for food as part of the "bushmeat" trade—the largest single factor contributing to the recent decrease in African ape populations. Given the evolutionary proximity of humans and African apes, this practice comes perilously close to cannibalism. We can engage in endless debate over the morality of the bushmeat trade, but its impact on the health of local human populations is beyond dispute. Outbreaks of Ebola hemorrhagic fever—one of the deadliest and most contagious diseases known to modern medicine—have been noted to occur in close proximity to infected African ape populations. Handling and eating contaminated ape meat is the most likely method of transmitting the Ebola virus to humans.

Even when there is no obvious connection between our failure to come to grips with our own evolutionary history and the human condition per se, a strong case can be made for adopting a more scientifically defensible perspective on humanity's rightful place in nature. The biblical command for human populations to multiply as we progressively subdue the earth and all of its living creatures made sense at the end of the Neolithic, but it hardly seems applicable today. With more than six billion humans currently inhabiting the planet, our species has probably earned the right to declare that particular mission accomplished. In any case, along with rights come responsibilities. Being granted dominion over the earth and its biological diversity therefore entails responsible stewardship. Yet so long as we regard ourselves as being separate and apart from nature, there is no compelling reason to conserve natural resources and preserve endangered species. Unfortunately, paleoanthropology in the strictest sense—with its fixation on charting our noble ascent from humble beginnings among the primates—hardly rectifies this discrepancy.

If Gregory's diagnosis of pithecophobia correctly identifies part of the problem, the solution must include a dramatic shift in outlook. Confronted with a patient suffering from delusions of grandeur, any well-trained psychotherapist would try to help the afflicted person develop a more realistic appraisal of his or her lot in life. We must do the same for humanity as a whole. The widespread view that humans are categorically distinct from other animals suffers from a lack of objectivity. "If man had not been his own classifier, he would never have thought of founding a separate order for his own reception," Darwin himself observed. How can we achieve a more balanced understanding of our place in nature?

Let's begin by returning to one of the discredited evolutionary scenarios we discussed earlier in this chapter, the tarsioid theory of human origins developed by Frederic Wood Jones. Tarsiers and humans make a compelling comparison because they are both so weird. Any impartial zoologist from outer space would immediately be intrigued by both species because of their unique biology, but he would have great difficulty choosing which of the two is more distinctive. At first, the alien zoologist would be impressed by the fact that humans stand out among primates in walking upright on their highly modified hindlimbs. Being utterly objective, he would also note that tarsiers are equally bizarre, using their grotesquely elongated ankles to leap across distances spanning many times their own body length. Upon evaluating these uniquely specialized modes of locomotion, the Martian zoologist would have no reason to rank humans ahead of tarsiers (or vice versa) on his scale of biological distinctiveness. Our alien scientist would then proceed to analyze their divergent sensory adaptations. Humans obviously differ from other primates in our legendary powers of cognition. The physical manifestation of this difference—our enormously enlarged brains—would be plain for the Martian zoologist to see. Yet he would also note that the nervous system of tarsiers stands out in an entirely different, yet no less obvious, way—tarsier eyeballs have swollen to encompass the same volume as a tarsier brain. Once again, the Martian expert would find both tarsiers and humans engagingly unique, but he would not necessarily rule that humans warrant special biological scrutiny.

The take-home message of this imaginary episode is that biological distinctiveness, like beauty, lies in the eyes of the beholder. As long as humans are the only judges at the pageant, the odds are stacked distinctly in our favor. An unbiased arbitrator—like an alien zoologist from beyond our solar system—might be considerably less smitten by our natural charms.

A more objective measure of biological divergence is the amount of time that has elapsed since a given species—or a closely related group of species like tarsiers—has separated from its nearest living relatives. The fossil record demonstrates that tarsiers have been evolving independently for roughly fifty-five million years, almost ten times as long as humans have been separated from chimpanzees. Evidence of this enormous chronological disparity is literally written into our genetic code. When tarsier DNA is sequenced and compared with that of other mammals, it sometimes fails to group unambiguously with primates.[29] Human DNA always falls neatly within the primate range of variation, being especially close to that of chimps and gorillas. If evolutionary isolation accurately gauges biological distinctiveness, tarsiers are exceptional primates indeed. By this same criterion, humans are merely run-of-the-mill.

It also helps put things in perspective to engage in some simple thought experiments. Imagine that a lucky primatologist discovers a small, isolated population of living robust australopithecines—the group of early hominids exemplified by Louis and Mary Leakey's *Zinjanthropus* skull from Olduvai Gorge—deep in the forest of central Africa. The species would immediately become the object of dramatic and intensive conservation efforts, and rightly so. Such close human relatives, with so much insight to offer us about our own evolution, would deserve all the resources at our disposal. Philosophers and legal experts might even debate whether the new species should be assigned certain rights normally reserved for humans. At the very least, a broad consensus would emerge that the robust australopithecines should not be killed and eaten. Genocide—in this case the eradication of an entire species—would be completely out of the question.

Now imagine yourself living a century in the future. The global human population has finally stabilized at a sustainable level, after expanding to more than ten billion people during the first half of the twenty-first century. Yet the dramatic expansion and contraction of human populations has taken its toll. Apes are entirely extinct in the wild. Zoos house the few living examples of chimps, gorillas, orangutans, and gibbons. Yet the loss of biological diversity is hardly limited to our nearest anthropoid kin. After succeeding for more than fifty-five million years in their daily struggle for existence, tarsiers finally encountered an evolutionary obstacle—in the form of extensive habitat destruction caused by human population growth—that they simply could not overcome. What was formerly a thriving and bushy branch of the primate family tree wilts and dies along with them.

Should these highly plausible events come to pass, I suspect that future generations of humans would mourn the loss. They would also hold us accountable, just as we look back on the racist attitudes of Henry Fairfield Osborn and like-minded Victorians with a mixture of pity and disgust. Our knowledge of the hominid fossil record demonstrates that all living humans, regardless of our skin color or ethnicity, share a very recent common ancestry. Racism is incompatible with such knowledge. By the same token, our close kinship with other primates requires us to view them in an entirely different light. For the vast majority of our long evolutionary journey, our fate has been bound by common ancestry with that of our primate relatives. Although it is tempting to emphasize the differences between humans and other primates by focusing on how humans evolved from apes, an unbiased account of our evolutionary history must make the opposite point. The deep evolutionary roots that we share with other anthropoids go back some fifty-five million years, while the human lineage itself is nearly an order of magnitude younger. Humanity as a whole is embedded within a rich biological tapestry. The living legacy of that common evolutionary journey deserves to be celebrated rather than despised. Pithecophobia in all of its manifestations conflicts with our own deep roots.

Notes

1. MISSING LINKS AND DAWN MONKEYS

1. As in many modern ecosystems, small muroid rodents are particularly abundant at many middle Eocene and younger fossil localities in China (Dawson and Tong 1998).

2. There is a great deal of literature on this subject. Among the more important primary references in recent years are White et al. 1994, M. G. Leakey et al. 1995, Haile-Selassie 2001, and Brunet et al. 2002.

3. Wilf and Labandeira 1999 and Wilf 2000 provide excellent empirical studies of the coevolution of plants and insects during this time in western North America, where the fossil record is exceptionally good.

4. Niemitz 1984a, 65, reports: "Under semiwild conditions this small swift flying bird [*Ceyx erithacus rufidorsus*, the forest kingfisher], which is very common in that area, was caught in the air, flying past the Bornean tarsier. The tarsier caught it like the goal keeper of a soccer game catches the ball with both hands in midleap. Both animals fell to the ground, where the tarsier killed the bird with a fast sequence of bites into the neck region."

5. Stephan 1984, 330.

6. Pollock and Mullin 1987.

7. Schmitz et al. 2001.

8. Beard 1998.

9. Despite its name, *Afrotarsius chatrathi* from the Fayum region of Egypt is more likely to be a very primitive species of African anthropoid than a fossil tarsier (Beard 1998).

10. Darwin 1874, 171.

11. Pilgrim 1927, 15.

12. B. Brown 1925 and esp. L. Brown 1950 provide engaging firsthand accounts of the Browns' 1923 expedition to Myanmar.

13. Colbert 1937.

14. Tsubamoto et al. 2002.

15. The significant role that these Burmese Eocene primates continued to play in the debate regarding anthropoid origins is shown by the fact that the new discoveries were published in *Nature* and *Science,* despite the fragmentary condition of the new fossils (Ba Maw et al., 1979; Ciochon et al., 1985).

16. Analysis of what was then known regarding the paleogeography and paleobiogeography of Africa during the early Cenozoic led P. A. Holroyd and M. C. Maas 1994 to support an endemic African origin for anthropoids. In the same edited volume, R. L. Ciochon, previously a strong proponent of the view that Burmese *Pondaungia* and *Amphipithecus* were the world's oldest anthropoids, changed his mind and argued instead that *Pondaungia* was related to North American notharctids (a well-known group of lemurlike adapiform primates from the Eocene), while the evolutionary position of *Amphipithecus* was left up in the air (Ciochon and Holroyd 1994).

17. Beard et al. 1994.

18. In a growing list of fossils relevant to this issue, those described by Ji et al. 1998, 2001, are among the most significant.

19. Le Gros Clark 1960, 349.

20. D. T. Rasmussen 1994, fig. 7, sketched out one of the most explicit recent attempts to tie anthropoids in a simple ancestor-descendant sequence with earlier prosimians. His ladderlike phylogeny derives anthropoids directly from lemurlike adapiform primates, while relegating tarsiers to a distant side branch of the primate family tree, thus disagreeing with Le Gros Clark's original version of the ladder.

21. This philosophical link between a relatively recent time of origin for anthropoids and their direct derivation from fossil prosimians is nicely illustrated by the following quotation from Rasmussen and Simons 1992, 503: "We predict that as more is learned about Old World primate faunas during the latest Eocene, more 'protoanthropoid' prosimians and prosimian-like anthropoids will be discovered, which will not be easily and unambiguously classified into either suborder."

22. The many problems involved in proposing direct ancestor-descendant relationships in the fossil record are explored in much greater detail by Gee 1999.

2. TOWARD EGYPT'S SACRED BULL

1. Thomas Jefferson's statement here dates from 1781–82 (Simpson 1942, 150).

2. An English translation of Cuvier's original paper on the subject, which was read at the public session of the French Institut National on April 4, 1796, is provided by Rudwick 1997, 18–24.

3. English translation in Rudwick 1997, 52, of an address given by Cuvier on November 17, 1800.

4. Rudwick 1997, 260–61.

5. An English translation of Cuvier's original paper is provided by Rudwick 1997, 69–73.

6. Delfortrie 1873; Filhol 1874.

7. Beard et al. 1988.

8. Godinot 1998, table 1.

9. Although there are valid scientific reasons for enlarging the concept of hominids to include all living and fossil great apes (orangutans, gorillas, and chimps), I use the term in its more restricted, traditional sense throughout this book, simply to avoid unnecessary confusion.

10. These include the fifteen European species assigned by Godinot 1998, table 1, to the subfamily Adapinae, as well as the primitive Chinese adapid *Adapoides troglodytes,* which Godinot 1998 refers to the subfamily Caenopithecinae.

11. Dagosto 1983.

12. Godinot and Jouffroy 1984.

13. Kay and Covert 1984.

14. Ravosa and Hylander 1994.

15. Rosenberger and Strasser 1985.

16. Gingerich and Martin 1981.

17. Beard et al. 1994.

18. Gingerich 1981a.

19. Godinot 1998, fig. 10, shows that the stratigraphic ranges of *Adapis* and *Leptadapis* overlap for most of their duration. Even the earliest members of the *Leptadapis* lineage are depicted as a distinct branch of adapids that was not ancestral to *Adapis.*

20. Franzen 1987, fig. 17.

21. Warren 1998.

22. Leidy 1869.

23. "Marsh and his assistants appeared on the scene at Fort Bridger in 1870, imperiously declaring that only they should work these fossil beds, despite Leidy's equal claim, if not obvious priority. Almost immediately there began an intense competition for fossils with Leidy's collector, Ferdinand Hayden, who wrote [to Leidy], 'Marsh is more ambitious than Cope ever was. He is raging ambitious . . . and feels disposed to be unfriendly'" (Warren 1998, 186).

24. Leidy 1870, 114.

25. Marsh 1872, 405–6.

26. Leidy 1873, 86.

27. Leidy 1873, 89–90.

28. Dagosto 1993.

29. Covert 1986, 350–51.

30. Gingerich 1981a.

31. Godinot 1992.

32. Godinot 1991; Bacon and Godinot 1998.

33. Krishtalka et al. 1990 describe the evidence for a high degree of canine

sexual dimorphism in *Notharctus venticolus*. Gingerich 1995 suggests that fossils of a much earlier notharctid species, *Cantius torresi,* also exhibit evidence of canine dimorphism. Although Gingerich's interpretation seems likely, the fossils in question do not preserve the actual crowns of the canines. As such, current evidence tentatively points toward of a high degree of canine sexual dimorphism in early Eocene *Cantius torresi.*

34. Beard and Dawson 1999.

35. Gingerich and Simons 1977; Gingerich and Haskin 1981.

36. Primitive Eocene sivaladapids are reported by Qi and Beard 1998, while some of the geologically youngest sivaladapids are described by Gingerich and Sahni 1984. Gebo et al. 1999 describe the only known postcranial element of a sivaladapid—a fragmentary first metatarsal of *Hoanghonius stehlini.*

37. Godinot 1998, table 1, lists twenty-nine species of European cercamonines. The only North American representative of this group is *Mahgarita stevensi* from the late middle Eocene of west Texas (Wilson and Szalay 1976). Several small cercamoniines, as well as the larger species known as *Aframonius dieides,* have been found in North Africa and Oman (Hartenberger and Marandat 1992; Gheerbrant et al. 1993; Simons et al. 1995; Simons 1997a; Simons and Miller 1997).

38. Dickens et al. 1997.

3. A GEM FROM THE WILLWOOD

1. Much of my account of Wortman's early exploration of the Wind River and Bighorn basins is adapted from Gingerich 1980a.

2. Wortman 1899, 140. Gingerich 1980a questions whether Wortman actually faced such dire threats to his personal safety.

3. Cope 1882a.

4. Cope 1882b, 154.

5. I first heard *Tetonius* referred to as the "gem of the Willwood" as a graduate student collecting fossils in the Bighorn Basin under the supervision of Dr. Kenneth D. Rose at Johns Hopkins University. Both Ken and his colleague, Dr. Thomas M. Bown, then of the U.S. Geological Survey, regarded *Tetonius* that way.

6. No single reference documents the amazing diversity of North American omomyids. Important primary sources of information include Szalay 1976; Bown and Rose 1987; Beard et al. 1992a; and Gunnell 1995.

7. Gingerich 1984.

8. Niemitz 1984a provides information on the diet of tarsiers, which are unique among living primates in that they eat only live animal prey. For a discussion of the importance of gums in the diets of small primates, see Nash 1986 and references therein.

9. Strait 2001.

10. Dagosto 1993 provides a comprehensive overview of omomyids' postcranial anatomy and their inferred locomotor behavior. Aspects of omomyid skeletal anatomy are documented more thoroughly by Dagosto 1985, 1988; Gebo 1988; Covert and Hamrick 1993; Dagosto and Schmid 1996; Dagosto et al. 1999; and Anemone and Covert 2000.

11. Dagosto et al. 1999 describe the anatomy of the tibia of *Shoshonius*

cooperi, while Covert and Hamrick 1993 describe the same element in *Absarokius abbotti.* Dagosto 1985 discusses the functional anatomy of this region across a broader spectrum of mammals, including *Hemiacodon gracilis.*

12. In his initial brief paper announcing Wortman's discovery of the skull of *Tetonius,* Cope 1882a, 74, notes that "[t]he orbits [of *Tetonius*] are large, approaching those of *Tarsius.*" Apparently from this fact alone, Cope went on to conclude that "[t]he animal was nocturnal in its habits."

13. Several complete skulls of *Shoshonius cooperi* reveal that this species possesses orbits at least as large as those of *Tetonius.* These specimens are treated in greater detail in chapter 6. Covert and Hamrick 1993 report that the omomyid *Absarokius abbotti* also possesses enlarged orbits, on the basis of upper jaw fragments preserving its lower margin. On the other hand, the North American primate *Rooneyia viejaensis,* which is often regarded as a late-occurring member of the Omomyidae, has relatively small orbits suggesting diurnal habits. For this and other reasons, *Rooneyia* is not regarded as an omomyid here (see also Beard and MacPhee 1994). However, the primitive Asian omomyid *Teilhardina asiatica* also has small orbits, suggesting that the earliest omomyids may have been diurnal (Ni et al. 2004).

14. Radinsky 1967.

15. Niemitz 1984b.

16. The positive correlation between body size and home range size among primates is documented by Milton and May 1976. Niemitz 1984b and Gursky 2000 document the areal extent of home ranges in tarsiers; Tan 1999 provides similar data for certain lemurs.

17. Species of omomyid primates from the Washakie Basin have been described by Gazin 1962; Savage and Waters 1978; and Williams and Covert 1994.

18. *Tetonius* provides the most prominent example of the practice of naming omomyids after mountain ranges, but it must be pointed out that the mountain range itself is substantially younger than its eponymous primate. The Teton Range had not yet been uplifted from the surrounding terrain of Jackson Hole when *Tetonius* roamed Wyoming in the early Eocene.

19. Wing 2001; Rose 2001.

20. For a great deal of additional information on omomyid evolution in the Bighorn Basin, see the monograph published by Bown and Rose 1987. The partial summary provided here is based entirely on their work.

21. Rose and Bown 1991.

22. Chinese omomyids are discussed by Beard and Wang 1991, Beard et al. 1994, and Ni et al. 2004.

23. Schmid 1982; Godinot et al. 1992.

24. For further discussion of hindlimb anatomy in microchoerids, see Schmid 1979; Godinot and Dagosto 1983; Dagosto 1985; Dagosto and Schmid 1996.

25. Schmid 1983.

26. Bown 1976; Gingerich 1977a.

27. Wortman in Cope 1882b, 139–40.

28. Apparently, the growing animosity between Wortman and Cope was known to their scientific peers. Years later, W. J. Holland noted: "No opportunity was afforded him [Wortman] to describe or write upon the material which

he had collected and prepared. Nevertheless so high were his attainments in comparative anatomy that Professor Joseph Leidy often requested him to take charge of his classes in comparative anatomy [at the University of Pennsylvania] when Leidy for various reasons was unable to meet them" (Holland 1927, 199).

29. After noting that Cope had prevented Wortman from publishing on the fossils he found, Osborn says: "As soon as he [Wortman] came under the writer's direction, he was given a generous share of the fossils of his own finding and in several instances the best finds of the season were placed in his hands" (Osborn 1926c, 653).

30. In Wortman's long treatise on the Ganodonta, a group he regarded as close relatives of living South American edentates (sloths, anteaters, and armadillos), he went out of his way to observe: "Cope at one time held that his group Taeniodonta [more or less equivalent to Wortman's Ganodonta] exhibits affinities with the Edentata, but this he subsequently abandoned, and he, too, utterly failed to grasp the fundamental facts of the problem" (Wortman, 1897, 106).

31. Rea 2001, 72.

32. The story of Wortman's role in the discovery of *Diplodocus carnegii* is recounted in much greater detail by Rea 2001.

33. Holland 1927, 200.

4. THE FOREST IN THE SAHARA

1. Bown and Kraus 1988; Kappelman 1992.

2. An earlier trip to Egypt, in the autumn of 1960, served mainly to lay the groundwork for the series of expeditions that began in 1961.

3. Leakey 1959. Today, *Zinjanthropus boisei* is regarded as a valid species of the group of hominids known as robust australopithecines. It is usually classified as either *Australopithecus boisei* or *Paranthropus boisei*.

4. As one example of the level of interest it garnered in the popular press, *Zinjanthropus* occupied the entire front page of the September 12, 1959, edition of the *Illustrated London News,* where the fossil skull is called a "stupendous" discovery. In the accompanying article, Louis Leakey is quoted as saying that *Zinjanthropus* represents "the oldest well-established stone toolmaker ever found anywhere" (217).

5. Osborn 1908.

6. At the time, the museum in Stuttgart that purchased Markgraf's Fayum specimens was known as the Naturalienkabinett. Today, the same institution is called the Staatliches Museum für Naturkunde. In 1906, Osborn himself described a fossil hyrax from the Fayum that had been collected by Markgraf and sold to the Naturalienkabinett. This was roughly a year before Osborn organized his own expedition to the Fayum.

7. Simons 1995a, 200.

8. Schlosser 1911.

9. Much has been written about Marsh, Cope, and their contemporaries. Lanham 1973 provides a useful introduction to these colorful, highly important figures in American paleontology.

10. Simons 1962.

11. Simons explains the curious failure of earlier researchers to follow up on the success of Osborn, Markgraf, and others in the Fayum as follows: "Another impediment to renewed work there [in the Fayum] may have been the belief that the workers at the turn of the century had already recovered almost all fossils of any importance . . . i.e., that the Fayum localities were 'worked out'" (Simons 1995a, 204).

12. The graduate student in question was G. Edward Lewis, who originally named *Ramapithecus brevirostris,* and who reached many of the same conclusions that would be echoed three decades later by Simons 1961. However, Lewis failed to garner the scientific attention that Simons received, for reasons that are discussed at length by Lewin 1997.

13. Simons 1964, 535.

14. Pilbeam and Simons 1965.

15. Leakey 1962.

16. Within a year of Leakey's description of *Kenyapithecus wickeri,* Simons 1963 began arguing that it was the same species as *Ramapithecus brevirostris,* an opinion that he would maintain throughout the 1960s (Simons 1969).

17. Leakey 1967.

18. At first, Simons 1963 merely hinted that *Proconsul* might be synonymous with *Dryopithecus.* He took the formal step of subsuming the three known species of *Proconsul* within the genus *Dryopithecus* two years later (Simons and Pilbeam 1965).

19. Lewin 1997 places the conflict between Simons and Leakey in the broader context of paleoanthropological disputes of the time. The encounter between Pilbeam and Leakey took place at a workshop sponsored by the Wenner-Gren Foundation for Anthropological Research in Chicago in 1964 (Lewin 1997, 93).

20. Simons as quoted in Lewin 1997, 93–94.

21. An article about Louis Leakey's research on *Kenyapithecus* in the February 3, 1967, edition of *Time* magazine says: "Leakey feels that he has little chance of finding the common ancestor of both man and the apes—a creature he believes may have lived some 40 million years ago, in the Oligocene epoch. Yale paleontologist Elwyn Simons is working in Egypt's Fayum province, Leakey notes, an area rich in material from the Oligocene. His somewhat sad prediction: 'He [Simons] will be the man who gets the common link [between apes and humans].'" According to modern geological timescales, the Oligocene epoch spanned about ten million years, from thirty-four to twenty-four million years ago. Hence, Leakey's estimate of forty million years ago for the dichotomy between apes and humans was too great, even by his own criteria.

22. *Aeolopithecus chirobates* was originally proposed by Simons 1965. While the species remains valid, most modern researchers consider *Aeolopithecus* to be the same as *Propliopithecus* (Kay et al. 1981). Hence, *Propliopithecus chirobates* now appears to be a slightly larger—and younger—relative of *Propliopithecus haeckeli.*

23. The Greek roots of the name translate roughly as "Egyptian ape that joins or connects." This reflects Simons's (1965) original view that *Aegyptopithecus*

was an evolutionary link between *Propliopithecus* and more advanced Miocene apes such as *Proconsul* and *Dryopithecus.*

24. These tentative evolutionary relationships were proposed by Simons 1965 when he originally named *Aegyptopithecus* and *Aeolopithecus.* Their reconstructed positions on the primate family tree have varied significantly and frequently through time.

25. Simons 1967, 33.

26. Simons as quoted in *Time,* November 24, 1967, 62.

27. Kortlandt 1980.

28. Bown et al. 1982.

29. Olson and Rasmussen 1986.

30. Rasmussen and Simons 2000.

31. For further discussion of the functional anatomy of the forelimb among primates, see Larson 1993 and Rose 1993.

32. Fleagle and Simons 1982a, 1995.

33. Hamrick et al. 1995.

34. Fleagle and Simons 1982b.

35. Kirk and Simons 2001.

36. Simons 1986.

37. Fleagle et al. 1980.

38. Lewin 1997 provides an excellent account of the rise and fall of *Ramapithecus* as an early hominid and the pivotal role of Elwyn Simons in this debate.

5. RECEIVED WISDOM

1. Leidy 1873, 89–90.

2. Happily, Matthew's views on climate and evolution agreed with those of his boss at the American Museum of Natural History, Henry Fairfield Osborn. Both men were convinced that central Asia was the most likely birthplace of the human lineage, although partly for different reasons. This possibility played a major role in launching the renowned Central Asiatic Expeditions of the 1920s, which were organized and financed by Osborn and Matthew's home institution, the American Museum of Natural History. Although the expeditions failed to uncover the remains of early humans, they did reveal an astonishing variety of paleontological treasures, mainly in the form of early mammals and Cretaceous dinosaurs.

3. The Miocene monkey *Homunculus patagonicus* was first described in 1891 by the prolific Argentinian paleontologist Florentino Ameghino. It appears to be closely related to South American titi monkeys belonging to the genus *Callicebus* (Fleagle and Tejedor 2002).

4. In Matthew's own words: "To the argument so often advanced that the transportation of a species across a wide stretch of sea and its survival and success in colonizing a new country in this way [by rafting] is an exceedingly improbable accident, it may be answered that, if we multiply the almost infinitesimal chance of this occurrence during the few centuries of scientific record by the almost infinite duration of geological epochs and periods, we obtain a finite and quite probable chance, which it is perfectly fair to invoke, where the evidence against land invasion is so strong" (Matthew 1915, 205–6).

5. Cartmill 1982 provides an excellent historical review of the arboreal theory and its relation to contemporary ideas in primate and human evolution.

6. For example, in his influential review of North American Eocene primates, the prominent Smithsonian paleontologist C. L. Gazin wrote: "One of the most impressive lessons learned from a study of primates is the very deep-seated effect that an arboreal adaptation has on the morphology of the animal. . . . Much has been written regarding the modifications of the foot for an arboreal existence, but emphasis should also be placed on the modification or morphological change related to the forward position of the eyes that must surely be related to this habitat. Undoubtedly the advantages of binocular vision are keenly appreciated where depth of vision is so important as it is among branches of trees. . . . The effect of such greatly modifying adaptive factors in both the skull and feet would be to bring about marked convergence in forms of unrelated or remotely related origins, and to result in strikingly parallel development in related groups. . . . The second of the above conditions cited, that of parallel development, is surely exemplified in the similarity between the platyrrhine [South American] and catarrhine [Old World] monkeys. . . . I am strongly convinced . . . that they have evolved quite separately from distinct Eocene families" (Gazin 1958, 98–99).

7. Luckett 1975.

8. Houle 1999.

9. Ciochon and Chiarelli 1980, 476.

10. Censky et al. (1998, 556) document the successful dispersal of iguanas on mats of floating vegetation in the Caribbean Sea, for example. After several powerful hurricanes in the region, "at least 15 individuals of the green iguana, *Iguana iguana*, appeared on the eastern beaches of Anguilla in the Caribbean. This species did not previously occur on the island. They arrived on a mat of logs and uprooted trees, some of which were more than 30 feet [nine meters] long and had large root masses. Local fishermen say the mat was extensive and took two days to pile up on shore. They reported seeing iguanas on both the beach and on logs in the bay." The most likely source for the iguanas that washed up on the beaches of Anguilla is the island of Guadeloupe, more than 200 kilometers (125 miles) away. Follow-up studies revealed that the invasive iguanas survived on Anguilla for at least twenty-nine months, at which time one of the female lizards was thought to be pregnant.

11. Houle 1999, table 3.

12. The exact evolutionary position of *Plesiadapis* and its close relatives is disputed. Their skulls have been shown to be less similar to those of primates than was long believed, and their postcranial skeletons share many features with modern flying lemurs, the only living members of the mammalian order Dermoptera (Beard 1993a, 1993b; Bloch and Silcox 2001).

13. Gingerich 1976, fig. 42. Gingerich's idea that anthropoids evolved from adapiforms recalls the earlier views of Leidy 1873, for example. Although Gingerich eventually abandoned his notion that tarsiers and omomyids evolved from archaic primates like *Plesiadapis,* he went on to elaborate his thesis that adapiforms gave rise to anthropoids (Gingerich 1977b, 1980b; Gingerich and Schoeninger 1977).

14. Gingerich 1981b.

15. According to Gingerich 1977b, 177: "Undoubted higher primates are to be found first in the Quarry G level of the Fayum where the genera *Apidium* and *Propliopithecus* are first found. Before the time represented by Quarry G, several primates are known from the late Eocene and early Oligocene that are equivocally placed in either the Adapidae [equivalent to Adapiformes in current usage] or the anthropoid Simiiformes, for example *Oligopithecus* and *Hoanghonius, Amphipithecus,* and *Cercamonius*. These forms all share some combination of molar morphology, symphyseal fusion, canine honing, premolar crowding, and deep mandibular rami that make some advanced Adapidae virtually indistinguishable from primitive anthropoids . . . the intermediate morphological characteristics and intermediate stratigraphic position of these equivocal late Eocene and early Oligocene genera make them very strong evidence that higher primates evolved from Adapidae."

16. Pocock 1918.

17. Szalay 1975.

18. F. S. Szalay reconstructed the front dentition of *Washakius, Uintanius,* and several other omomyids as having small incisors in his influential monograph on the Omomyidae (Szalay 1976). Fifteen years later, Covert and Williams 1991 confirmed Szalay's reconstruction of *Washakius,* based on a specimen in which the crowns of all front teeth were preserved.

19. Dagosto 1985.

20. Cartmill and Kay 1978.

21. Cartmill 1980.

22. MacPhee and Cartmill 1986.

23. Rosenberger 1985.

24. Simons and Rasmussen 1989.

25. Gingerich 1979.

26. Patterson 1981.

27. Cachel 1979.

6. THE BIRTH OF A GHOST LINEAGE

1. Some authors consider the North American primate *Rooneyia* and the European microchoerids *Necrolemur* and *Microchoerus* as members of the Omomyidae. If so, Wortman's discovery of *Tetonius* is not quite so unique. However, *Rooneyia* appears to be only distantly related to *Tetonius* and other omomyids (Beard and MacPhee 1994). By the same token, all European primates for which skulls have been described are either adapiforms or microchoerids.

2. Krishtalka 1993, 338.

3. Beard et al. 1992b.

4. Stucky 1984.

5. Stucky et al. 1990.

6. Beard et al. 1992a.

7. Beard and MacPhee 1994 provide further details on the cranial anatomy of *Shoshonius* and other living and fossil primates.

8. Dagosto et al. 1999.

9. Strait 2001.

10. Alternatively, one could argue that certain anatomical features were lost through a process known as evolutionary reversal. Hence, it could be that *Shoshonius* evolved from ancestors that possessed a postorbital septum and an anterior accessory cavity, or that anthropoids evolved from *Shoshonius*-like primates with enormous eyes and dual bony flanges over each auditory bulla. In this case, I strongly prefer the hypothesis of convergence (evolutionary acquisition of similar traits) over reversal (evolutionary loss of similar traits).

11. Cartmill 1994, 556.

12. Simons and Rasmussen 1989.

13. Beard et al. 1991; Beard and MacPhee 1994; Dagosto et al. 1999.

14. Norell 1992.

7. INITIAL HINTS FROM DEEP TIME

1. Godinot and Mahboubi 1992.

2. Coiffait et al. 1984.

3. Although the third lower molars of early anthropoids generally have reduced hypoconulid lobes, their first and second lower molars tend to have enlarged hypoconulids compared to Eocene prosimians.

4. Bonis et al. 1988.

5. Kappelman 1992.

6. Simons and Rasmussen 1994.

7. Simons 1989.

8. I suspect that *Catopithecus* and *Oligopithecus* are so close that they should be classified in the same genus.

9. Recall that the front part, or trigonid, of mammalian lower molars typically projects above the level of the back part, or talonid, of the tooth. In primates, the talonid is broad and generally basinlike. It functions as a mortar, occluding with the pestlelike lingual cusp (or protocone) on the upper molars. In most anthropoids, the lower molar trigonids are reduced in height, so that trigonid and talonid lie at roughly the same level.

10. Simons 1990.

11. Simons 1997b.

12. Simons 1992; Simons et al. 2002. *Abuqatrania basiodontos* is so similar to *Qatrania wingi* that the two species may actually be one and the same. In any case, they should be referred to the same genus.

13. Simons 1990, 1997b; Simons and Seiffert 1999; Simons et al. 1999; Seiffert et al. 2000; Seiffert and Simons 2001; Kirk and Simons 2001.

14. Exactly how *Stockia* relates to various other North American omomyids is currently controversial. However, several *Stockia*-like omomyids are known. These include *Utahia kayi, Ageitodendron matthewi, Chipetaia lamporea,* and *Ourayia uintensis,* all of which are known from middle Eocene localities from a small region of northeastern Utah and adjacent southwestern Wyoming (Honey 1990; Gunnell 1995; Rasmussen 1996).

15. Beard and Wang 1991.

16. Qi et al. 1991.

17. Wang and Dawson 1994; Dawson and Wang 2001.

8. GHOST BUSTERS

1. Mebrouk et al. 1997.
2. Beard and MacPhee 1994.
3. Gebo 1986, 1988; Beard et al. 1988; Dagosto 1988; Covert and Williams 1994.
4. Ford 1988, 1994.
5. Dagosto 1990; Dagosto and Gebo 1994.
6. Elizabeth Culotta, a science journalist who covered the Duke "Anthropoid Origins" conference for *Science* magazine, quoted my comments regarding the anthropoid status of *Eosimias* (Culotta 1992, 1517).
7. Gingerich's opinion that *Eosimias* is a hedgehog was cited by Culotta 1992, 1517.
8. Rasmussen et al. 1992; Simons et al. 1994; Simons and Rasmussen 1994.
9. Sudre 1979; Godinot 1994, 238–39.
10. Jaeger et al. 1985.
11. Vianey-Liaud et al. 1994.
12. Beard et al. 1994.
13. Godinot 1994; Godinot and Mahboubi 1994.
14. "It is now clear that cranial features must be known for definite placement of a species in Anthropoidea," according to Simons and Rasmussen 1994, 135–36.
15. Simons 1990.
16. Divergent views on the affinities of *Oligopithecus* are expressed by Gingerich 1977b; Delson and Rosenberger 1980; and Rasmussen and Simons 1988.
17. See the opposing viewpoints of Michel Brunet and his colleagues, who discovered *Sahelanthropus* and interpret it as the oldest known hominid (Brunet et al. 2002), and M. H. Wolpoff and his co-authors, who view it as an ape (Wolpoff et al. 2002).
18. Simons 1990, 1995b, refers *Catopithecus* to the Propliopithecidae, which also includes *Aegyptopithecus* and *Propliopithecus*. Propliopithecids are widely acknowledged as early catarrhines, whose modern relatives include Old World monkeys, apes, and humans. By definition, early catarrhines occur several levels up the evolutionary tree from the origin of anthropoids.
19. Seiffert et al. 2000; Seiffert and Simons 2001.

9. RESURRECTING THE GHOST

1. My account of Andersson's discovery of the Eocene fossil site now known as "Locality 1" in the Yuanqu Basin is adapted from Andersson's own version of the events in question (Andersson 1923).
2. Andersson's account conveys the excitement he felt: "At the time I sent my material to this expert I was under the impression that the fossils were most likely of early Pleistocene age, but, to my great and pleasant surprise, Dr. Odhner reported that the collection was of much higher interest, being, in fact, the first indication of Eocene deposits in China. . . . Dr. Odhner has identified 8 species of fresh-water mollusks, 6 of which are characteristic of the Eocene deposits of France and Western Germany, whereas the two remaining species are new. . . .

There was every probability that a much larger number of mollusk species remained to be discovered, and the occurrence in the first small material of a tiny fragment of a vertebrate gave some hope of still more interesting possibilities" (Andersson 1923, 25).

3. Gillin 1967, 63.

4. Mateer and Lucas 1985.

5. Triumphantly, Osborn noted in the foreword to Andrews's popular book *On the Trail of Ancient Man* that he had predicted decades earlier that Asia was the birthplace of the primate ancestors of modern humans: "the home of the anthropoid apes, the chimpanzee, the orang, the gibbon and the gorilla, was placed in southern Asia . . . but the home of the more remote ancestors of man, *Primates,* was placed in northern Asia, where our Expedition went to work" (Andrews 1926, ix).

6. According to Andersson, "In most courteous and pleasant manner Dr. [Roy Chapman] Andrews has acceded to the desire expressed by Dr. Ting [Andersson's direct superior at the Geological Survey of China] and myself that a regional division of the field of research be made in order that duplication of work may be avoided and the most useful co-operation assured" (Andersson 1923, 3).

7. Morgan and Lucas 2002, 20.

8. At the time Andersson found the first fossil primate specimens in the Yuanqu Basin, Asian Eocene primates had never been formally reported in the scientific literature. However, the British paleontologist G. D. P. Cotter had discovered equally scrappy Eocene primate specimens in Myanmar as early as 1913, although they were not formally described as *Pondaungia cotteri* until many years later (Pilgrim 1927).

9. Intriguingly, Andersson and Granger rapidly developed a cordial relationship, possibly because Andersson acknowledged Granger's vast expertise as a paleontologist. Soon after Granger arrived in Beijing in June 1921 (almost exactly corresponding to Zdansky's arrival there), Andersson showed him the Eocene mammals he had just collected in the Yuanqu Basin. Granger readily indentified Andersson's primate specimens as belonging to either a lemuroid or an insectivore (Andersson 1923, 34). Given enough time to make the appropriately detailed comparisons, Granger almost certainly would have eliminated the latter possibility.

10. Zdansky 1930.

11. Gingerich 1977b; Rasmussen and Simons 1988; Rasmussen 1994.

12. The generic name *Xanthorhysis* comes from the Greek words *xanthos* (yellow) and *rhysis* (river) (Beard 1998). The etymology pays homage to Zdansky's fossil primate *Hoanghonius* (also named after the Yellow River, but on the basis of the local dialect of Mandarin). It also reflects the common practice of naming fossil tarsiiform primates (omomyids, microchoerids, and true tarsiids) after major rivers (see chapter 3).

13. The ten-million-year discrepancy in the anthropoid ghost lineage caused by using *Shoshonius* as opposed to *Xanthorhysis* as a baseline for assessing anthropoid origins is cut in half if we consider the isolated teeth of *Tarsius eocaenus* from the Shanghuang fissures, which date to nearly forty-five million years ago (Beard et al. 1994).

14. For example, the upper molars lacked any trace of the postprotocingulum, or "*Nannopithex* fold," that is so common among Eocene adapiforms and omomyids. Instead, a well-developed postprotocrista ran directly from the protocone toward the metacone, as is the case in such primitive North African anthropoids as *Algeripithecus, Catopithecus,* and *Proteopithecus.*

15. Beard and Wang 2004.

16. The two exceptions are *Altanius orlovi* from the earliest Eocene of Mongolia and *Ekgmowechashala philotau* from the late Oligocene of South Dakota and Oregon. The affinities of both of these fossil primates are controversial. Here, they are regarded as exceedingly primitive primates that cannot be linked unambiguously with any of the major living or extinct primate clades.

17. The sivaladapid affinities of *Hoanghonius* are also substantiated by *Guangxilemur*, a slightly younger fossil that is morphologically intermediate between *Hoanghonius* and Miocene sivaladapids (Qi and Beard 1998).

18. Beard et al. 1996.

19. Gebo et al. 2000a.

10. INTO THE AFRICAN MELTING POT

1. Kuhn 1962.

2. Begun 2002.

3. For information on the age, anatomy, and possible evolutionary relationships of *Branisella*, see Takai et al. 2000b and additional references cited therein.

4. In the Linnean system of classification, orders are fairly high-level taxonomic units, being interposed between the ranks of class (such as Mammalia) and family (such as Camelidae, the group of even-toed ungulates that includes camels and llamas, but not pigs, hippos, deer, and cattle). Generally speaking, living mammalian orders segregate the major types of mammals into different groups. Thus, the major groups of mammals that are native to Africa belong to the orders Proboscidea (elephants), Hyracoidea (hyraxes), Macroscelidea (elephant shrews), Tubulidentata (aardvarks), Sirenia (dugongs and manatees), and Afrosoricida (golden moles and tenrecs).

5. Springer et al. 1997; Madsen et al. 2001; Murphy et al. 2001a, 2001b.

6. Murphy et al. 2001b.

7. Court and Mahboubi 1993; Gheerbrant et al. 1998a; Tabuce et al. 2001.

8. The late Paleocene site of Adrar Mgorn 1 in Morocco has yielded abundant material of small insectivorous mammals, but the earliest African rodents date to the early or middle Eocene (Vianey-Liaud et al. 1994; Gheerbrant et al. 1998b).

9. Simons and Bown 1995.

10. Notions of how primates are related to other groups of mammals have varied substantially through the years. Among the more lasting and influential of these hypotheses was that espoused by the great American comparative anatomist William King Gregory (1910), who proposed the name Archonta for a large segment of living mammals including primates, tree shrews, flying lemurs, bats, and elephant shrews. Subsequent research in molecular evolutionary systematics has shown that elephant shrews are afrotheres and that bats are only

distantly related to the other "archontan" groups. Accordingly, Waddell et al. 1999 proposed the new term Euarchonta, which literally means "true Archonta," for what remains of Gregory's original Archonta—namely primates, tree shrews, and flying lemurs.

11. Murphy et al. 2001b have coined the term Euarchontoglires for the group that includes Euarchonta, rodents, and lagomorphs.

12. Based on analyses of DNA sequences of living primates using the molecular clock, Meireles et al. 2003 calculate that tarsiers and anthropoids have been segregated since the late Paleocene, about fifty-eight million years ago. This conclusion complements similar work published by Schmitz et al. 2001; Murphy et al. 2001a; and Yoder 2003.

13. Beard 1998.

14. The generic name is based on the nearby village of Bahin, while the species name comes from the Pondaung Formation, the stratigraphic unit that yielded the fossils (Jaeger et al. 1999).

15. Tsubamoto et al. 2002 obtained a fission-track date of 37.2 million years for a tuff in the Pondaung Formation that lies one meter above primate postcranial elements attributed to *Pondaungia* by Ciochon et al. 2001.

16. Chaimanee et al. 1997, 2000.

17. Takai et al. 2001.

18. Jaeger et al. 1998.

19. Ciochon et al. 2001.

20. Shigehara et al. 2002.

21. Takai et al. 2000a, 2003; Gunnell et al. 2002.

22. Cachel 1979.

23. Ford 1980.

24. Atsalis et al. 1996.

25. Dagosto and Terranova 1992.

26. Gebo et al. 2000b.

27. Rowley 1996.

28. Meireles et al. 2003.

29. In their original description of *Altiatlasius,* Bernard Sigé and his colleagues (1990) referred to it as an omomyid, although they recognized that it was probably related to anthropoids at that time. Subsequent discoveries of middle Eocene anthropoids such as *Eosimias* and *Algeripithecus* have closed the gap between *Altiatlasius* and other anthropoids, indicating that Sigé et al. were correct in regarding *Altiatlasius* as a basal anthropoid. The generic name refers to the High Atlas Mountains, which are located near the fossil site. The species name *koulchii* is transliterated from an Arabic term for an entity whose meaning is vague.

30. The primitive adapiform *Donrussellia* was originally described as an omomyid (Godinot 1978). Kenneth D. Rose and his colleagues have commented extensively on the numerous similarities and the subtle distinctions between primitive adapiforms and omomyids (Rose and Bown 1991; Rose et al. 1994).

31. Indeed, this is precisely the stance that my colleague Marc Godinot has adopted. Godinot 1994 accepts the anthropoid affinities of *Altiatlasius* and views its late Paleocene age as further evidence that anthropoids originated in Africa.

32. Yoder et al. 1996.
33. Tattersall 1982.
34. Hartwig and Meldrum 2002.

11. PALEOANTHROPOLOGY AND PITHECOPHOBIA

1. "If man had not been his own classifier, he would never have thought of founding a separate order for his own reception," Darwin observed (1874, 515).

2. "As far as differences in certain important points of structure are concerned, man may no doubt rightly claim the rank of a Sub-order; and this rank is too low, if we look chiefly to his mental faculties. Nevertheless, from a genealogical point of view it appears that this rank is too high, and that man ought to form merely a Family, or possibly even only a Sub-family" (Darwin, 1874, 517).

3. Darwin 1874, 519–20.
4. Simons 1963, 886.
5. Wood Jones 1918.
6. Elliot Smith 1919.
7. Chalmers Mitchell 1919, 497.
8. Cartmill 1982.
9. Osborn 1928, 202.
10. Lewin 1997.
11. Osborn 1930, 6.
12. Osborn 1928, 191.

13. Ironically, in his initial description of the Taung skull, Dart 1925 agreed with Osborn that open-country living played a critical role in stimulating early hominids to develop their advanced humanlike traits. However, Dart believed that the most likely locale for human origins was the open veld savannas of southern Africa, not the central Asian steppes that Osborn preferred.

14. Lewin 1997 provides a particularly thorough account of how and why the Piltdown forgery became accepted by Osborn and the leading British anthropologists of that era.

15. Osborn 1922.
16. Osborn 1927, fig. 1.
17. Osborn 1925.
18. Gregory 1927c.
19. Shipman 2001.
20. Osborn 1930, 4.
21. Osborn 1926b, 266.
22. Ibid.
23. Osborn 1926a, 6.

24. In Osborn's own words, "the evolution of man is arrested or retrogressive in every region where the natural food supply is abundant and accessible without effort; in tropical and semi-tropical regions where natural food fruits abound, human effort—individual and racial—immediately ceases" (Osborn 1926b, 266–67).

25. Gregory was clearly aware that he was placed in a difficult political position by his scientific disagreement with Osborn: "But, like the slave in the clas-

sical story whose unpleasant and doubtless risky duty it was to remind royalty '*Memento te hominem esse*,' I conceive it as my hard duty to remind mankind that these poor [anthropoid] relations of ours, mute witnesses of the past, are still with us and that the evidence of our lowly origin can hardly be waved aside on the ground of the length and aloofness of our own lineage" (Gregory 1927a, 440).

26. "I have been wonderfully impressed with the movement that is called the rebirth of the Fatherland," Osborn proclaimed in a speech in Germany on September 24, 1934: "Among all classes of people, high and low, and especially among the youth, the eager boys and girls, I observed a spirit of determination and loyalty to the ideals of the new government which I believe is without parallel in any part of the world. In fact, I may say without exaggeration, and shall say when I return to America, that all the environmental influences of the youth of Germany at this time are superior to those of any country and far superior to those of most countries, including the United States of America" (quoted in Richter 1934, 438).

27. Gregory 1927b.

28. Walsh et al. 2003.

29. Adkins and Honeycutt 1993.

References Cited

Adkins, R. M., and Honeycutt, R. L. 1993. A molecular examination of archontan and chiropteran monophyly. In *Primates and Their Relatives in Phylogenetic Perspective*, ed. R. D. E. MacPhee, 227–49. New York: Plenum Press.

Andersson, J. G. 1923. Essays on the Cenozoic of northern China. *Memoirs of the Geological Survey of China*, ser. A, 3: 1–152.

Andrews, R. C. 1926. *On the Trail of Ancient Man*. Garden City, N.Y.: Garden City Publishing Co.

Anemone, R. L., and Covert, H. H. 2000. New skeletal remains of *Omomys* (Primates, Omomyidae): Functional morphology of the hindlimb and locomotor behavior of a middle Eocene primate. *Journal of Human Evolution* 38: 607–33.

Atsalis, S., Schmid, J., and Kappeler, P. M. 1996. Metrical comparisons of three species of mouse lemur. *Journal of Human Evolution* 31: 61–68.

Bacon, A.-M., and Godinot, M. 1998. Analyse morphofonctionnelle des fémurs et des tibias des *"Adapis"* du Quercy: Mise en évidence de cinq types morphologiques. *Folia Primatologica* 69: 1–21.

Ba Maw, Ciochon, R. L., and Savage, D. E. 1979. Late Eocene of Burma yields earliest anthropoid primate, *Pondaungia cotteri*. *Nature* 282: 65–67.

Beard, K. C. 1993a. Phylogenetic systematics of the Primatomorpha, with special reference to Dermoptera. In *Mammal Phylogeny: Placentals*, ed. F. S. Szalay, M. J. Novacek, and M. C. McKenna, 129–50. New York: Springer.

———. 1993b. Origin and evolution of gliding in early Cenozoic Dermoptera

(Mammalia, Primatomorpha). In *Primates and Their Relatives in Phylogenetic Perspective*, ed. R. D. E. MacPhee, 63–90. New York: Plenum Press.

———. 1998. A new genus of Tarsiidae (Mammalia: Primates) from the middle Eocene of Shanxi Province, China, with notes on the historical biogeography of tarsiers. In *The Dawn of the Age of Mammals in Asia*, ed. K. C. Beard and M. R. Dawson, 260–77. Pittsburgh: Carnegie Museum of Natural History.

Beard, K. C., and Dawson, M. R., eds. 1998. *The Dawn of the Age of Mammals in Asia*. Bulletin of The Carnegie Museum of Natural History, no. 34. Pittsburgh: Carnegie Museum of Natural History.

———. 1999. Intercontinental dispersal of Holarctic land mammals near the Paleocene/Eocene boundary: Paleogeographic, paleoclimatic and biostratigraphic implications. *Bulletin de la Société Géologique de France* 170: 697–706.

Beard, K. C., and MacPhee, R. D. E. 1994. Cranial anatomy of *Shoshonius* and the antiquity of Anthropoidea. In *Anthropoid Origins*, ed. J. G. Fleagle and R. F. Kay, 55–97. New York: Plenum Press.

Beard, K. C., and Wang, B. 1991. Phylogenetic and biogeographic significance of the tarsiiform primate *Asiomomys changbaicus* from the Eocene of Jilin Province, People's Republic of China. *American Journal of Physical Anthropology* 85: 159–66.

Beard, K. C., and Wang, J. 2004. The eosimiid primates (Anthropoidea) of the Heti Formation, Yuanqu Basin, Shanxi and Henan provinces, People's Republic of China. *Journal of Human Evolution* 46:401–32.

Beard et al. 1988. [Beard, K. C., Dagosto, M., Gebo, D. L., and Godinot, M.] Interrelationships among primate higher taxa. *Nature* 331: 712–14.

———. 1991. [Beard, K. C., Krishtalka, L., and Stucky, R. K.] First skulls of the early Eocene primate *Shoshonius cooperi* and the anthropoid-tarsier dichotomy. *Nature* 349: 64–67.

———. 1992a. [Beard, K. C., Krishtalka, L., and Stucky, R. K.] Revision of the Wind River faunas, early Eocene of central Wyoming, part 12: New species of omomyid primates (Mammalia: Primates: Omomyidae) and omomyid taxonomic composition across the early-middle Eocene boundary. *Annals of the Carnegie Museum of Natural History* 61: 39–62.

———. 1992b. [Beard, K. C., Sigé, B., and Krishtalka, L.] A primitive vespertilionoid bat from the early Eocene of central Wyoming. *Comptes Rendus de l'Académie des Sciences* (Paris), 2d ser., 314: 735–41.

———. 1994. [Beard, K. C., Qi, T., Dawson, M. R., Wang, B., and Li, C.] A diverse new primate fauna from middle Eocene fissure-fillings in southeastern China. *Nature* 368: 604–9.

———. 1996.[Beard, K. C., Tong, Y., Dawson, M. R., Wang, J., and Huang, X.] Earliest complete dentition of an anthropoid primate from the late middle Eocene of Shanxi Province, China. *Science* 272: 82–85.

Begun, D. R. 2002. The Pliopithecoidea. In *The Primate Fossil Record*, ed. W. C. Hartwig, 221–40. Cambridge: Cambridge University Press.

Bloch J. I., and Silcox, M. T. 2001. New basicrania of Paleocene-Eocene *Ignacius*: Re-evaluation of the plesiadapifom-dermopteran link. *American Journal of Physical Anthropology* 116: 184–98.

Bonis, L. de, Jaeger, J.-J., Coiffat, B., and Coiffat, P.-E. 1988. Découverte du plus ancien primate catarrhinien connu dans l'Éocène supérieur d'Afrique du Nord. *Comptes Rendus de l'Académie des Sciences* (Paris), 2d ser., 306: 929–34.

Bown, T. M. 1976. Affinities of *Teilhardina* (Primates, Omomyidae) with description of a new species from North America. *Folia Primatologica* 25: 62–72.

Bown, T. M., and Kraus, M. J. 1988. Geology and paleoenvironment of the Oligocene Jebel Qatrani Formation and adjacent rocks, Fayum Depression, Egypt. *U.S. Geological Survey Professional Papers* 1452: 1–60.

Bown, T. M., and Rose, K. D. 1987. Patterns of dental evolution in early Eocene anaptomorphine primates (Omomyidae) from the Bighorn Basin, Wyoming. *Paleontological Society Memoirs* 23: 1–162.

———, eds. 1990. *Dawn of the Age of Mammals in the Northern Part of the Rocky Mountain Interior, North America.* Geological Society of America Special Paper no. 243. Boulder, Colo.: Geological Society of America.

Bown et al. 1982. [Bown, T. M., Kraus, M. J., Wing, S. L., Fleagle, J. G., Tiffney, B. H., Simons, E. L., and Vondra, C. F.] The Fayum primate forest revisited. *Journal of Human Evolution* 11: 603–32.

Brown, B. 1925. Byways and highways in Burma. *Natural History* 25: 294–308.

Brown, L. 1950. *I Married a Dinosaur.* New York: Dodd, Mead.

Brunet, M., et al. 2002. A new hominid from the upper Miocene of Chad, central Africa. *Nature* 418: 145–51.

Cachel, S. 1979. A paleoecological model for the origin of higher primates. *Journal of Human Evolution* 8: 351–59.

Cartmill, M. 1980. Morphology, function, and evolution of the anthropoid postorbital septum. In *Evolutionary Biology of the New World Monkeys and Continental Drift,* ed. R. L. Ciochon and A. B. Chiarelli, 243–74. New York: Plenum Press.

———. 1982. Basic primatology and prosimian evolution. In *A History of American Physical Anthropology, 1930–1980,* ed. F. Spencer, 147–86. New York: Academic Press.

———. 1994. Anatomy, antinomies, and the problem of anthropoid origins. In *Anthropoid Origins,* ed. J. G. Fleagle and R. F. Kay, 549–66. New York: Plenum Press.

Cartmill, M., and Kay, R. F. 1978. Cranio-dental morphology, tarsier affinities, and primate suborders. In *Recent Advances in Primatology,* vol. 3, *Evolution,* ed. D. J. Chivers and K. A. Joysey, 205–14. London: Academic Press.

Censky, E. J., Hodge, K., and Dudley, J. 1998. Over-water dispersal of lizards due to hurricanes. *Nature* 395: 556.

Chaimanee et al. 1997. [Chaimanee, Y., Suteethorn, V., Jaeger, J.-J., and Ducrocq, S.] A new late Eocene anthropoid primate from Thailand. *Nature* 385: 429–31.

———. 2000. [Chaimanee, Y., Khansuba, S., and Jaeger, J.-J.] A new lower jaw of *Siamopithecus eocaenus* from the late Eocene of Thailand. *Comptes Rendus de l'Académie des Sciences* (Paris), 2d ser., *Sciences de la Vie,* 323: 235–41.

Chalmers Mitchell, P. 1919. Discussion on the zoological position and affinities of *Tarsius. Proceedings of the Zoological Society of London* 1919: 496–97.

Ciochon, R. L., and Chiarelli, A. B. 1980. Paleobiogeographic perspectives on the origin of Platyrrhini. In *Evolutionary Biology of the New World Monkeys and Continental Drift,* ed. R. L. Ciochon and A. B. Chiarelli, 459–93. New York: Plenum Press.

Ciochon, R. L., and Holroyd, P. A. 1994. The Asian origin of Anthropoidea revisited. In *Anthropoid Origins,* ed. J. G. Fleagle and R. F. Kay, 143–62. New York: Plenum Press.

Ciochon et al. 1985. [Ciochon, R. L., Savage, D. E., Thaw Tint, and Ba Maw.] Anthropoid origins in Asia? New discovery of *Amphipithecus* from the Eocene of Burma. *Science* 229: 756–59.

———. 2001. [Ciochon, R. L., Gingerich, P. D., Gunnell, G. F., and Simons, E. L.] Primate postcrania from the late middle Eocene of Myanmar. *Proceedings of the National Academy of Sciences* 98: 7672–77.

Coiffat, P.-E., Coiffat, B., Jaeger, J.-J., and Mahboubi, M. 1984. Un nouveau gisement à mammifères fossils d'âge Éocène supérieur sur le versant sud des Nementcha (Algérie orientale): Découverte des plus anciens rongeurs d'Afrique. *Comptes Rendus de l'Académie des Sciences* (Paris), 2d ser., 299: 893–98.

Colbert, E. H. 1937. A new primate from the upper Eocene Pondaung Formation of Burma. *American Museum Novitates* 951: 1–18.

Cope, E. D. 1882a. An anthropomorphous lemur. *American Naturalist* 16: 73–74.

———. 1882b. Contributions to the history of Vertebrata of the lower Eocene of Wyoming and New Mexico, made during 1881. *Proceedings of the American Philosophical Society* 20: 139–97.

Court, N., and Mahboubi, M. 1993. Reassessment of lower Eocene *Seggeurius amourensis:* Aspects of primitive dental morphology in the mammalian order Hyracoidea. *Journal of Paleontology* 67: 889–93.

Covert, H. H. 1986. Biology of early Cenozoic primates. In *Comparative Primate Biology,* vol. 1, *Systematics, Evolution, and Anatomy,* ed. D. R. Swindler and J. Erwin, 335–59. New York: Alan R. Liss.

Covert, H. H., and Hamrick, M. W. 1993. Description of new skeletal remains of the early Eocene anaptomorphine primate *Absarokius* (Omomyidae) and a discussion about its adaptive profile. *Journal of Human Evolution* 25: 351–62.

Covert, H. H., and Williams, B. A. 1991. The anterior lower dentition of *Washakius insignis* and adapid-anthropoid affinities. *Journal of Human Evolution* 21: 463–67.

———. 1994. Recently recovered specimens of North American Eocene omomyids and adapids and their bearing on debates about anthropoid origins. In *Anthropoid Origins,* ed. J. G. Fleagle and R. F. Kay, 29–54. New York: Plenum Press.

Culotta, E. 1992. A new take on anthropoid origins. *Science* 256: 1516–17.

Dagosto, M. 1983. Postcranium of *Adapis parisiensis* and *Leptadapis magnus* (Adapiformes, Primates): Adaptational and phylogenetic significance. *Folia Primatologica* 41: 49–101.

———. 1985. The distal tibia of primates with special reference to the Omomyidae. *International Journal of Primatology* 6: 45–75.

———. 1988. Implications of postcranial evidence for the origin of euprimates. *Journal of Human Evolution* 17: 35–56.

———. 1990. Models for the origin of the anthropoid postcranium. *Journal of Human Evolution* 19: 121–39.

———. 1993. Postcranial anatomy and locomotor behavior in Eocene primates. In *Postcranial Adaptation in Nonhuman Primates,* ed. D. L. Gebo, 199–234. DeKalb: Northern Illinois University Press.

Dagosto, M., and Gebo, D. L. 1994. Postcranial anatomy and the origin of the Anthropoidea. In *Anthropoid Origins,* ed. J. G. Fleagle and R. F. Kay, 567–93. New York: Plenum Press.

Dagosto, M., and Schmid, P. 1996. Proximal femoral anatomy of omomyiform primates. *Journal of Human Evolution* 30: 29–56.

Dagosto, M., and Terranova, C. J. 1992. Estimating the body size of Eocene primates: A comparison of results from dental and postcranial variables. *International Journal of Primatology* 13: 307–44.

Dagosto et al. 1999. [Dagosto, M., Gebo, D. L., and Beard, K. C.] Revision of the Wind River faunas, early Eocene of central Wyoming, part 14: Postcranium of *Shoshonius cooperi* (Mammalia: Primates). *Annals of the Carnegie Museum of Natural History* 68: 175–211.

Dart, R. A. 1925. *Australopithecus africanus:* The man-ape of South Africa. *Nature* 115: 195–99.

Darwin, C. 1874 [1871]. *The Descent of Man and Selection in Relation to Sex.* 2d ed. Philadelphia: David McKay.

Dawson, M. R., and Tong, Y. 1998. New material of *Pappocricetodon schaubi* (Mammalia: Cricetidae) from the Yuanqu Basin, Shanxi Province, China. In *Dawn of the Age of Mammals in Asia,* ed. K. C. Beard and M. R. Dawson, 278–85. Pittsburgh: Carnegie Museum of Natural History.

Dawson, M. R., and Wang, B. 2001. Middle Eocene Ischyromyidae (Mammalia: Rodentia) from the Shanghuang fissures, southeastern China. *Annals of the Carnegie Museum* 70: 221–30.

Delfortrie, M. 1873. Un singe de la famille des lémuriens dans les phosphates de chaux quaternaires du Département du Lot. *Actes de la Société Linnéenne de Bordeaux* 29: 87–95.

Delson, E., and Rosenberger, A. L. 1980. Phyletic perspectives on platyrrhine origins and anthropoid relationships. In *Evolutionary Biology of the New World Monkeys and Continental Drift,* ed. R. L. Ciochon and A. B. Chiarelli, 445–58. New York: Plenum Press.

Dickens, G. R., Castillo, M. M., and Walker, J. C. G. 1997. A blast of gas in the latest Paleocene: Simulating first-order effects of massive dissociation of oceanic methane hydrate. *Geology* 25: 259–62.

Elliot Smith, G. 1919. Discussion on the zoological position and affinities of *Tarsius. Proceedings of the Zoological Society of London* 1919: 465–75.

Filhol, M. H. 1874. Nouvelles observations sur les mammifères des gisements de phosphates de chaux (lémuriens et pachylémuriens). *Annales des Sciences Géologiques* 5: 1–36.

Fleagle, J. G., and Simons, E. L. 1982a. The humerus of *Aegyptopithecus:* A primitive anthropoid. *American Journal of Physical Anthropology* 59: 175–93.

―――. 1982b. Skeletal remains of *Propliopithecus chirobates* from the Egyptian Oligocene. *Folia Primatologica* 39: 161–77.

―――. 1995. Limb skeleton and locomotor adaptations of *Apidium phiomense*, an Oligocene anthropoid from Egypt. *American Journal of Physical Anthropology* 97: 235–89.

Fleagle, J. G., and Tejedor, M. F. 2002. Early platyrrhines of southern South America. In *The Primate Fossil Record*, ed. W. C. Hartwig, 161–73. Cambridge: Cambridge University Press.

Fleagle et al. 1980. [Fleagle, J. G., Kay, R. F., and Simons, E. L.] Sexual dimorphism in early anthropoids. *Nature* 287: 328–30.

Ford, S. M. 1980. Callitrichids as phyletic dwarfs, and the place of Callitrichidae in Platyrrhini. *Primates* 21: 31–43.

―――. 1988. Postcranial adaptations of the earliest platyrrhine. *Journal of Human Evolution* 17: 155–92.

―――. 1994. Primitive platyrrhines? Perspectives on anthropoid origins from platyrrhine, parapithecid, and preanthropoid postcrania. In *Anthropoid Origins*, ed. J. G. Fleagle and R. F. Kay, 595–673. New York: Plenum Press.

Franzen, J. L. 1987. Ein neuer Primate aus dem Mitteleozän der Grube Messel (Deutschland, S Hessen). *Courier Forschungsinstitut Senckenberg* (Frankfurt a/M) 91: 151–87.

Gazin, C. L. 1958. A review of the middle and upper Eocene primates of North America. *Smithsonian Miscellaneous Collections* 136: 1–112.

―――. 1962. A further study of the lower Eocene mammalian faunas of southwestern Wyoming. *Smithsonian Miscellaneous Collections* 144: 1–98.

Gebo, D. L. 1986. Anthropoid origins—the foot evidence. *Journal of Human Evolution* 15: 421–30.

―――. 1988. Foot morphology and locomotor adaptation in Eocene primates. *Folia Primatologica* 50: 3–41.

Gebo et al. 1999. [Gebo, D. L., Dagosto, M., Beard, K. C., and Wang, J.] A first metatarsal of *Hoanghonius stehlini* from the late middle Eocene of Shanxi Province, China. *Journal of Human Evolution* 37: 801–6.

―――. 2000a. [Gebo, D. L., Dagosto, M., Beard, K. C., Qi, T., and Wang, J.] The oldest known anthropoid postcranial fossils and the early evolution of higher primates. *Nature* 404: 276–78.

―――. 2000b. [Gebo, D. L., Dagosto, M., Beard, K. C., and Qi, T.] The smallest primates. *Journal of Human Evolution* 38: 585–94.

Gee, H. 1999. *In Search of Deep Time: Beyond the Fossil Record to a New History of Life*. New York: Free Press.

Gheerbrant, E., Sudre, J., Capetta, H., and Bignot, G. 1998a. *Phosphatherium escuilliei* du Thanétien du Bassin des Ouled Abdoun (Maroc), plus ancien proboscidien (Mammalia) d'Afrique. *Geobios* 30: 247–69.

Gheerbrant, E., Sudre, J., Sen, S., Abrial, C., Marandat, B., Sigé, B., and Vianey-Liaud, M. 1998b. Nouvelles données sur les mammifères du Thanétien et de l'Yprésien du Bassin d'Ouarzazate (Maroc) et leur contexte stratigraphique. *Palaeovertebrata* 27: 155–202.

Gheerbrant, E., Thomas, H., Roger, J., Sen, S., and Al-Sulaimani, Z. 1993. Deux nouveaux primates dans l'Oligocène inférieur de Taqah (Sultanat d'Oman):

Premiers Adapiformes (? Anchonomyini) de la péninsule arabique? *Palaeovertebrata* 22: 141–96.

Gillin, D. 1967. *Warlord: Yen Hsi-shan in Shansi Province, 1911–1949.* Princeton: Princeton University Press.

Gingerich, P. D. 1976. Cranial anatomy and evolution of early Tertiary Plesiadapidae (Mammalia, Primates). *University of Michigan Papers on Paleontology* 15: 1–141.

———. 1977a. Dental variation in early Eocene *Teilhardina belgica,* with notes on the anterior dentition of some early Tarsiiformes. *Folia Primatologica* 28: 144–53.

———. 1977b. Radiation of Eocene Adapidae in Europe. *Geobios, Mémoire Special* 1: 165–82.

———. 1979. The stratophenetic approach to phylogeny reconstruction in vertebrate paleontology. In *Phylogenetic Analysis and Paleontology,* ed. J. Cracraft and N. Eldredge, 41–77. New York: Columbia University Press.

———. 1980a. History of early Cenozoic vertebrate paleontology in the Bighorn Basin. In *Early Cenozoic Paleontology and Stratigraphy of the Bighorn Basin, Wyoming,* ed. id., 7–24. University of Michigan Papers on Paleontology, no. 24. Ann Arbor: University of Michigan Museum of Paleontology.

———. 1980b. Eocene Adapidae, paleobiogeography, and the origin of South American Platyrrhini. In *Evolutionary Biology of the New World Monkeys and Continental Drift,* ed. R. L. Ciochon and A. B. Chiarelli, 123–38. New York: Plenum Press.

———. 1981a. Cranial morphology and adaptations in Eocene Adapidae, I: Sexual dimorphism in *Adapis magnus* and *Adapis parisiensis. American Journal of Physical Anthropology* 56: 217–34.

———. 1981b. Early Cenozoic Omomyidae and the evolutionary history of tarsiiform primates. *Journal of Human Evolution* 10: 345–74.

———. 1984. Paleobiology of tarsiiform primates. In *Biology of Tarsiers,* ed. C. Niemitz, 33–44. Stuttgart: Gustav Fischer.

———. 1995. Sexual dimorphism in earliest Eocene *Cantius torresi* (Mammalia, Primates, Adapoidea). *Contributions from the Museum of Paleontology, University of Michigan* 29: 185–99.

Gingerich, P. D., and Haskin, R. A. 1981. Dentition of early Eocene *Pelycodus jarrovii* (Mammalia, Primates) and the generic attribution of species formerly referred to *Pelycodus. Contributions from the Museum of Paleontology, University of Michigan* 25: 327–37.

Gingerich, P. D., and Martin, R. D. 1981. Cranial morphology and adaptations in Eocene Adapidae, II: The Cambridge skull of *Adapis parisiensis. American Journal of Physical Anthropology* 56: 235–57.

Gingerich, P. D., and Sahni, A. 1984. Dentition of *Sivaladapis nagrii* (Adapidae) from the late Miocene of India. *International Journal of Primatology* 5: 63–79.

Gingerich, P. D., and Schoeninger, M. 1977. The fossil record and primate phylogeny. *Journal of Human Evolution* 6: 483–505.

Gingerich, P. D., and Simons, E. L. 1977. Systematics, phylogeny, and evolution of early Eocene Adapidae (Mammalia, Primates) in North America. *Contributions from the Museum of Paleontology, University of Michigan* 24: 245–79.

Godinot, M. 1978. Un nouvel Adapidé (Primate) de l'Éocène inférieur de Provence. *Comptes Rendus de l'Académie des Sciences* (Paris), ser. D, 286: 1869–72.

———. 1991. Toward the locomotion of two contemporaneous *Adapis* species. *Zeitschrift für Morphologie und Anthropologie* 78: 387–405.

———. 1992. Apport à la systématique de quatre genres d'Adapiformes (Primates, Éocène). *Comptes Rendus de l'Académie des Sciences* (Paris), 2d ser., 314: 237–42.

———. 1994. Early North African primates and their significance for the origin of Simiiformes (= Anthropoidea). In *Anthropoid Origins*, ed. J. G. Fleagle and R. F. Kay, 235–95. New York: Plenum Press.

———. 1998. A summary of adapiform systematics and evolution. *Folia Primatologica* 69, suppl. 1: 218–49.

Godinot, M., and Dagosto, M. 1983. The astragalus of *Necrolemur* (Primates, Microchoerinae). *Journal of Paleontology* 57: 1321–24.

Godinot, M., and Jouffroy, F. K. 1984. La main d'*Adapis* (Primate, Adapidé). In *Actes du Symposium Paléontologique Georges Cuvier*, ed. E. Buffetaut, J. M. Mazin, and E. Salmon, 221–42. Montbéliard: Le Serpentaire.

Godinot, M., and Mahboubi, M. 1992. Earliest known simian primate found in Algeria. *Nature* 357: 324–26.

———. 1994. Les petits primates simiiformes de Glib Zegdou (Éocène inférieur à moyen d'Algérie). *Comptes Rendus de l'Académie des Sciences* (Paris), 2d ser., 319: 357–64.

Godinot, M., Russell, D. E., and Louis, P. 1992. Oldest known *Nannopithex* (Primates, Omomyiformes) from the early Eocene of France. *Folia Primatologica* 58: 32–40.

Gregory, W. K. 1910. The orders of mammals. *Bulletin of the American Museum of Natural History* 27: 1–524.

———. 1927a. The origin of man from the anthropoid stem—when and where? *Proceedings of the American Philosophical Society* 66: 439–63.

———. 1927b. Two views of the origin of man. *Science* 65: 601–5.

———. 1927c. *Hesperopithecus* apparently not an ape nor a man. *Science* 66: 579–81.

Gunnell, G. F. 1995. Omomyid primates (Tarsiiformes) from the Bridger Formation, middle Eocene, southern Green River Basin, Wyoming. *Journal of Human Evolution* 28: 147–87.

Gunnell, G. F., Ciochon, R. L., Gingerich, P. D., and Holroyd, P. A. 2002. New assessment of *Pondaungia* and *Amphipithecus* (Primates) from the late middle Eocene of Myanmar, with a comment on "Amphipithecidae." *Contributions from the Museum of Paleontology, University of Michigan* 30: 337–72.

Gursky, S. 2000. Effect of seasonality on the behavior of an insectivorous primate, *Tarsius spectrum*. *International Journal of Primatology* 21: 477–95.

Haile-Selassie, Y. 2001. Late Miocene hominids from the middle Awash, Ethiopia. *Nature* 412: 178–81.

Hamrick, M. W., Meldrum, D. J., and Simons, E. L. 1995. Anthropoid phalanges from the Oligocene of Egypt. *Journal of Human Evolution* 28: 121–45.

Hartenberger, J.-L., and Marandat, B. 1992. A new genus and species of an early Eocene primate from North Africa. *Human Evolution* 7: 9–16.

Hartwig, W. C., and Meldrum, D. J. 2002. Miocene platyrrhines of the northern Neotropics. In *The Primate Fossil Record,* ed. W. C. Hartwig, 175–88. Cambridge: Cambridge University Press.

Holland, W. J. 1927. Obituary: Dr. Jacob L. Wortman. *Annals of the Carnegie Museum* 17: 199–201.

Holroyd, P. A., and Maas, M. C. 1994. Paleogeography, paleobiogeography, and anthropoid origins. In *Anthropoid Origins,* ed. J. G. Fleagle and R. F. Kay, 297–334. New York: Plenum Press.

Honey, J. G. 1990. New washakiin primates (Omomyidae) from the Eocene of Wyoming and Colorado, and comments on the evolution of the Washakiini. *Journal of Vertebrate Paleontology* 10: 206–21.

Houle, A. 1999. The origin of platyrrhines: An evaluation of the Antarctic scenario and the floating island model. *American Journal of Physical Anthropology* 109: 541–59.

Jaeger, J.-J., Denys, C., and Coiffat, B. 1985. New Phiomorpha and Anomaluridae from the late Eocene of north-west Africa: Phylogenetic implications. In *Evolutionary Relationships Among Rodents: A Multidisciplinary Analysis,* ed. W. P. Luckett and J.-L. Hartenbeger, 567–88. New York: Plenum Press.

Jaeger, J.-J., Soe, A. N., Aung, A. K., Benammi, M., Chaimanee, Y., Ducrocq, R. M., Tun, T., Thein, T., and Ducrocq, S. 1998. New Myanmar middle Eocene anthropoids. An Asian origin for catarrhines? *Comptes Rendus de l'Académie des Sciences* (Paris), 3d ser., *Sciences de la Vie,* 321: 953–59.

Jaeger, J.-J., Thein, T., Benammi, M., Chaimanee, Y., Soe, A. N., Lwin, T., Tun, T., Wai, S., and Ducrocq, S. 1999. A new primate from the middle Eocene of Myanmar and the Asian early origin of anthropoids. *Science* 286: 528–30.

Ji, Q., Currie, P. J., Ji, S., and Norell, M. A. 1998. Two feathered dinosaurs from northeastern China. *Nature* 393: 753–61.

Ji, Q., Norell, M. A., Gao, K., Ji, S., and Ren, D. 2001. The distribution of integumentary structures in a feathered dinosaur. *Nature* 410: 1084–88.

Kappelman, J. 1992. The age of the Fayum primates as determined by paleomagnetic reversal stratigraphy. *Journal of Human Evolution* 22: 495–503.

Kay, R. F., and Covert, H. H. 1984. Anatomy and behaviour of extinct primates. In *Food Acquisition and Processing in Primates,* ed. D. J. Chivers, B. A. Wood, and A. Bilsborough, 467–508. New York: Plenum Press.

Kay, R. F., Fleagle, J. G., and Simons, E. L. 1981. A revision of the Oligocene apes of the Fayum Province, Egypt. *American Journal of Physical Anthropology* 55: 293–322.

Kirk, E. C., and Simons, E. L. 2001. Diets of fossil primates from the Fayum Depression of Egypt: A quantitative analysis of molar shearing. *Journal of Human Evolution* 40: 203–29.

Kortlandt, A. 1980. The Fayum primate forest: Did it exist? *Journal of Human Evolution* 9: 277–97.

Krishtalka, L. 1993. Anagenetic angst: Species boundaries in Eocene primates. In *Species, Species Concepts, and Primate Evolution,* ed. W. H. Kimbel and L. B. Martin, 331–44. New York: Plenum Press.

Krishtalka, L., Stucky, R. K., and Beard, K. C. 1990. The earliest fossil evidence for sexual dimorphism in primates. *Proceedings of the National Academy of Sciences* 87: 5223–26.

Kuhn, T. S. 1962. *The Structure of Scientific Revolutions*. Chicago: University of Chicago Press.

Lanham, U. 1973. *The Bone Hunters*. New York: Columbia University Press.

Larson, S. G. 1993. Functional morphology of the shoulder in primates. In *Postcranial Adaptation in Nonhuman Primates,* ed. D. L. Gebo, 45–69. DeKalb: Northern Illinois University Press.

Leakey, L. S. B. 1959. A new fossil skull from Olduvai. *Nature* 184: 491–93.

———. 1962. A new lower Pliocene fossil primate from Kenya. *Annals and Magazine of Natural History,* 13th ser., 4: 689–96.

———. 1967. An early Miocene member of Hominidae. *Nature* 213: 155–63.

Leakey, M. G., Feibel, C. S., McDougall, I., and Walker, A. 1995. New four-million-year-old hominid species from Kanapoi and Allia Bay, Kenya. *Nature* 376: 565–71.

Le Gros Clark, W. E. 1960 [1959]. *The Antecedents of Man: An Introduction to the Evolution of the Primates*. Chicago: Quadrangle Books.

Leidy, J. 1869. Notice of some extinct vertebrates from Wyoming and Dakota. *Proceedings of the Academy of Natural Sciences of Philadelphia* 21: 63–67.

———. 1870. Descriptions of *Palaeosyops paludosus, Microsus cuspidatus,* and *Notharctus tenebrosus. Proceedings of the Academy of Natural Sciences of Philadelphia* 22: 113–14.

———. 1873. Contributions to the extinct vertebrate fauna of the western territories. *Report of the United States Geological Survey* (F. V. Hayden) 1: 1–358.

Lewin, R. 1997. *Bones of Contention: Controversies in the Search for Human Origins*. 2d ed. Chicago: University of Chicago Press.

Luckett, W. P. 1975. Ontogeny of the fetal membranes and placenta: Their bearing on primate phylogeny. In *Phylogeny of the Primates: A Multidisciplinary Approach,* ed. id. and F. S. Szalay, 157–82. New York: Plenum Press.

MacPhee, R. D. E., and Cartmill, M. 1986. Basicranial structures and primate systematics. In *Comparative Primate Biology,* vol. 1, *Systematics, Evolution, and Anatomy,* ed. D. R. Swindler and J. Erwin, 219–75. New York: Alan R. Liss.

Madsen, O., Scally, M., Douady, C. J., Kao, D. J., DeBry, R. W., Adkins, R., Amrine, H. M., Stanhope, M. J., de Jong, W. W., and Springer, M. S. 2001. Parallel adaptive radiations in two major clades of placental mammals. *Nature* 409: 610–14.

Marsh, O. C. 1872. Discovery of fossil Quadrumana in the Eocene of Wyoming. *American Journal of Science and Arts,* 3d ser., 4: 405–6.

Mateer, N. J., and Lucas, S. G. 1985. Swedish vertebrate palaeontology in China: A history of the Lagrelius collection. *Bulletin of the Geological Institutions of the University of Uppsala* 11: 1–24.

Matthew, W. D. 1915. Climate and evolution. *Annals of the New York Academy of Sciences* 24: 171–318.

Matthew, W. D., and Granger, W. 1915. A revision of the lower Eocene Wasatch and Wind River faunas, part IV: Entelonychia, Primates, Insectivora (Part). *Bulletin of the American Museum of Natural History* 34: 429–83.

Mebrouk, F., Mahboubi, M., Bessedik, M., and Feist, M. 1997. L'apport des charophytes à la stratigraphie des formations continentales Paléogènes de l'Algérie. *Geobios* 30: 171–77.

Meireles, C. M., Czelusniak, J., Page, S. L., Wildman, D. E., and Goodman, M. 2003. Phylogenetic position of tarsiers within the order Primates: Evidence from γ-Globin DNA sequences. In *Tarsiers: Past, Present, and Future,* ed. P. C. Wright, E. L. Simons, and S. Gursky, 145–60. New Brunswick: Rutgers University Press.

Milton, K., and May, M. L. 1976. Body weight, diet, and home range area in primates. *Nature* 259: 459–62.

Morgan, V. L., and Lucas, S. G. 2002. Walter Granger, 1872–1941, paleontologist. *New Mexico Museum of Natural History and Science Bulletin* 19: 1–58.

Murphy, W. J., Eizirik, E., Johnson, W. E., Zhang, Y. P., Ryder, O. A., and O'Brien, S. J. 2001a. Molecular phylogenetics and the origins of placental mammals. *Nature* 409: 614–18.

Murphy, W. J., Eizirik, E., O'Brien, S. J., Madsen, O., Scally, M., Douady, C. J., Teeling, E., Ryder, O. A., Stanhope, M. J., de Jong, W. W., and Springer, M. S. 2001b. Resolution of the early placental mammal radiation using Bayesian phylogenetics. *Science* 294: 2348–51.

Nash, L. T. 1986. Dietary, behavioral, and morphological aspects of gummivory in primates. *Yearbook of Physical Anthropology* 29: 113–37.

Ni, X., Wang, Y., Hu, Y., and Li, C. 2004. A euprimate skull from the early Eocene of China. *Nature* 427: 65–68.

Niemitz, C. 1984a. Synecological relationships and feeding behaviour of the genus *Tarsius*. In *Biology of Tarsiers,* ed. id., 59–75. Stuttgart: Gustav Fischer.

———. 1984b. An investigation and review of the territorial behaviour and social organisation of the genus *Tarsius*. In *Biology of Tarsiers,* ed. id., 117–27. Stuttgart: Gustav Fischer.

Norell, M. A. 1992. Taxic origin and temporal diversity: The effect of phylogeny. In *Extinction and Phylogeny,* ed. M. J. Novacek and Q. D. Wheeler, 89–118. New York: Columbia University Press.

Olson, S. L., and Rasmussen, D. T. 1986. Paleoenvironment of the earliest hominoids: New evidence from the Oligocene avifauna of Egypt. *Science* 233: 1202–4.

Osborn, H. F. 1906. Milk dentition of the hyracoid *Saghatherium* from the upper Eocene of Egypt. *Bulletin of the American Museum of Natural History* 22: 263–66.

———. 1908. New fossil mammals from the Fayûm Oligocene, Egypt. *Bulletin of the American Museum of Natural History* 24: 265–72.

———. 1922. *Hesperopithecus*, the first anthropoid primate found in America. *American Museum Novitates* 37: 1–5.

———. 1925. *The Earth Speaks to Bryan*. New York: C. Scribner's Sons.

———. 1926a. The evolution of human races. *Natural History* 26: 3–13.

———. 1926b. Why central Asia? *Natural History* 26: 263–69.

———. 1926c. J. L. Wortman—A biographical sketch. *Natural History* 26: 652–53.

————. 1927. Recent discoveries relating to the origin and antiquity of man. *Science* 65: 481–88.

————. 1928. Recent discoveries relating to the origin and antiquity of man. *Palaeobiologica* 1: 189–202.

————. 1930. The discovery of Tertiary man. *Science* 71: 1–7.

Patterson, C. 1981. Significance of fossils in determining evolutionary relationships. *Annual Review of Ecology and Systematics* 12: 195–223.

Pilbeam, D. R., and Simons, E. L. 1965. Some problems of hominid classification. *American Scientist* 53: 237–59.

Pilgrim, G. E. 1927. A *Sivapithecus* palate and other primate fossils from India. *Memoirs of the Geological Survey of India, Palaeontologia Indica*, n.s., 14: 1–26.

Pocock, R. I. 1918. On the external characters of the lemurs and of *Tarsius*. *Proceedings of the Zoological Society of London* 1918: 19–53.

Pollock, J. I., and Mullin, R. J. 1987. Vitamin C biosynthesis in prosimians: Evidence for the anthropoid affinity of *Tarsius*. *American Journal of Physical Anthropology* 73: 65–70.

Qi, T., and Beard, K. C. 1998. Late Eocene sivaladapid primate from Guangxi Zhuang Autonomous Region, People's Republic of China. *Journal of Human Evolution* 35: 211–20.

Qi, T., Zong, G., and Wang, Y. 1991. Discovery of *Lushilagus* and *Miacis* in Jiangsu and its zoogeographical significance. *Vertebrata PalAsiatica* 29: 59–63.

Radinsky, L. B. 1967. The oldest primate endocast. *American Journal of Physical Anthropology* 27: 385–88.

Rasmussen, D. T. 1994. The different meanings of a tarsioid-anthropoid clade and a new model of anthropoid origin. In *Anthropoid Origins*, ed. J. G. Fleagle and R. F. Kay, 335–60. New York: Plenum Press.

————. 1996. A new middle Eocene omomyine primate from the Uinta Basin, Utah. *Journal of Human Evolution* 31: 75–87.

Rasmussen, D. T., and Simons, E. L. 1988. New specimens of *Oligopithecus savagei*, early Oligocene primate from the Fayum, Egypt. *Folia Primatologica* 51: 182–208.

————. 1992. Paleobiology of the oligopithecines, the earliest known anthropoid primates. *International Journal of Primatology* 13: 477–508.

————. 2000. Ecomorphological diversity among Paleogene hyracoids (Mammalia): A new cursorial browser from the Fayum, Egypt. *Journal of Vertebrate Paleontology* 20: 167–76.

Rasmussen et al. 1992. [Rasmussen, D. T., Bown, T. M., and Simons, E. L.] The Eocene-Oligocene transition in continental Africa. In *Eocene-Oligocene Climatic and Biotic Evolution*, ed. D. R. Prothero and W. A. Berggren, 548–66. Princeton: Princeton University Press.

Ravosa, M. J., and Hylander, W. L. 1994. Function and fusion of the mandibular symphysis in primates: Stiffness or strength? In *Anthropoid Origins*, ed. J. G. Fleagle and R. F. Kay, 447–68. New York: Plenum Press.

Rea, T. 2001. *Bone Wars: The Excavation and Celebrity of Andrew Carnegie's Dinosaur*. Pittsburgh: University of Pittsburgh Press.

Richter, R. 1934. Henry Fairfield Osborn und "Senckenberg." *Natur und Volk* 64: 435–39.

Rose, K. D. 2001. Wyoming's Garden of Eden. *Natural History* 110, no. 3: 55–59.

Rose, K. D., and Bown, T. M. 1991. Additional fossil evidence on the differentiation of the earliest euprimates. *Proceedings of the National Academy of Sciences* 88: 98–101.

Rose, K. D., Godinot, M., and Bown, T. M. 1994. The early radiation of euprimates and the initial diversification of Omomyidae. In *Anthropoid Origins*, ed. J. G. Fleagle and R. F. Kay, 1–28. New York: Plenum Press.

Rose, M. D. 1993. Functional anatomy of the elbow and forearm in primates. In *Postcranial Adaptation in Nonhuman Primates*, ed. D. L. Gebo, 70–95. DeKalb: Northern Illinois University Press.

Rosenberger, A. L. 1985. In favor of the necrolemur-tarsier hypothesis. *Folia Primatologica* 45: 179–94.

Rosenberger, A. L., and Strasser, E. 1985. Toothcomb origins: Support for the grooming hypothesis. *Primates* 26: 73–84.

Rowley, D. B. 1996. Age of initiation of collision between India and Asia: A review of stratigraphic data. *Earth and Planetary Science Letters* 145: 1–13.

Rudwick, M. J. S. 1997. *Georges Cuvier, Fossil Bones, and Geological Catastrophes*. Chicago: University of Chicago Press.

Savage, D. E., and Waters, B. T. 1978. A new omomyid primate from the Wasatch Formation of southern Wyoming. *Folia Primatologica* 30: 1–29.

Schlosser, M. 1911. Beiträge zur Kenntnis der Oligozänen Landsäugetiere aus dem Fayum (Ägypten). *Beiträge zur Paläontologie und Geologie Österreich-Ungarns und des Orients* 24: 51–167.

Schmid, P. 1979. Evidence of microchoerine evolution from Dielsdorf (Zürich region, Switzerland)—a preliminary report. *Folia Primatologica* 31: 301–11.

———. 1982. *Die systematische Revision der europäischen Microchoeridae Lydekker, 1887 (Omomyiformes, Primates)*. Ph.D. diss., Universität Zürich. Zurich: Peter Schmid.

———. 1983. Front dentition of the Omomyiformes (Primates). *Folia Primatologica* 40: 1–10.

Schmitz, J., Ohme, M., and Zischler, H. 2001. SINE insertions in cladistic analyses and the phylogenetic affiliations of *Tarsius bancanus* to other primates. *Genetics* 157: 777–84.

Seiffert, E. R., and Simons, E. L. 2001. Astragalar morphology of late Eocene anthropoids from the Fayum Depression (Egypt) and the origin of catarrhine primates. *Journal of Human Evolution* 41: 577–606.

Seiffert, E. R., Simons, E. L., and Fleagle, J. G. 2000. Anthropoid humeri from the late Eocene of Egypt. *Proceedings of the National Academy of Sciences* 97: 10062–67.

Shigehara, N., Takai, M., Kay, R. F., Aung, A. K., Soe, A. N., Tun, S. T., Tsubamoto, T., and Thein, T. 2002. The upper dentition and face of *Pondaungia cotteri* from central Myanmar. *Journal of Human Evolution* 43: 143–66.

Shipman, P. 2001. *The Man Who Found the Missing Link: Eugène Dubois and His Lifelong Quest to Prove Darwin Right*. New York: Simon & Schuster.

Sigé, B., Jaeger, J.-J., Sudre, J., and Vianey-Liaud, M. 1990. *Altiatlasius koulchii*

n. gen. et sp., primate omomyidé du Paléocène supérieur du Maroc, et les origines des euprimates. *Palaeontographica*, sec. A, 214: 31–56.

Simons, E. L. 1961. The phyletic position of *Ramapithecus*. *Postilla* 57: 1–9.

———. 1962. Two new primate species from the African Oligocene. *Postilla* 64: 1–12.

———. 1963. Some fallacies in the study of hominid phylogeny. *Science* 141: 879–89.

———. 1964. On the mandible of *Ramapithecus*. *Proceedings of the National Academy of Sciences* 51: 528–35.

———. 1965. New fossil apes from Egypt and the initial differentiation of Hominoidea. *Nature* 205: 135–39.

———. 1967. The earliest apes. *Scientific American* 217, no. 6: 28–35.

———. 1969. Late Miocene hominid from Fort Ternan, Kenya. *Nature* 221: 448–51.

———. 1986. *Parapithecus grangeri* of the African Oligocene: An archaic catarrhine without lower incisors. *Journal of Human Evolution* 15: 205–13.

———. 1989. Description of two new genera and species of late Eocene Anthropoidea from Egypt. *Proceedings of the National Academy of Sciences* 86: 9956–60.

———. 1990. Discovery of the oldest known anthropoidean skull from the Paleogene of Egypt. *Science* 247: 1567–69.

———. 1992. Diversity in the early Tertiary anthropoidean radiation in Africa. *Proceedings of the National Academy of Sciences* 89: 10743–47.

———. 1995a. Egyptian Oligocene primates: A review. *Yearbook of Physical Anthropology* 38: 199–238.

———. 1995b. Skulls and anterior teeth of *Catopithecus* (Primates: Anthropoidea) from the Eocene and anthropoid origins. *Science* 268: 1885–88.

———. 1997a. Discovery of the smallest Fayum Egyptian primates (Anchomomyini, Adapidae). *Proceedings of the National Academy of Sciences* 94: 180–84.

———. 1997b. Preliminary description of the cranium of *Proteopithecus sylviae*, an Egyptian late Eocene anthropoidean primate. *Proceedings of the National Academy of Sciences* 94: 14970–75.

Simons, E. L., and Bown, T. M. 1995. Ptolemaiida, a new order of Mammalia—with description of the cranium of *Ptolemaia grangeri*. *Proceedings of the National Academy of Sciences* 92: 3269–73.

Simons, E. L., and Miller, E. R. 1997. An upper dentition of *Aframonius dieides* (Primates) from the Fayum, Egyptian Eocene. *Proceedings of the National Academy of Sciences* 94: 7993–96.

Simons, E. L., and Pilbeam, D. R. 1965. Preliminary revision of the Dryopithecinae (Pongidae, Anthropoidea). *Folia Primatologica* 3: 81–152.

Simons, E. L., and Rasmussen, D. T. 1989. Cranial morphology of *Aegyptopithecus* and *Tarsius* and the question of the tarsier-anthropoid clade. *American Journal of Physical Anthropology* 79: 1–23.

———. 1994. A whole new world of ancestors: Eocene anthropoideans from Africa. *Evolutionary Anthropology* 3: 128–39.

Simons, E. L., and Seiffert, E. R. 1999. A partial skeleton of *Proteopithecus sylviae* (Primates, Anthropoidea): First associated dental and postcranial remains of an Eocene anthropoidean. *Comptes Rendus de l'Académie des Sciences, Paris (Sciences de la terre et des planètes)* 329: 921–27.

Simons et al. 1994. [Simons, E. L., Rasmussen, D. T., Bown, T. M., and Chatrath, P. S.] The Eocene origin of anthropoid primates: Adaptation, evolution, and diversity. In *Anthropoid Origins*, ed. J. G. Fleagle and R. F. Kay, 179–201. New York: Plenum Press.

———. 1995. [Simons, E. L., Rasmussen, D. T., and Gingerich, P. D.] New cercamoniine adapid from Fayum, Egypt. *Journal of Human Evolution* 29: 577–89.

———. 1999. [Simons, E. L., Plavcan, J. M., and Fleagle, J. G.] Canine sexual dimorphism in Egyptian Eocene anthropoid primates: *Catopithecus* and *Proteopithecus*. *Proceedings of the National Academy of Sciences* 96: 2559–62.

———. 2002. [Simons, E. L., Seiffert, E. R., Chatrath, P. S., and Attia, Y.] Earliest record of a parapithecid anthropoid from the Jebel Qatrani Formation, northern Egypt. *Folia Primatologica* 72: 316–31.

Simpson, G. G. 1942. The beginnings of vertebrate paleontology in North America. *Proceedings of the American Philosophical Society* 86: 130–88.

Springer, M. S., Cleven, G. C., Madsen, O., de Jong, W. W., Waddell, V. G., Amrine, H. M., and Stanhope, M. J. 1997. Endemic African mammals shake the phylogenetic tree. *Nature* 388: 61–64.

Stephan, H. 1984. Morphology of the brain in *Tarsius*. In *Biology of Tarsiers*, ed. C. Niemitz, 319–44. Stuttgart: Gustav Fischer.

Strait, S. G. 2001. Dietary reconstruction of small-bodied omomyoid primates. *Journal of Vertebrate Paleontology* 21: 322–34.

Stucky, R. K. 1984. Revision of the Wind River faunas, early Eocene of central Wyoming, part 5: Geology and biostratigraphy of the upper part of the Wind River Formation, northeastern Wind River Basin. *Annals of the Carnegie Museum* 53: 231–94.

Stucky, R. K., Krishtalka, L., and Redline, A. D. 1990. Geology, vertebrate fauna, and paleoecology of the Buck Spring Quarries (early Eocene, Wind River Formation), Wyoming. In *The Dawn of the Age of Mammals in the Northern Part of the Rocky Mountain Interior, North America*, ed. T. M. Bown and K. D. Rose, 169–86. Boulder, Colo.: Geological Society of America.

Sudre, J. 1979. Nouveaux mammifères Éocènes du Sahara occidental. *Palaeovertebrata* 9: 83–115.

Szalay, F. S. 1975. Phylogeny of primate higher taxa: The basicranial evidence. In *Phylogeny of the Primates: A Multidisciplinary Approach*, ed. W. P. Luckett and F. S. Szalay, 91–125. New York: Plenum Press.

———. 1976. Systematics of the Omomyidae (Tarsiiformes, Primates): Taxonomy, phylogeny, and adaptations. *Bulletin of the American Museum of Natural History* 156: 157–450.

Szalay, F. S., and Eric Delson. 1979. *Evolutionary History of the Primates*. New York: Academic Press.

Tabuce, R., Coiffat, B., Coiffat, P.-E., Mahboubi, M., and Jaeger, J.-J. 2001. A new genus of Macroscelidea (Mammalia) from the Eocene of Algeria: A possible origin for elephant-shrews. *Journal of Vertebrate Paleontology* 21: 535–46.

Takai et al. 2000a. [Takai, M., Shigehara, N., Tsubamoto, T., Egi, N., Aung, A. K., Thein, T., Soe, A. N., and Tun, S. T.] The latest middle Eocene primate fauna in Pondaung area, Myanmar. *Asian Paleoprimatology* 1: 7–28.

———. 2000b. [Takai, M., Anaya, F., Shigehara, N., and Setoguchi, T.] New fossil materials of the earliest New World monkey, *Branisella boliviana*, and the problem of platyrrhine origins. *American Journal of Physical Anthropology* 111: 263–81.

———. 2001. [Takai, M., Shigehara, N., Aung, A. K., Tun, S. T., Soe, A. N., Tsubamoto, T., and Thein, T.] A new anthropoid from the latest middle Eocene of Pondaung, central Myanmar. *Journal of Human Evolution* 40: 393–409.

———. 2003. [Takai, M., Shigehara, N., Egi, N., and Tsubamoto, T.] Endocranial cast and morphology of the olfactory bulb of *Amphipithecus mogaungensis* (latest middle Eocene of Myanmar). *Primates* 44: 137–44.

Tan, C. L. 1999. Group composition, home range size, and diet of three sympatric bamboo lemur species (genus *Hapalemur*) in Ranomafana National Park, Madagascar. *International Journal of Primatology* 20: 547–66.

Tattersall, I. 1982. *The Primates of Madagascar.* New York: Columbia University Press.

Tsubamoto, T., Takai, M., Shigehara, N., Egi, N., Tun, S. T., Aung, A. K., Maung, M., Danhara, T., and Suzuki, H. 2002. Fission-track zircon age of the Eocene Pondaung Formation, Myanmar. *Journal of Human Evolution* 42: 361–69.

Vianey-Liaud, M., Jaeger, J.-J., Hartenberger, J.-L., and Mahboubi, M. 1994. Les rongeurs de l'Éocène d'Afrique nord-occidentale [Glib Zegdou (Algérie) et Chambi (Tunisie)] et l'origine des Anomaluridae. *Palaeovertebrata* 23: 93–118.

Waddell, P. J., Okada, N., and Hasegawa, M. 1999. Towards resolving the interordinal relationships of placental mammals. *Systematic Biology* 48: 1–5.

Walsh, P. D., et al. 2003. Catastrophic ape decline in western equatorial Africa. *Nature* 422: 611–14.

Wang, B., and Dawson, M. R. 1994. A primitive cricetid (Mammalia: Rodentia) from the middle Eocene of Jiangsu Province, China. *Annals of the Carnegie Museum* 63: 239–56.

Warren, L. 1998. *Joseph Leidy: The Last Man Who Knew Everything.* New Haven: Yale University Press.

White, T. D., Suwa, G., and Asfaw, B. 1994. *Australopithecus ramidus*, a new species of early hominid from Aramis, Ethiopia. *Nature* 371: 306–12.

Wilf, P. 2000. Late Paleocene-early Eocene climate changes in southwestern Wyoming: Paleobotanical analysis. *Geological Society of America Bulletin* 112: 292–307.

Wilf, P., and Labandeira, C. C. 1999. Response of plant-insect associations to Paleocene-Eocene warming. *Science* 284: 2153–56.

Williams, B. A., and Covert, H. H. 1994. New early Eocene anaptomorphine primate (Omomyidae) from the Washakie Basin, Wyoming, with comments on the phylogeny and paleobiology of anaptomorphines. *American Journal of Physical Anthropology* 93: 323–40.

Wilson, J. A., and Szalay, F. S. 1976. New adapid primate of European affinities from Texas. *Folia Primatologica* 25: 294–312.

Wing, S. L. 2001. Hot times in the Bighorn Basin. *Natural History* 110, no. 3: 48–54.

Wolpoff, M. H., Senut, B., Pickford, M., and Hawks, J. 2002. *Sahelanthropus* or *"Sahelpithecus"*? *Nature* 419: 581–82.

Wood Jones, F. 1918. *The Problem of Man's Ancestry.* London: Society for Promoting Christian Knowledge.

Wortman, J. L. 1897. The Ganodonta and their relationship to the Edentata. *Bulletin of the American Museum of Natural History* 9: 59–110.

———. 1899. Restoration of *Oxyaena lupina* Cope, with descriptions of certain new species of Eocene creodonts. *Bulletin of the American Museum of Natural History* 12: 139–48.

Yoder, A. D. 2003. The phylogenetic position of genus *Tarsius:* Whose side are you on? In *Tarsiers: Past, Present, and Future,* ed. P. C. Wright, E. L. Simons, and S. Gursky, 161–75. New Brunswick: Rutgers University Press.

Yoder, A. D., Cartmill, M., Ruvolo, M., Smith, K., and Vilgalys, R. 1996. Ancient single origin for Malagasy primates. *Proceedings of the National Academy of Sciences* 93: 5122–26.

Zdansky, O. 1930. Die alttertiären Säugetiere Chinas nebst stratigraphischen Bemerkungen. *Palaeontologia Sinica,* ser. C, 6, no. 2: 1–87.

Index

Compositor:	Integrated Composition Systems
Text:	10/13 Sabon
Display:	Akzidenz
Printer and binder:	Edwards Brothers